Synthetic Biology Brings New Opportunity for Antibiotics Discovery

Synthetic Biology Brings New Opportunity for Antibiotics Discovery

Editors

Yaojun Tong
Zixin Deng
Linquan Bai

MDPI • Basel • Beijing • Wuhan • Barcelona • Belgrade • Manchester • Tokyo • Cluj • Tianjin

MDPI

Editors

Yaojun Tong
State Key Laboratory of Microbial Metabolism, School of Life Sciences and Biotechnology
Shanghai Jiao Tong University
Shanghai
China

Zixin Deng
State Key Laboratory of Microbial Metabolism, School of Life Sciences and Biotechnology
Shanghai Jiao Tong University
Shanghai
China

Linquan Bai
State Key Laboratory of Microbial Metabolism, School of Life Sciences and Biotechnology
Shanghai Jiao Tong University
Shanghai
China

Editorial Office
MDPI
St. Alban-Anlage 66
4052 Basel, Switzerland

This is a reprint of articles from the Special Issue published online in the open access journal *Antibiotics* (ISSN 2079-6382) (available at: www.mdpi.com/journal/antibiotics/special_issues/Synthetic).

For citation purposes, cite each article independently as indicated on the article page online and as indicated below:

LastName, A.A.; LastName, B.B.; LastName, C.C. Article Title. *Journal Name* **Year**, *Volume Number*, Page Range.

ISBN 978-3-0365-7261-1 (Hbk)
ISBN 978-3-0365-7260-4 (PDF)

Contents

Preface to "Synthetic Biology Brings New Opportunity for Antibiotics Discovery"

Though society is developing much more rapidly today than it did in the past, we are facing a complicated public health issue in the form of global drug shortages, and antimicrobials are unfortunately on that list. These shortages are caused by various issues, e.g., technological limitations, which have led to a drying up of the pipeline of antimicrobial development. Indeed, it has been decades since antibiotics with a completely novel mode of action were last delivered to the clinic. The limited availability of antibiotics, in addition to the rapid emergence of multidrug resistance, are putting us in a very dangerous position. Luckily, the advances of new technologies, especially biotechnologies, such as DNA sequencing, precise genome editing, system metabolic engineering, multiomics, synthetic biology, big data processing, and artificial intelligence are offering a great opportunity to develop better and efficient bioactive molecules, including next-generation antibiotics. Currently, various achievements and new discoveries from both academia and industry are taking place in the broad field of antibiotics, which reflects the fact that a new era is coming. As one of the most important driving forces, academia plays a critical role in drug discovery for a safer society, and, thus, we would like to bring academia, industry, and clinic together in this Special Issue to report recent progress in antibiotics discovery.

Yaojun Tong, Zixin Deng, and Linquan Bai
Editors

antibiotics

Editorial

Synthetic Biology Facilitates Antimicrobials Discovery

Linquan Bai *, Zixin Deng * and Yaojun Tong *

State Key Laboratory of Microbial Metabolism, Joint International Research Laboratory of Metabolic and Developmental Sciences, School of Life Sciences and Technology, Shanghai Jiao Tong University, Shanghai 200240, China

* Correspondence: bailq@sjtu.edu.cn (L.B.); zxdeng@sjtu.edu.cn (Z.D.); yaojun.tong@sjtu.edu.cn (Y.T.)

We are currently facing two big global challenges: antibiotics shortage and multidrug resistance. Since the "Golden Age" of natural products-based antibiotic discovery in the mid-20th century, we have only discovered a very limited number of new antibiotics. This indicates that the traditional antibiotic discovery pipeline is drying up. Moreover, the misuse and overuse of limited classes of antibiotics have caused the emergence and spread of multidrug resistance. Fortunately, the advance of synthetic biology brings us new opportunities for antibiotic discovery. A starting place could be the genomic (gene) and related enzymatic information with the "bottom-up" strategy [1], empowered by "cutting-edge" biotechnology such as CRISPR-based genome editing [2], to facilitate antibiotic discoveries from natural resources.

Several important studies [3–7] on synthetic biology-assisted antibiotic research have been published recently, and we strongly feel that it is about time to re-attract researchers' attention to antibiotic discovery using cutting-edge concepts such as synthetic biology. This idea encourages us to organize this Special Issue on "Synthetic Biology Brings New Opportunity for Antibiotics Discovery". This Special Issue covers a broad range of antibiotic research, comprising three review articles and seven original research papers.

In the review article written by Huang and colleagues [8], the biosynthesis and working mechanisms of specific plant-derived antimicrobial agents such as artemisinin, oleanolic acid, berberine, colchicine, and baicalin are summarized and discussed with a goal to provide insights for the future discovery and development of such agents.

The review by Wang and colleagues [9] summarizes the diverse secondary metabolites that are produced by a popular edible and medicinal mushroom, *Inonotus hispidus*. Its secondary metabolites are mainly polyphenols and triterpenoids, with multiple bioactivities, including anticancer, immunomodulatory, anti-inflammatory, antioxidant, antimicrobial, and enzyme inhibitory activities. *I. hispidus* is a promising source of bioactive compounds, including antibiotics for health promotion and functional food development.

To address the challenge of the shortage of antifungals, Zhong and colleagues have written a review [10] that focuses on how to use synthetic biology to facilitate the development of novel antifungal drugs that originate from traditional Chinese medicine derived from natural products.

In the study conducted by Geng and colleagues, the mechanisms of antibiotic resistance and genes associated with antibiotic resistance in *Zymomonas mobilis* are deeply investigated using bioinformatics and CRISPR/Cas12a-based genome-editing technology. Six ampicillin-resistant genes are identified, and their related mutants are constructed. It shows that ZMO0103 is the key to ampicillin resistance in *Z. mobilis* and how other ampicillin-resistant genes may have a synergistic effect [11]. This study could lay the foundation for further studies of other antibiotic resistance mechanisms.

The research article conducted by Yang and colleagues reports a novel global regulatory protein, SspH, which is widespread in *Streptomyces*, and this was found to play an important role in controlling different types of antibiotic production. This study provides

Citation: Bai, L.; Deng, Z.; Tong, Y. Synthetic Biology Facilitates Antimicrobials Discovery. *Antibiotics* **2023**, *12*, 578. https://doi.org/10.3390/antibiotics12030578

Received: 20 February 2023
Accepted: 9 March 2023
Published: 15 March 2023

insight into the regulation of antibiotic biosynthesis and the potential targets for future antibiotic discovery and production [12].

Liu and colleagues, in their research paper, aimed to identify the genes that are involved in the export of the anti-coccidiosis agent salinomycin produced by *Streptomyces albus* BK3-25. By analyzing transcriptomes, eight putative transporter genes were found and proven to positively impact salinomycin exports. The overexpression of these genes will lead to an increase in the titer of salinomycin and improved self-resistance. The study provides a transcriptome-based strategy for improving the titer of salinomycin and identifying transporter genes for other antibiotics [13].

In their research article, Xue and colleagues found that L-kynurenine (Kyn) is a building block for the biosynthesis of bioactive natural products. They used a genome-mining approach and discovered a biosynthetic gene cluster (BGC) from *Neosartorya pseudofischeri* that can produce pseudofisnins: novel 1-benzazepine-containing compounds. A methyltransferase named PseC was identified as a crucial enzyme in this BGC, catalyzing di-methylation in an amine group [14].

Using a combinatory strategy of genome-mining and OSMAC (one strain of many compounds), Zhang and colleagues identified two new and seven known cyclodipeptide derivatives from *Streptomyces* sp. 26D9-414. The new compound 2 showed similar cytotoxicity to cisplatin in treating various cancer cells [15], which is promising for further investigations.

In the study conducted by Al-Thubaiti and colleagues, metal cefotaxime complexes of Ca(II), Cr(III), Zn(II), Cu(II), and Se(VI) were synthesized and characterized. They then investigated the effects of cefotaxime and cefotaxime metal complexes on oxidative stress and their activity against cancer cells (HepG-2) and bacteria (*Bacillus subtilis* and *Escherichia coli*). The results showed that cefotaxime metal complexes with Zn and Se had high antioxidant activity against HepG-2 cells, and they also presented potent antibacterial activities at low concentrations [16].

El-Megharbel and colleagues, in their research article, investigated the formation of Mg(II), Fe(III), Cu(II), Zn(II), and Se(IV) complexes of the antibiotic ceftriaxone (CFX). Furthermore, they also investigated the effect of CFX and its metal complexes on oxidative stress and tissue injury in rats. It showed that CFX and its metal complexes, especially CFX/Zn, had high antibacterial activity [17].

I hope that readers find the articles in this Special Issue interesting and inspiring for their own research. Joining together, let us bring society another "Golden Age" of antibiotic discovery!

Acknowledgments: We would like to thank all authors and reviewers, without your contribution, it is impossible to form such an impactful Special Issue. We also sincerely thank the staff from the Editorial Office who provided great assistance.

Conflicts of Interest: The authors declare no conflict of interest.

References

1. Tong, Y.; Deng, Z. An aurora of natural products-based drug discovery is coming. *Synth. Syst. Biotechnol.* **2020**, *5*, 92–96. [CrossRef] [PubMed]
2. Tong, Y.; Weber, T.; Lee, S.Y. CRISPR/Cas-based genome engineering in natural product discovery. *Nat. Prod. Rep.* **2019**, *36*, 1262–1280. [CrossRef] [PubMed]
3. Wang, Z.Q.; Koirala, B.; Hernandez, Y.; Zimmerman, M.; Park, S.; Perlin, D.S.; Brady, S.F. A naturally inspired antibiotic to target multidrug-resistant pathogens. *Nature* **2022**, *601*, 606–611. [CrossRef] [PubMed]
4. Wang, Z.Q.; Koirala, B.; Hernandez, Y.; Zimmerman, M.; Brady, S.F. Bioinformatic prospecting and synthesis of a bifunctional lipopeptide antibiotic that evades resistance. *Science* **2022**, *376*, 991–996. [CrossRef] [PubMed]
5. Mitcheltree, M.J.; Pisipati, A.; Syroegin, E.A.; Silvestre, K.J.; Klepacki, D.; Mason, J.D.; Terwilliger, D.W.; Testolin, G.; Pote, A.R.; Wu, K.J.Y.; et al. A synthetic antibiotic class overcoming bacterial multidrug resistance. *Nature* **2021**, *599*, 507–512. [CrossRef] [PubMed]

6. Durand-Reville, T.F.; Miller, A.A.; O'Donnell, J.P.; Wu, X.Y.; Sylvester, M.A.; Guler, S.; Iyer, R.; Shapiro, A.B.; Carter, N.M.; Velez-Vega, C.; et al. Rational design of a new antibiotic class for drug-resistant infections. *Nature* **2021**, *597*, 698–702. [CrossRef] [PubMed]
7. Imai, Y.; Meyer, K.J.; Iinishi, A.; Favre-Godal, Q.; Green, R.; Manuse, S.; Caboni, M.; Mori, M.; Niles, S.; Ghiglieri, M.; et al. A new antibiotic selectively kills Gram-negative pathogens. *Nature* **2019**, *576*, 459–464. [CrossRef] [PubMed]
8. Huang, W.Q.; Wang, Y.X.; Tian, W.S.; Cui, X.X.; Tu, P.F.; Li, J.; Shi, S.P.; Liu, X. Biosynthesis investigations of terpenoid, alkaloid, and flavonoid antimicrobial agents derived from medicinal plants. *Antibiotics* **2022**, *11*, 1380. [CrossRef] [PubMed]
9. Wang, Z.X.; Feng, X.L.; Liu, C.W.; Gao, J.M.; Qi, J.Z. Diverse metabolites and pharmacological effects from the Basidiomycetes *Inonotus hispidus*. *Antibiotics* **2022**, *11*, 1097. [CrossRef] [PubMed]
10. Zhong, H.; Han, L.; Lu, R.Y.; Wang, Y. Antifungal and immunomodulatory ingredients from traditional Chinese medicine. *Antibiotics* **2023**, *12*, 48. [CrossRef] [PubMed]
11. Geng, B.N.; Huang, X.Y.; Wu, Y.L.; He, Q.N.; Yang, S.H. Identification and characterization of genes related to ampicillin antibiotic resistance in *Zymomonas mobilis*. *Antibiotics* **2022**, *11*, 1476. [CrossRef] [PubMed]
12. Yang, X.; Zhang, Y.Y.; Li, S.S.; Ye, L.; Wang, X.J.; Xiang, W.S. SspH, a novel HATPase family regulator, controls antibiotic biosynthesis in *Streptomyces*. *Antibiotics* **2022**, *11*, 538. [CrossRef] [PubMed]
13. Liu, X.; Wu, Y.T.; Zhang, X.J.; Kang, Q.J.; Yan, Y.S.; Bai, L.Q. Comparative transcriptome-based mining of genes involved in the export of polyether antibiotics for titer improvement. *Antibiotics* **2022**, *11*, 600. [CrossRef] [PubMed]
14. Xue, X.X.; Chen, L.; Tang, M.C. Genome mining discovery of a new benzazepine alkaloid pseudofisnin A from the marine fungus *Neosartorya pseudofischeri* F27-1. *Antibiotics* **2022**, *11*, 1444. [CrossRef] [PubMed]
15. Zhang, D.S.; Wang, J.B.; Qiao, Y.J.; Lin, B.X.; Deng, Z.X.; Kong, L.X.; You, D.L. Genome mining and metabolic profiling reveal cytotoxic cyclodipeptides in *Streptomyces hygrospinosus* var. Beijingensis. *Antibiotics* **2022**, *11*, 1463. [CrossRef] [PubMed]
16. Al-Thubaiti, E.H.; El-Megharbel, S.M.; Albogami, B.; Hamza, R.Z. Synthesis, spectroscopic, chemical characterizations, anticancer capacities against HepG-2, antibacterial and antioxidant activities of cefotaxime metal complexes with Ca(II), Cr(III), Zn(II), Cu(II) and Se(IV). *Antibiotics* **2022**, *11*, 967. [CrossRef] [PubMed]
17. El-Megharbel, S.M.; Qahl, S.H.; Alaryani, F.S.; Hamza, R.Z. Synthesis, spectroscopic studies for five new Mg (II), Fe (III), Cu (II), Zn (II) and Se (IV) ceftriaxone antibiotic drug complexes and their possible hepatoprotective and antioxidant capacities. *Antibiotics* **2022**, *11*, 547. [CrossRef] [PubMed]

antibiotics

Review

Biosynthesis Investigations of Terpenoid, Alkaloid, and Flavonoid Antimicrobial Agents Derived from Medicinal Plants

Wenqian Huang [1,†], Yingxia Wang [1,†], Weisheng Tian [1,†], Xiaoxue Cui [1], Pengfei Tu [1,2], Jun Li [1], Shepo Shi [1,*] and Xiao Liu [1,2,*]

1 Modern Research Center for Traditional Chinese Medicine, Beijing University of Chinese Medicine, Beijing 100029, China
2 State Key Laboratory of Natural and Biomimetic Drugs, School of Pharmaceutical Sciences, Peking University, Beijing 100191, China
* Correspondence: shishepo@bucm.edu.cn (S.S.); liuxiao@bucm.edu.cn (X.L.)
† These authors contributed equally to this work.

Abstract: The overuse of antibiotics in the past decades has led to the emergence of a large number of drug-resistant microorganisms. In recent years, the infection rate caused by multidrug-resistant microorganisms has been increasing, which has become one of the most challenging problems in modern medicine. Plant-derived secondary metabolites and their derivatives have been identified to display significant antimicrobial abilities with good tolerance and less adverse side effects, potentially having different action mechanisms with antibiotics of microbial origin. Thus, these phyto-antimicrobials have a good prospect in the treatment of multidrug-resistant microorganisms. Terpenoids, alkaloids, and flavonoids made up the predominant part of the currently reported phytochemicals with antimicrobial activities. Synthetic biology research around these compounds is one of the hotspot fields in recent years, which not only has illuminated the biosynthesis pathways of these phyto-antimicrobials but has also offered new methods for their production. In this review, we discuss the biosynthesis investigations of terpenoid, alkaloid, and flavonoid antimicrobial agents—using artemisinin and oleanolic acid (terpenoids), berberine and colchicine (alkaloids), and baicalin (flavonoids) as examples—around their antimicrobial action mechanisms, biosynthesis pathway elucidation, key enzyme identification, and heterologous production, in order to provide useful hints for plant-derived antimicrobial agent discovery and development.

Keywords: phytochemicals; antimicrobial agents; biosynthetic pathway; secondary metabolites

Citation: Huang, W.; Wang, Y.; Tian, W.; Cui, X.; Tu, P.; Li, J.; Shi, S.; Liu, X. Biosynthesis Investigations of Terpenoid, Alkaloid, and Flavonoid Antimicrobial Agents Derived from Medicinal Plants. *Antibiotics* **2022**, *11*, 1380. https://doi.org/10.3390/antibiotics11101380

Academic Editor: Bernhard Witulski

Received: 10 September 2022
Accepted: 7 October 2022
Published: 9 October 2022

1. Introduction

Infectious diseases caused by pathogenic microorganisms are becoming one of the major causes of death worldwide [1]. Antibiotic refers to a chemical substance, with an organic chemical of natural or synthetic origin, that has the capacity to inhibit the growth of and even kill pathogenic bacteria and other micro-organisms [2]. The discovery and development of antibiotics during the 20th century substantially reduced the threat of infectious diseases [3,4]. However, it has been decades since antibiotics with a completely novel mode of action were last delivered to the clinic. Specifically, in the first decade of the 21st century, with the emergence of resistant strains of several important microbials, including *Pneumococci*, *Enterococci*, *Staphylococci*, *Plasmodium falciparum*, and *Mycobacterium tuberculosis* [5], people were faced with this continuing threat on a wider scale than ever before. Multidrug-resistant pathogens are expected to kill about 300 million people prematurely and will have costed the global economy up to USD 100 trillion by 2050 [6,7]. Several factors are involved in the rise of antibiotic resistance, including the existence of efflux pumps, the lack of sensitive antibiotic targets, induction of a stress response of bacterial cells (SOS reaction and RPOS regulation), the transport of drug-resistant genes through the horizontal gene transfer (HGT) mechanism, and the inactivation of antibiotics by hydrolysis or modification, but the main reason is the overuse and misuse of antibiotics in human and animal health and the lack of development of new antibiotics [8–10]. Thus, there is an urgent need for new compounds with different mechanisms that can limit antibiotic resistance.

In recent years, people have focused more attention on plants for new antibiotics discovery and development since many experiments have proven that compounds from plants have significant and potentially different antimicrobial effects compared with antibiotics of microbial origin [11–13]. The antimicrobial compounds from medicinal plants may inhibit the growth of bacteria, fungi, viruses, and protozoa by different mechanisms than those of presently used antimicrobials and may have a significant clinical value in the treatment of resistant microbial strains [5,14,15]. These agents could perform direct bactericidal action by blocking bacterial DNA synthesis; inhibiting ATPase activity; inhibiting biofilm formation, membrane integrity, or permeability; and resisting the quorum sensing effect [16]. Moreover, many phytochemicals also showed effective antibiotic drug resistance reversal activity, mainly through enzyme modification, plasmid curing, or drug efflux pump [9]. Although some of them do not hold substantial antibacterial potential on their own, their application along with other drugs may considerably augment the antibiotic potential of the drug against which the pathogen was resilient [17]. Moreover, compared with synthetic drugs, plant-derived antibiotics usually have fewer side effects and a lower possibility of drug resistance [10]. On the basis of these advantages, exploring plant-based metabolites is a promising choice to identify new bioactive compounds, which can be used to develop new and effective antimicrobial agents or multidrug-resistant reversal agents.

Secondary plant metabolites are molecules indirectly necessary for the life of plants, which can serve as structural elements or as important tools for plants to adapt to their environment and play a crucial role in many aspects of plant life activities [18]. According to the different chemical structure skeletons and natural origins, plant-derived natural products can be divided into diverse categories. Among them, there is no doubt that terpenoids, alkaloids, and flavonoids compose the dominant part of phytochemicals in the plant kingdom. Moreover, according to a large number of pharmaceutical reports, these three kinds of natural products were also the major source of bioactive antimicrobial candidates' discovery. This review aims to focus on terpenoid, alkaloid, and flavonoid compounds with antimicrobial activities (including antibacterial, antifungal, antiviral, and/or antiparasitic activities) from medicinal plants, mainly discussing their action mechanisms, biosynthesis pathway elucidation, and biosynthesis key enzyme identification, as well as engineering strain construction.

2. Terpenoids

Terpenoids, also known as isoprenoids, are one of the largest natural product families, constituting more than 40,000 primary and secondary metabolites, including monoterpenes (53%), diterpenoids (1%), sesquiterpenes (28%), and others (18%). The basic unit of terpenes is the isoprene unit (C_5H_8), which is a simple hydrocarbon. It is the main precursor and could be post-modified through the cytosolic mevalonate (MVA) pathway or the plastid methyl erythritol phosphate (MEP) pathway. Terpenoids are a major source of bioactive natural products. Especially because of their lipophilic characteristics, terpenoids have become one of the major kinds of antimicrobial agents against various microorganisms [19].

2.1. The Antimicrobial Mechanisms of Terpenoids

There are mainly five mechanisms through which terpenoids exhibit antimicrobial action according to previous reports.

- Cell membrane destruction: Terpenoids mainly use their lipophilicity to destroy the cell membrane of bacteria. Terpenoids can pass through the phospholipid bilayer of bacteria and diffuse inward, showing antibacterial or bactericidal effects [20]. Since the integrity of the cell membrane is very important for the normal physiological activities of bacteria, the damage of terpenoids to the membrane will affect the bacteria's basic physiological activities, and the important substances such as proteins and important enzymes in the cell will be lost, finally achieving the antimicrobial effect [21]. It is reported that 1,8-cineole (Table 1), a monoterpene substance extracted from *Eucalyptus globulus Labill*, showed antibacterial effect against *Acinetobacter baumannii*, *Candida albicans*, a methicillin-resistant *Staphylococcus aureus* (MRSA) strain, and *Escherichia coli* by destroying the cell membrane [22]. In another study, the researchers exposed *Salmonella typhimurium*, *E. coli* O157: H7, *Pseudomonas fluorescence*, *Brochotrix thermophacta*, and *Staphylococcus aureus* cells to cinnamaldehyde (Table 1), carvacrol (Table 1), thymol (Table 1), eugenol (Table 1), and limonene (Table 1), and observed their membrane damage through scanning electron microscopy. These results found that terpenoids can achieve bacteriostatic effects by destroying the membrane structure [23]. The mechanism of action and target sites on microbial cells are graphically illustrated in [20,21].

- Anti-quorum sensing (QS) action: The QS system is an intercellular communication system [20]. It is a communication mode for bacteria to coordinate the interaction between bacteria and other organisms, which is also the main reason for the emergence of antibiotic resistance [19]. The group sensing signal loop of Gram-positive and Gram-negative bacteria has been introduced and illustrated in the literature [24]. Studies have shown that a low concentration of cinnamaldehyde can effectively inhibit the QS effect between bacteria [25]. Low concentrations of carvacrol and thymol can effectively inhibit the self-inducer of bacteria, namely, acyl homoserine lactone (AHL), thus achieving the inhibition of QS [26]. The action mechanism of cinnamaldehyde inhibiting the acyl homoserine lactones and other autoinducers involved in the quorum sensing is illustrated in [27].
- Inhibition of ATP and its enzyme: ATP is the most direct energy source in organisms, and it is also a necessary element for microorganisms to maintain normal operation and work. Terpenoids can act on the cell membrane, resulting in the difference in ATP concentration inside and outside the cell, leading to the disorder of the cell membrane, thus conducting the antibacterial activity [20]. For example, terpenoid eugenol and thymol could target the cell membrane to show fungicidal activity against *C. albicans* by inhibiting H^+-ATPase, which will lead to intracellular acidification and cell death [28]. In another study, the researchers treated the target pathogen with the MIC of carvacrol. The extracellular ATP concentrations of the samples were measured with the help of a luminometer (Biotek). On the basis of absorbance analysis at 260 nm, this study observed that carvacrol disrupted the *E. coli* membrane, while the release of potassium ions and ATP was also detected [29].
- Inhibition of protein synthesis: The physiological activity of bacteria is inseparable from protein synthesis. Terpenoids, as inhibitors of protein synthesis, can achieve an antibacterial effect by blocking any process of the protein synthesis pathway. Some studies have shown that cinnamaldehyde can reduce the in vitro assembly reaction and the binding reaction of FtsZ (filamenting temperature-sensitive mutant Z)-type protein, a prokaryotic homolog of tubulin that regulates cell division. In addition, cinnamaldehyde can inhibit the hydrolysis of GTP and bind to FtsZ, as well as interfere with the formation of z-loop of cell dynamics, thus showing antibacterial activity against bacteria [30]. In the latest research, the researchers used calculations, biochemistry, and in-vivo-based assays to verify that cinnamaldehyde is a potential inhibitor of *S. typhimurium* (stFtsZ), and its inhibition rate of stFtsZ GTPase activity and polymerization is up to 70% [31].
- The synergistic effect: For example, the synergistic antibacterial effect of eugenol with carvacrol and thymol is duc to thc ability of carvacrol and thymol to penetrate the extracellular membrane, thus making it easier for eugenol to enter the cytoplasmic membrane or increasing the number, size, and duration of pores to bind to membrane proteins for better antibacterial activity [32]. The reaction mechanism is shown in the literature [27].

Table 1. Summary of the antimicrobial effects of some plant-derived terpenoids, alkaloids, and flavonoids.

	Compounds	Chemical Structures	Target Microorganisms	Antimicrobial Effects	Reference
Terpenoids	1,8-cineole	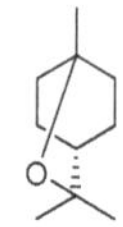	*A. baumannii* *C. albicans* MRSA strain *E. coli*	Cell membrane destruction	[22]
	cinnamaldehyde	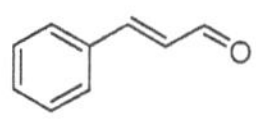	*S. typhimurium* *E. coli* O157: H7 *P. fluorescence* *B. thermophacta* *S. aureus*	1. Cell membrane destruction 2. Anti-quorum sensing action 3. Inhibition of protein synthesis	[23,25,30,31]
	carvacrol	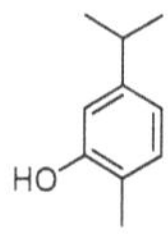	*S. typhimurium* *E. coli* O157: H7 *P. fluorescence* *B. thermophacta* *S. aureus* *P. fluorescens* KM121	1. Cell membrane destruction 2. Anti-quorum sensing action 3. Inhibition of nucleic acid synthesis 4. The synergistic effect 5. Inhibits cell movement and bacterial invasion	[23,26,27,29,32]
	thymol	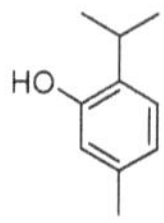	*S. typhimurium* *E. coli* O157: H7 *P. fluorescence* *B. thermophacta* *S. aureus* *P. fluorescens* KM121	1. Cell membrane destruction 2. Anti-quorum sensing action 3. Inhibition of nucleic acid synthesis 4. The synergistic effect	[23,26–28,32]
	eugenol	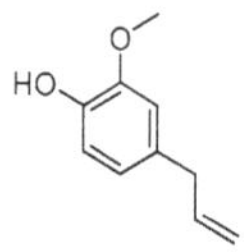	*S. typhimurium* *E. coli* O157: H7 *P. fluorescence* *B. thermophacta* *S. aureus*	1. Cell membrane destruction 2. Inhibition of nucleic acid synthesis 3. The synergistic effect	[23,27,28,32]
	limonene	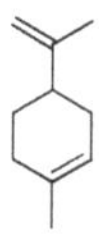	*A. baumannii* *C. albicans* MRSA strain *E. coli*	Cell membrane destruction	[23]
	oleanolic acid	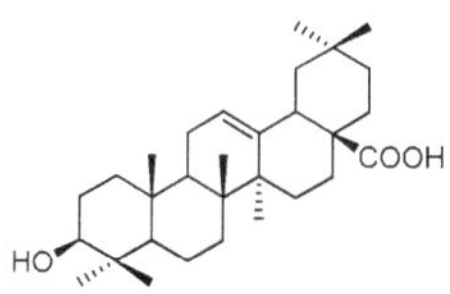	*E. coli* *S. aureus* *Enterococcus faecalis* *P. aeruginosa*	Antibacterial	[33]
Alkaloids	piperine	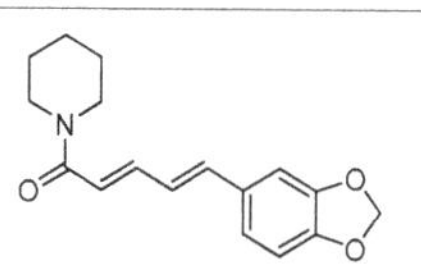	*S. aureus* *B. subtilis* *Salmonella* sp. *E. coli*	Efflux pump inhibition	[34,35]
	reserpine	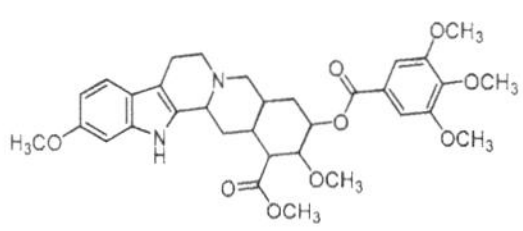	*E. coli*	Efflux pump inhibition	[36]

Table 1. *Cont.*

Compounds	Chemical Structures	Target Microorganisms	Antimicrobial Effects	Reference
berberine	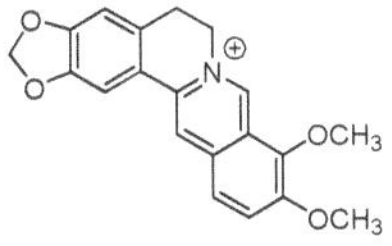	*E. coli* *Micrococcus luteus* *P. aeruginosa* *Prevotella intermedia* *Fusobacterium nucleatum* MRSA strain	1. Efflux pump inhibition 2. DNA-intercalating 3. Growth inhibition	[37–39]
L-ephedrine	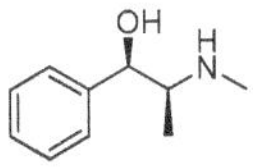	Influenza A virus	DNA-intercalating	[40]
D-pseudoephedrine	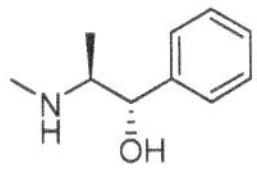	Influenza A virus	DNA-intercalating	[40]
L-methylephedrine	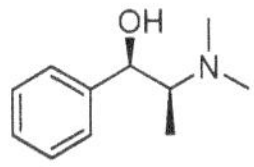	Influenza A virus	DNA-intercalating	[40]
chelerythrine	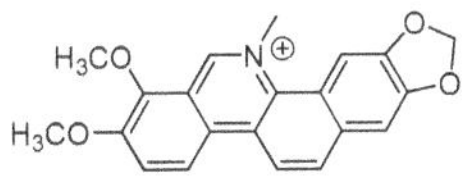	*S. aureus* MRSA strain ESBLs-SA	1. Nucleic acid synthesis and repair inhibition 2. Growth inhibition	[41]
8-hydroxy quinoline	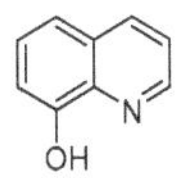	*S. aureus* *H. influenza* *S. pneumoniae*	Permeability change of membrane	[42,43]
michellamine b	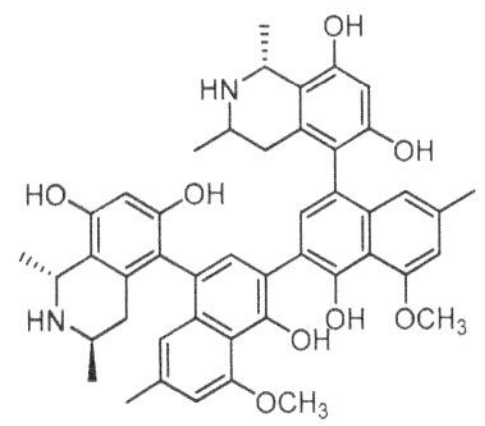	HIV	Protein activity inhibition	[44]
sanguinarine	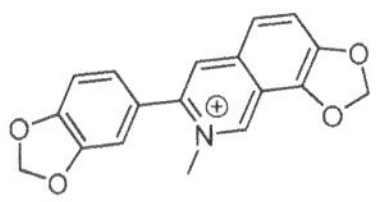	*K. pneumoniae* MRSA strain *P. aeruginosa* *Streptococcus pyogenes*	1. DNA-intercalating 2. Growth inhibition	[45,46]
roemerine	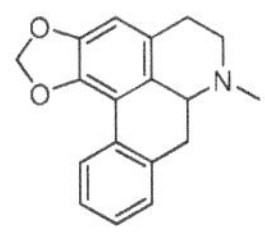	*S. aureus* *B. subtilis*	1. Efflux pump inhibition 2. Permeability change of membrane	[47,48]
dihydrochelerythrine	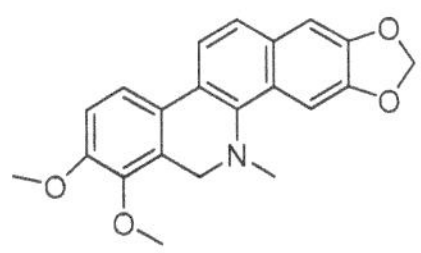	*S. aureus* MRSA strain	Growth inhibition	[49]
evodiamine	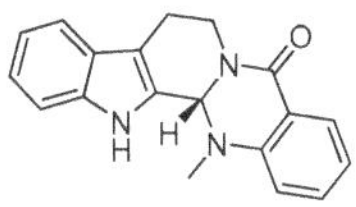	*M. tubercolosis*	Peptidoglycan biosynthesis inhibitor	[50,51]

Table 1. *Cont.*

	Compounds	Chemical Structures	Target Microorganisms	Antimicrobial Effects	Reference
Flavonoids	hesperidin	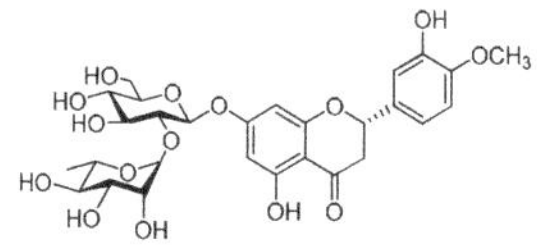	*S. aureus* *L. monocytogenes*	Inhibit bacterial growth by modulating the expression of virulence factors	[52] [53]
	oroxylin a	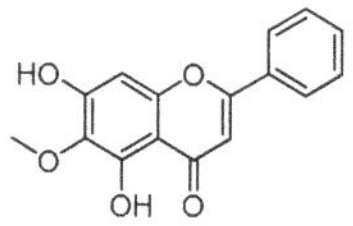	*B. subtilis* *S. aureus*	/	[54]
	apigenin	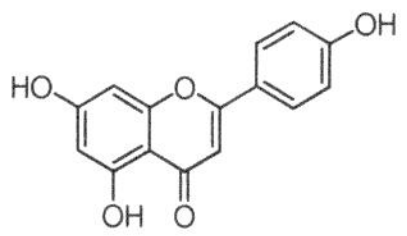	*S. aureus* *B. subtilis* *E. coli* *P. aeruginosa.*	1. Inhibits peptidoglycan synthesis 2. Increases cell membrane permeability	[55]
	morin	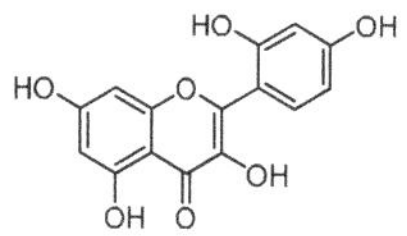	*E. coli*	Inhibition of ATP synthetase	[56]
	silymarin	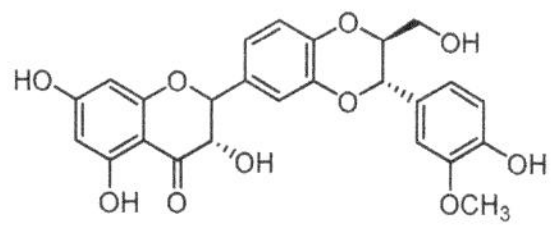	*E. coli*	Inhibition of ATP synthetase	[56]
	epigallocatechin gallate	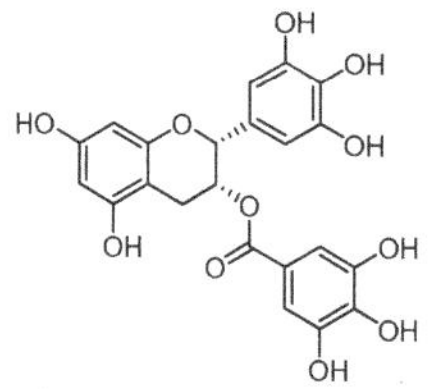	*S. maltophilia*	Inhibits dihydrofolate reductase	[57]
	quercetin	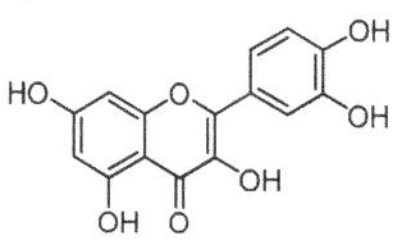	*P. aeruginosa*	1. Inhibits viral polymerase and viral nucleic acid 2. Inhibits the formation of its biofilm	[58]
	galangin	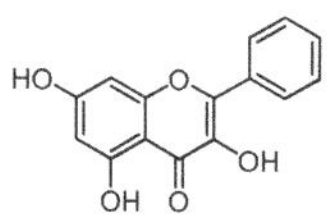	*S. aureus*	1. Destroys the plasma membrane 2. Weakens the cell wall	[59]
	catechin	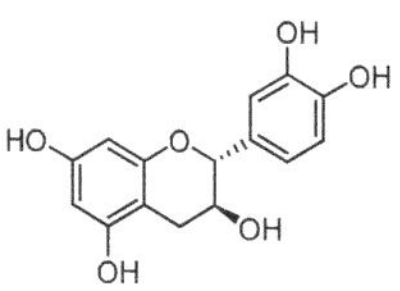	*B. subtilis* *E. coli*	Inhibits the bacterial DNA gyrase	[60] [61]
	baicalin	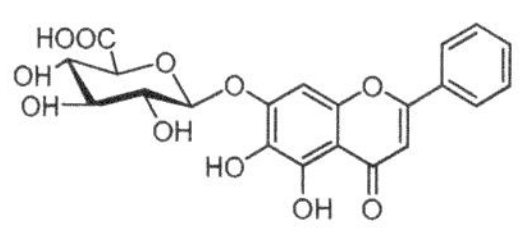	*Salmonella* spp. *Staphylococcus* spp.	Inhibits biofilm formation	[62] [63]

Table 1. *Cont.*

Compounds	Chemical Structures	Target Microorganisms	Antimicrobial Effects	Reference
phloretin	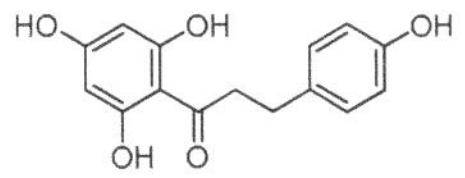	*C. albicans*	1. Inhibits the pathogenicity 2. Inhibits virulence factors	[64]
silybin	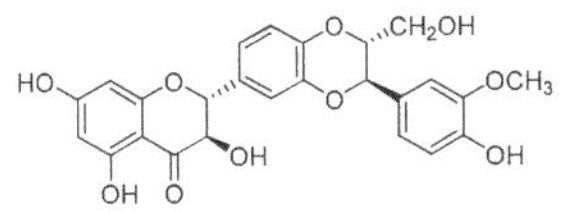	MRSA strain	Inhibits the efflux pump	[65]

2.2. Biosynthesis of Terpenoid Precursors

There are two important precursors for terpenoid biosynthesis, dimethylallyl pyrophosphates (DMAPP) and isopentenyl diphosphate (IPP), which can both be produced via the MVA or MEP pathways (Figure 1), depending on the organism. The MVA pathway is based on the formation of IPP and DMAPP from acetyl coenzyme A (CoA) through the precursor substance MVA, followed by further condensation of IPP and DMAPP to form sesquiterpenes, triterpenes, and sterols by the action of polyisoprene pyrophosphate synthase. The MEP pathway, on the other hand, is based on pyruvate and glyceraldehyde-3-phosphate in the presence of 1-deoxyxylulose-5-phosphate synthase (DXS) to form DXP. Then, under the catalysis of 1-deoxyxylulose-5-phosphate reductor isomerase (DXR), MEP was formed, followed by further phosphorylation and cyclization to produce IPP, which will be used in the downstream biosynthesis of monoterpenes, diterpenes, and other terpenoids. In plants, both pathways can occur, with the MVA pathway acting in the cytoplasm and the MEP pathway acting in the plastid. In bacteria, terpenoids are generally produced via the MEP pathway, whereas terpenoids are mostly synthesized via the MVA pathway in fungi. Although there are slight differences in the processes of these two pathways, the end products are both DMAPP and IPP [66]. In general, the MEP pathway provides C_5-pentenyl diphosphate for the synthesis of C_{10} monoterpenes, C_{20} diterpenes, and C_{40} tetraterpenes, while the MVA pathway provides the same generic precursors for the synthesis of C_{15} sesquiterpenes, $C_{27–29}$ sterols, C_{30} triterpenes, and their saponin derivatives [67].

2.3. Discovery, Biosynthesis Investigations, and Engineering Strain Construction of the Representative Terpenoid Antimicrobial Agent—Artemisinin

2.3.1. Discovery and Predicted Action Mechanism of Artemisinin

So far, there have been several reports about terpenoid compounds that displayed desired antimicrobial activities [68]. Among them, the most representative one is undoubtedly artemisinin (Figure 2). Artemisinin (Qinghaosu) is a sesquiterpene endoperoxide isolated from the leaves of the plant *Artemisia annua*, which has a long history of use in traditional Chinese medicine. Malaria, caused by *Plasmodium falciparum*, has been a life-threatening disease for thousands of years [69]. Nowadays, 40% of the world's population is at risk of malaria infection, and artemisinin is designated as the first-line antimalarial drug by the World Health Organization (WHO). Since the discovery of the antimalarial activities of artemisinin by Chinese scientists in 1971, it has saved millions of lives and represents one of the significant contributions of China to global health. On account of this, the 2015 Nobel Prize for Medicine was awarded to Professor Youyou Tu for her contributions to the discovery and recognition of artemisinin [70].

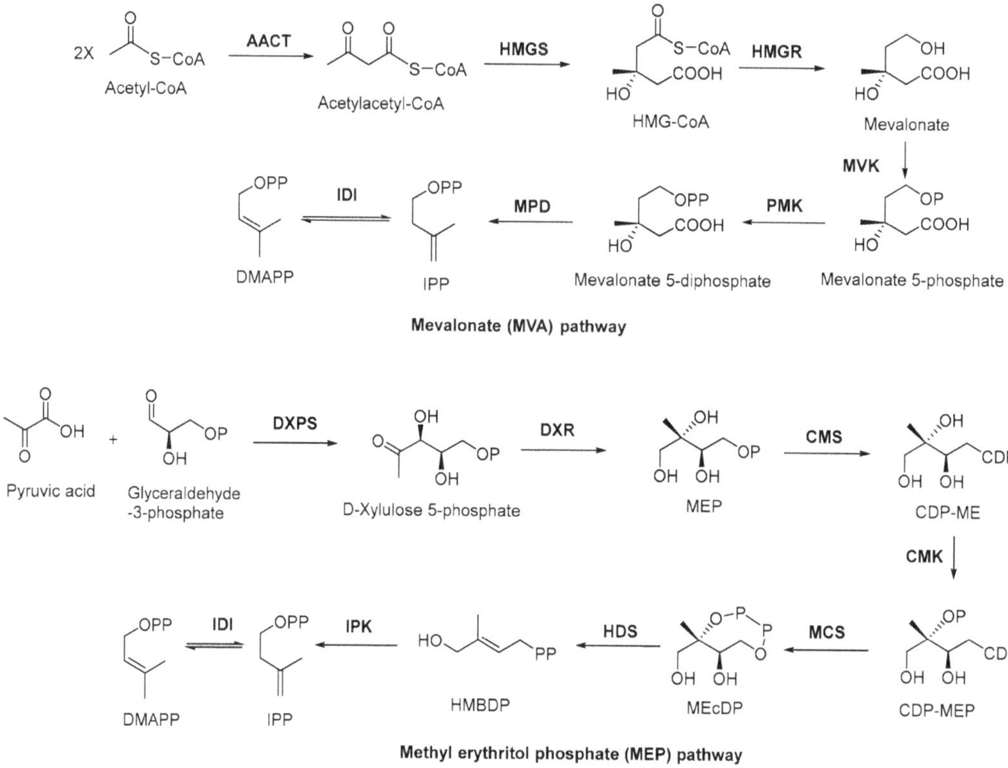

Figure 1. MVA and MEP pathways involved in terpenoid biosynthesis. AACT, acetoacetyl coenzyme A thiolase; HMGS, 3-hydroxy-3-methyl glutaryl coenzyme A synthetase; HMGR, 3-hydroxy-3-methyl glutaryl coenzyme A reductase; MVK, mevalonate kinase; PMK, phosphomevalonate kinase; MPD, mevalonate-5-pyrophosphate decarboxylase; IDI, isopentenyl diphosphate isomerase; DXPS, 1-deoxy-xylose-5-phosphate synthase; DXR, 1-deoxy-xylose-5-phosphate racemic enzyme; CMS, 4-diphoxphocyt-idyl-2-*C*-methyl-2-(E)-butenyl-4-diphosphate synthase; CDP, cytidine-4-diphosphate; CDP-ME, cytidine-4-diphosphate-2-*C*-methylerythritol; CMK, 4-diphoxphocyt-idyl-2-*C*-methyl-D-erythritol kinase; CDP-MEP, cytidine-4-diphosphate-2-*C*-methyl-D-erythritol-2-phosphate; MCS, 2-*C*-methyl-D-erythritol-2,4-cyclodiphosphats synthase; MEcDP, 2-*C*-methyl-D-erythritol-2,4-cyclophosphoric acid; HDS, 1-hydroxy-2-methyl-2-(E)-butenyl-4-diphosphate synthase; HMBDP, 1-hydroxy-2-methyl-2-(E)-butenyl-4-diphosphate; IPK, isopentenyl monophosphate kinase.

Although widespread investigations have been carried out, the mechanism of action of artemisinin is still incompletely clarified. It has been widely accepted that the antimalarial activity of artemisinin is largely dependent on the unusual endoperoxide since derivatives lacking the endoperoxide bridge are discovered to be devoid of antimalarial activity, and the activity could be enhanced by high oxygen tension and by the addition of other free-radical-generating compounds, while some radical scavengers could block the antimalarial activity [71]. Considerable evidence has proven that the killing parasite's ability of artemisinin-based combination therapies is mediated by free radicals, which are produced from the endoperoxide bridge [72]. The degradation of the endoperoxide bridge in a heme-dependent process could form carbon-centered radicals, which then alkylate multiple targets including heme and proteins at the pathogenic *Plasmodium* blood stage and lead to the conversion of heme to hemozoin and finally lead to the death of the parasite [73].

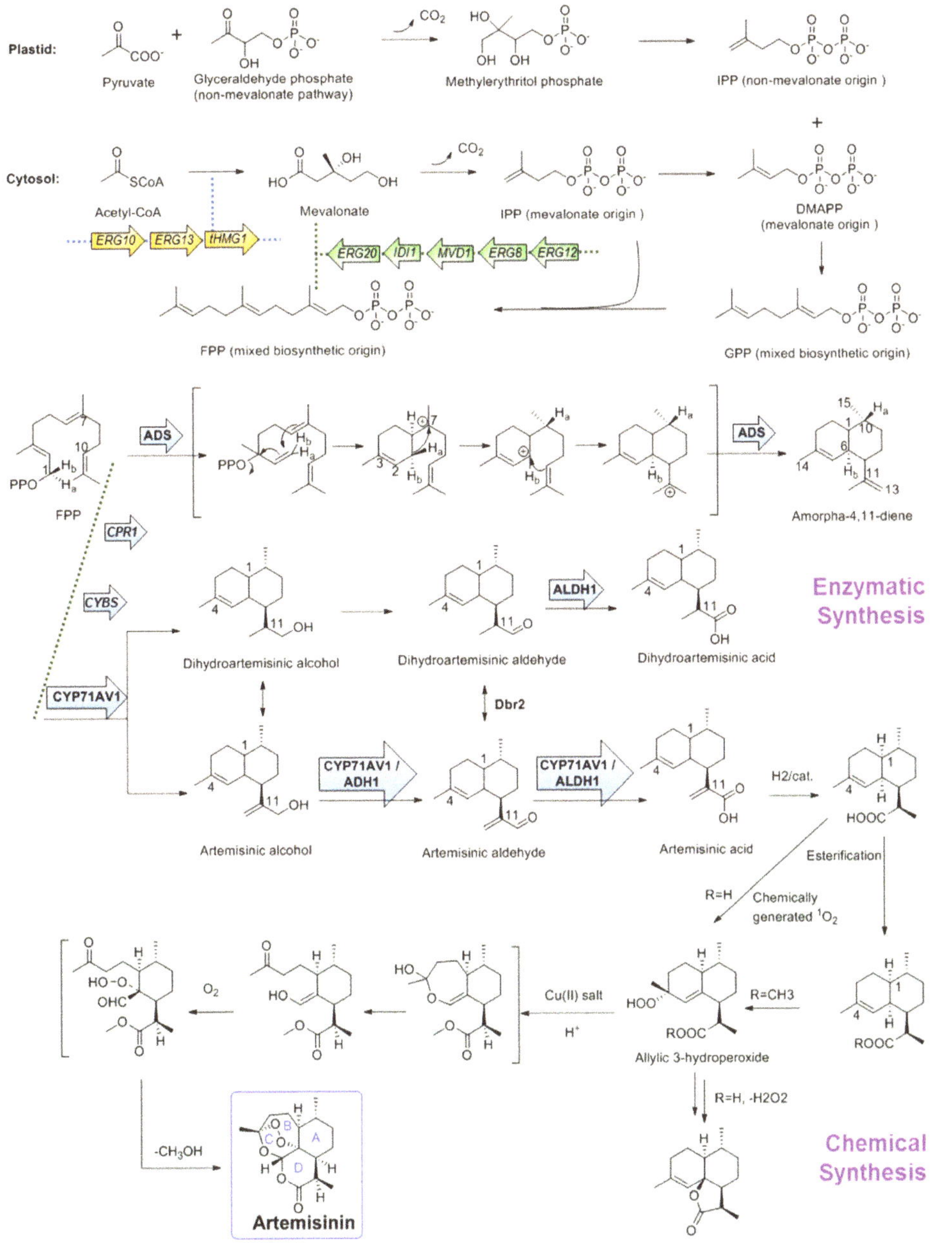

Figure 2. Chemo-enzymatic synthesis of artemisinin. Yellow region shows the biosynthesis pathways for artemisinic acid production. Green region shows the chemical conversion route of artemisinic acid to artemisinin. tHMGR, 3-hydroxy-3-methylglutaryl-coenzyme A reductase; ADS, amorphadiene synthase; CYP71AV1, amorpha-4; 11-diene monooxygenase; ADH1, artemisinic alcohol dehydrogenase; ALDH1, artemisinic aldehyde dehydrogenase 1; Dbr2, artemisinic aldehyde Δ11 (13) reductase; CPR1, cognate reductase of CYP71AV1. Arrows with frames showed the gene elements manipulated by Keasling's team for artemisinic acid production engineering strain construction.

2.3.2. Key Enzymes Involved in The Biosynthesis Pathway of Artemisinin

The large demand for artemisinin-based combination therapies has caused artemisinin to fall in short supply. To provide more alternative sources, the biosynthesis pathway of artemisinin has been investigated for many years and remarkable achievements have been obtained. Like other regular sesquiterpenes, artemisinin's biosynthesis precursor is farnesyl pyrophosphate (FPP), which is formed by the condensation of three IPP molecules by either the MVA or plastid non-MVA pathway, respectively [74]. To verify the origin of the precursors, a plant of *A. annua* was grown in an atmosphere containing labeled $^{13}CO_2$ for 100 min. Following a chase period of 10 days, artemisinin was isolated and analyzed by ^{13}C NMR spectroscopy. The result shows that the precursor IPP can be provided by both the MVA pathway and the non-MVA pathway. As shown in Figure 2, DMAPP was initially provided by MVA origin and then transferred to the plastid, where an IPP unit of non-MVA origin is used for elongation to form geranyl diphosphate (GPP). In the subsequent step, GPP is exported to the cytosolic compartment and converted into FPP using IPP from the MVA pathway [75] (Figure 2). After FPP is formed, the first committed step of artemisinin biosynthesis is the conversion of FPP to amorphadiene by the terpene synthase enzyme amorphadiene synthase (ADS). To explore the catalysis mechanism of ADS, deuterium-labeled FPP at H-1 position was used as the substrate to trace the H-1 hydrogen migration of FPP during cyclization. ^{1}H NMR results of amorphadiene showed that one of the hydrogen Ha-1 of FPP migrated to H-10 of amorphadiene, while the other hydrogen Hb-1 remained at its position to label amorphadiene H-6. These observations indicated that ADS may act through an initial formation of a bisabolyl cation intermediate through 1,6-ring closure and one 1,3-hydride shift. Bisabolyl carbocation intermediate would then undergo hydride shift through one direct suprafacial 1,3-shift of axial Ha-1 to C-7 (Figure 2), resulting in the correct *cis*-decalin configuration at C-1 and C-6 of amorphadiene [76–79]. Following the formation of amorpha-4,11-diene, a cytochrome P450, CYP71AV1, was cloned from *A. annua* and characterized by expression in *Saccharomyces cerevisiae*. CYP71AV1 could catalyze the multiple oxidation steps of amorpha-4,11-diene to produce artemisinic alcohol and artemisinic aldehyde, and finally yield artemisinic acid [80]. In addition, two genes encoding putative artemisinic alcohol dehydrogenase (ADH1) and artemisinic aldehyde dehydrogenase 1 (ALDH1) were characterized from *A. annua* glandular trichomes [81]. ADH1 is a NAD-dependent alcohol dehydrogenase of the medium-chain dehydrogenase/reductase superfamily, with specificity towards artemisinic alcohol. ALDH1 could effectively convert artemisinic aldehyde to artemisinic acid [82].

It is obvious that the Δ11 (13) double bond in amorpha-4,11-diene is reduced during the biosynthesis of artemisinin, which is assumed to occur in artemisinic aldehyde. A corresponding gene, *Dbr2*, was cloned and characterized from *A. annua* [83]. It could specifically reduce artemisinic aldehyde to produce dihydroartemisinic aldehyde, which could be then converted to dihydroartemisinic acid by ALDH1. Further study showed that ALDH1 could also catalyze the oxidation of artemisinic aldehyde as CYP71AV1 did [49]. Conversely, CYP71AV1 cannot catalyze the oxidation of dihydroartemisinic aldehyde. Meanwhile, experimental results showed that there was no direct enzymatic conversion of artemisinic acid into dihydroartemisinic acid. Therefore, there should be two branches that exist during artemisinin biosynthesis [84]. It is well accepted that the primary route is through dihydroartemisinic acid, and the route through artemisinic acid is a side pathway [85–87]. From dihydroartemisinic acid, biosynthesis of artemisinin still requires a photooxidative formation of the endoperoxide ring. However, the details of this process, such as the potential involvement of additional enzyme activities, are currently unclear. In 2004, there was a report that, through using the cell-free extracts of *A. annua*, realized the bioconversion of artemisinic acid to artemisinin, but the activity was not observed when using artemisinic acid as the only substrate [88]. Thus, the enzyme in charge of this reaction is still a question. One possibility is that artemisinic acid could be converted into several other compounds such as arteannuin B non-enzymatically, which is later transformed into artemisinin [89]. Another possibility is that dihydroartemisinic acid could undergo rapid plant pigment photosensitized oxidation, followed by subsequent spontaneous oxidation to form artemisinin [90].

2.3.3. Microbial Production of Artemisinic Acid

On the basis of the biosynthesis pathway elucidation, increasingly more attention on artemisinin is now shifting to its microbial production. Particularly represented by Dr. Jay D. Keasling, his team has made great achievements in this field [91]. They combined the biological synthesis of the earlier steps to produce the precursor artemisinic acid and the organic synthetic steps of artemisinic acid to produce artemisinin together and realized the industrial production of semi-synthetic artemisinin for commerce needs. They first constructed the biosynthesis pathway of amorphadiene in *E. coli*. Compared with the expression of DXP pathway genes, a dramatic increase in isoprenoid precursor production was observed when the *S. cerevisiae* MVA pathway was heterologously expressed in *E. coli*. Thus, two plasmids were correspondingly designed. One encoded the MevT operon (known as the 'top pathway'), which comprises three genes (*atoB*, *ERG13*, and *tHMG1*) that are needed for the conversion of acetyl-CoA to MVA. The second plasmid encoded the MevB operon (known as the 'bottom pathway') comprising five genes (*idi*, *ispA*, *MVD1*, *ERG8*, and *ERG12*) for the conversion of MVA to FPP. These two plasmids were subsequently expressed in *E. coli* with the codon-optimized amorphadiene synthase (ADS) gene together. Combined with the optimization of the fermentation conditions, the production of amorphadiene could reach 0.5 g per liter in *E. coli* [92–94]. Following this is the next stage: after the identification of CYP71AV1, this project meets a quandary that although the amorphadiene was produced with a higher yield in *E. coli* than in *S. cerevisiae*, *E. coli* is unsuitable for the expression of the P450 enzyme CYP71AV1, which is crucial for the following steps. Thus, in this stage, Keasling's team switched the expression system of artemisinin to *S. cerevisiae*. Following this, a series of gene manipulations were performed, including: (1) The *S. cerevisiae* strain was engineered to overexpress the MVA pathway, and all genes were integrated into the genome; (2) *ADS* and *CYP71AV1* genes were constructed as plasmid borne; (3) overexpression of a 3-hydroxy-3-methylglutaryl-CoA reductase (tHMGR) occurred to improve the production of amorphadiene; (4) downregulation of *ERG9* occurred, which encodes squalene synthase, catalyzing the first step in the sterol biosynthetic pathway to inhibit the flux from FPP to sterol; (5) a methionine repressible promoter P_{MET3} was used to increase amorphadiene production; (6) the *ADS* gene was expressed under the control of the *GAL1* promoter; (7) the *CYP71AV1* gene was expressed along with its cognate reductase (CPR1); (8) yeast strain CEN.PK2 was chosen as the host, which is capable of sporulating sufficiently; (9) every enzyme of the MVA pathway including *ERG20* (the final step for the production of FPP) was overexpressed in CEN.PK2 in an effort to increase the production of amorphadiene; (10) the *GAL80* gene was deleted to ensure constitutive expression of the overexpressed MVA pathway enzymes and the *A. annua*-derived genes; (11) the much cheaper glucose was used as the carbon source instead of galactose; (12) another two enzymes, aldehyde dehydrogenase (ALDH1) and artemisinic alcohol dehydrogenase (ADH1), were combinedly expressed with CYP71AV1, which resulted in the highest production yield of artemisinic acid. With all the above manipulations coupled with the development of the fermentation process, the production of artemisinic acid in the engineering yeast strain was finally as high as 25 g per liter [81,95,96].

2.3.4. Chemical Conversion to Produce Artemisinin

The final stage for artemisinin chemo-enzymatic synthesis is the chemical conversion of artemisinic acid to artemisinin (Figure 2). The chemical process involves a four-step conversion that begins with the reduction of artemisinic acid to dihydroartemisinic acid. Then, the esterification of the carboxylic acid moiety will be performed to block the subsequent formation of side products. The third step is an 'ene-type' reaction of the C4–C5 double bond with singlet oxygen ($^1O_2$) to produce an allylic 3-hydroperoxide. Moreover, in the final step, the allylic hydroperoxide undergoes an acid-catalyzed hock fragmentation and rearrangement to afford a ring-opened keto-aldehyde enol. Trapping of this enol with $^3O_2$ produces a vicinal hydroperoxide aldehyde, followed by a cascade reaction of acid-catalyzed cyclization that could form an endoperoxide bridge to provide artemisinin at last [81]. Finally, through the metabolic engineering of the earlier steps using multiple gene manipulations and following synthetic organic chemistry, the anti-malaria drug artemisinin production system was successfully established and effectively used for industrial production by Sanofi company as the worldwide supplement [91]. Artemisinin is by far the most successful and representative example of the perfect combination of biosynthetic pathway research and industrial production.

2.4. Biosynthesis Pathway Investigation of the Terpenoid Antimicrobial Agent—Oleanolic Acid

Oleanolic acid (Table 1) is a pentacyclic triterpenoid originating from a number of medicinal plants. It has desired antimicrobial activity against various bacterial pathogens and viruses [33,97–100]. Furthermore, the study on this antimicrobial agent is of importance because as a natural source product, there has been no resistance case toward oleanolic acid found yet [101]. The biosynthesis pathway of oleanolic acid has been relatively clear [102].

In plant cells, acetyl CoA generates DMAPP and IPP through the MVA pathway in the cytosol. IPP and DMAPP are isomerized into FPP under the action of farnesyl pyrophosphate synthase (FPS), and FPP is then converted into squalene under the action of squalene synthase (SQS). Squalene cyclooxygenase (SQE) then oxidizes squalene into a precursor molecule for primary sterol metabolism, 2,3-oxsqualene [103]. From this step, the different cyclizations of 2,3-oxidized squalene become a branching point between primary sterol and secondary triterpene metabolism. For the biosynthesis of plant sterols, the cyclization of 2,3-oxysqualene to the tetracyclic plant sterol precursor cycloartenol is mainly catalyzed by cycloartenol synthase (CAS) [104]. Conversely, the oleanolic acid biosynthetic pathway of our interest, 2,3-oxysqualene, is cyclized by β-amyrin synthase (BAS), which was first cloned from the medicinal plant ginseng and subsequently from a variety of other plants [104,105]. This pentacyclic carbon skeleton is assumed to be formed from (3*S*)-2,3-oxidosqualene folded in pre-chair–chair–chair conformation [106]. Opening of the epoxide ring followed by cation–π cyclization initially produces a tetracyclic dammarenyl cation. Following ring expansion and the formation of fifth ring, the lupenyl cation is formed [105]. Another ring expansion followed by a series of stereospecific 1,2-hydride shifts and the final abstraction of 12α proton produces β-amyrin [107] (Figure 3). The C-28 position of β-amyrin is then oxidized in three consecutive steps by a single cytochrome P450 enzyme, CYP716A12, to produce oleanolic acid. The key enzyme for this step—CYP716A12—was first identified in *Medicago truncatula*, and the study found that erythrodiol, oleanolic aldehyde, and oleanolic acid production were detected in the reaction solution catalyzed by this enzyme [108,109]. Thus, it is suggested that CYP716A12 is a C-28 oxidase of β-amyrin, catalyzing three sequential oxidation reactions of oleanane main chain C-28 rather than a one-step generation. The oleanolic acid biosynthetic pathway is shown in Figure 3.

With the development of synthetic biology, some conventional biosynthetic pathways were interfered with using genetic engineering to improve the target compound's production. For example, limonene, a cyclic monoterpene of plant origin, is antimicrobially sensitive to *Listeria monocytogenes* and can damage its cell integrity and wall structure [110]. The most classical biosynthetic pathway of limonene is the condensation of IPP and DMAPP to form GPP by the action of geranyl pyrophosphate synthase, and limonene synthase (LS) uses GPP as a substrate to synthesize limonene. However, GPP can also subsequently condense with a molecule of IPP to form FPP, and studies have shown that the synthesis of excessive FPP hinders the efficient synthesis of monoterpenes. According to a recent report, researchers have developed an FPPS mutant (F96W, N127W; $FPPS^{F96W,\ N127W}$) that can selectively produce GPP without further extension to FPP. In the yeast strain with high isoprene production, $fpps^{F96W,\ N127W}$ genes were combined with nine plant LS genes, and the *N*-terminal sequence of plasma-membrane-targeted transport peptide (TLS) was truncated. The best effect of 15.5 mg L^{-1} limonene on *Citrus lemon* tls1 (cltls1) was achieved.

Moreover, an orthogonal engineering pathway was constructed. In this pathway, limonene could be produced through the condensation of IPP and DMAPP by neryl pyrophosphate (NPP) synthase to form NPP, and limonene synthase can also use NPP as a substrate to synthesize limonene. The expression of *Solanum lycopersicum* nerolidyl diphosphate synthase (SlNDPS1) and *Citrus limon* tLS2 (CltLS2) genes in the same yeast strain made the limonene yield higher than that of traditional methods (28.9 mg L^{-1}). Under the action of glucose-induced promoter HXT1, the production of limonene can be increased to more than 900 mg L^{-1} by extensive pathway engineering using the FPPS competitive gene [111].

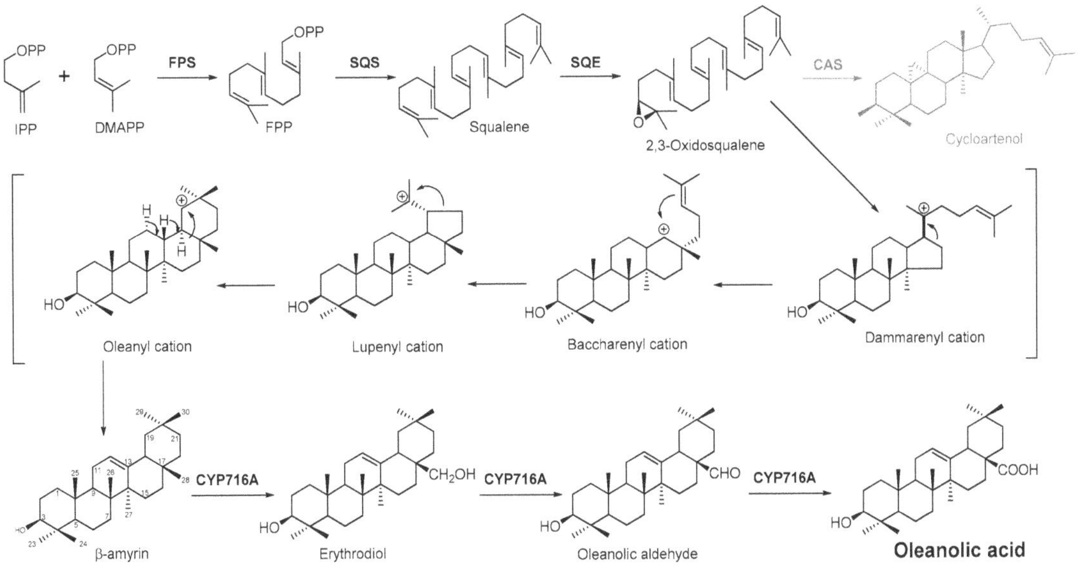

Figure 3. Biosynthesis pathway of terpenoid antimicrobial agent oleanolic acid. FPS, farnesyl diphosphate synthase; SQS, squalene synthase; SQE, squalene epoxidase; CAS, cycloartenol synthase.

3. Alkaloids

Alkaloids are a class of structurally diverse nitrogen-containing organic compounds, including more than 20,000 different molecules whose basic nitrogen atom can occur in the form of primary amine (RNH_2), a secondary amine (R_2NH), or a tertiary amine (R_3N) [112]. From the perspective of chemical structure or natural origin, alkaloids can be divided into two broad divisions. The first division contains the non-heterocyclic or atypical alkaloids, also known as protoalkaloids or biological amines, containing nitrogen in the side chain. The second division includes the heterocyclic or typical alkaloids (true alkaloids), containing nitrogen in the heterocycle. Because of their structural complexity, the second division can be further subdivided into 14 subgroups on the basis of the ring structure, as shown in Figure 4 [113].

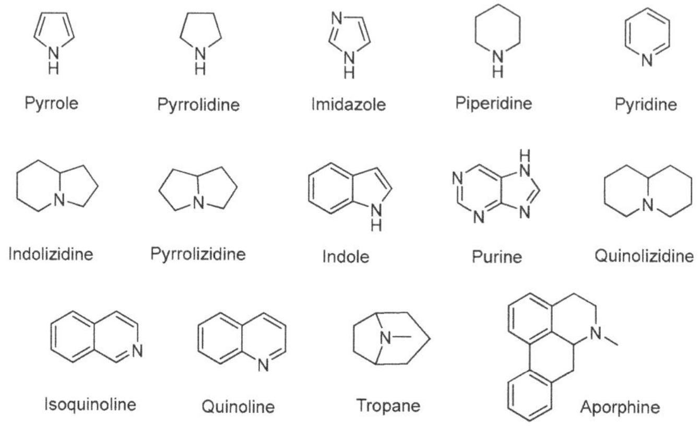

Figure 4. The 14 subgroups of alkaloids.

3.1. Plant-Originated Alkaloids with Antimicrobial Bioactivities

Because alkaloids have a proton-accepting nitrogen atom, and one or even more proton-donating amine hydrogen atoms in addition to functional groups, they can easily form hydrogen bonds with proteins, enzymes, and receptors [113]. As a result, alkaloids show a variety of pharmacological activities [8,114–116]. Nowadays, there are numerous reports on the antimicrobial activity of plant-derived alkaloids. They could inhibit the growth of fungi, bacteria, viruses, and protozoa through a variety of mechanisms, and may have important clinical value in the treatment of resistant microbial strains [117]. Most alkaloids act as efflux pump inhibitors (EPIs) to exert antimicrobial effects—for example, isoquinoline, protoberberine, quinoline, indole, monoterpene indole, and steroidal alkaloids are reported to be used as competitive inhibitors of efflux pumps in bacteria and fungi [118]. Piperine (Table 1)—a piperidine-type alkaloid—has strong antimicrobial activity against both Gram-positive and -negative bacteria [34], acting as an EPIs in *S. aureus* when combined with ciprofloxacin [35]. Reserpine (Table 1)—an indole alkaloid—is known to be a competitive inhibitor of both primary and secondary active transporter systems. In particular, regarding this latter function, reserpine acts mainly on resistance nodulation division (RND) and the major facilitator superfamily (MFS) [119,120]. In addition, reserpine could reverse Bmr-mediated multidrug resistance by inhibiting drug transport [36,121]. Berberine (Table 1) is a kind of isoquinoline alkaloid. It displays a synergistic effect with the carbapenem antibiotic to re-sensitize imipenem-resistant *Pseudomonas aeruginosa* by inhibiting the MexXY-OprM efflux pump system [37–39]. Berberine has shown antibacterial activity against selected oral pathogens and is more effective than saline as an endodontic irrigant against selected endodontic pathogens [122]. Some alkaloids play an antimicrobial role by inhibiting nucleic acid synthesis and repair—for instance, berberine is also an excellent DNA intercalator that accumulates under the drive of cell membrane potential [123]. L-Ephedrine (Table 1), D-pseudoephedrine (Table 1), and L-methylephedrine (Table 1) have antiviral effects on influenza A virus (IAV) in vitro by inhibiting viral replication and altering inflammatory response [40]. Chelerythrine (Table 1), an isoquinoline alkaloid, displays strong antibacterial activity against *S. aureus*, MRSA, and extended-spectrum β-lactamase *S. aureus* (ESBLs-SA) through inhibition of cellular division and nucleic acid synthesis [41]. Some alkaloids play an antimicrobial role by changing the permeability of the membrane. 8-Hydroxyquinoline (Table 1) is one of the oldest antibacterial agents [124]. Its high lipophilicity allows it to penetrate bacterial cell membranes to reach its target site of action [42], displaying activity against *S. aureus*, *Haemophilus influenzae*, and *Streptococcus pneumoniae* [43]. Some alkaloids conduct antimicrobial effects by inhibiting the activity of enzymes. For example, michellamine B (Table 1) obtained from the tropical plant *Ancistrocladus korupensis* showed anti-HIV activity by inhibiting the enzymatic activities of reverse transcriptases from both HIV types 1-2 as well as by inhibiting human DNA polymerases α and β [44]. Some alkaloids perform antimicrobial effects by inhibiting the growth of bacteria, such as benzophenanthridine alkaloid sanguinarine (Table 1). It could interfere with Z-ring assembly through inhibiting filamenting temperature-sensitive mutant Z (FtsZ) binding, thus preventing cytokinesis in both Gram-positive and Gram-negative bacteria [45]. Sanguinarine can also affect the binding of FtsZ protofilaments to have a bacteriostatic effect [46].

3.2. Biosynthesis Investigation of the Representative Antimicrobial Alkaloid Compound—Berberine

Alkaloids are biosynthetically derived from amino acids such as phenylalanine (Phe), tyrosine (Tyr), tryptophan, ornithine, and lysine. Building blocks from the acetate, shikimate, or deoxyxylulose phosphate pathways are also frequently incorporated into alkaloid structures. Nowadays, the synthetic pathways of multiple kinds of antimicrobial alkaloids have been analyzed and confirmed, such as berberine, colchicine, benzylisoquinoline alkaloids (BIAs), and tropane alkaloids (TAs).

Berberine is the main representative quaternary ammonium salt of protoberberines produced from *Berberis* spp. with various antimicrobial activities, especially against Gram-negative bacteria [125–128]. The generally accepted biosynthesis precursor of berberine is L-Tyr [129]. Biosynthesis from L-Tyr to berberine has 13 steps involving different enzymatic reactions, and all the enzymes involved in this pathway have been biochemically characterized, as shown in Figure 5 [130]. It begins with the formation of the first committed intermediate (*S*)-norcoclaurine, which is formed through the condensation of two Tyr derivatives, dopamine and 4-hydroxyphenylacetaldehyde (4-HPAA). Dopamine and 4-HPAA are synthesized by Tyr decarboxylase (TYDC) and Tyr/tyramine 3-hydroxylase (3OHase), or L-Tyr aminotransferase (TyrAT) and 4-hydroxyphenylpuruvate decarboxylase (4HPPDC), respectively, and they were further condensed by the formation of C-C bonds under the action of (*S*)-norcoclaurine synthase (NCS) to generate the basic 1-benzylisoquinoline core (*S*)-norcoclaurine [131–133]. (*S*)-Norcoclaurine continues to be methylated and oxidized to form (*S*)-reticuline, which is a key molecule to derive a series of alkaloids, through four steps under the action of three methyltransferases (*S*-adenosyl-L-methionine (SAM): (*S*)-norcoclaurine 6-*O*-methyltransferase (6OMT) [134,135], SAM: (*S*)-coclaurine-*N*-methyltransferase (CNMT) [136,137], SAM: 3′-hydroxy-*N*-methylcoclaurine 4′-*O*-methyltransferase (4′OMT) [138,139], and one cytochrome P450 enzyme [P450, (S)-*N*-methylcoclaurine 3′-hydroxylase (NMCH)) [140–142]. During this biosynthesis, 6OMT catalyzes *O*-methylation at C6 on (*S*)-norcoclaurine to yield (*S*)-coclaurine [135], and (*S*)-coclaurine further undergoes *N*-methylation under the action of CNMT to generate (*S*)-*N*-methylcoclaurine [137]. Then, the P450 enzyme NMCH can convert (*S*)-*N*-methylcoclaurine to (*S*)-3′-hydroxy-*N*-methylcoclaurine [140], and finally it catalyzes the transfer of the *S*-methyl group of SAM to the previous product through 4′OMT to form an important intermediate (*S*)-reticuline base for the synthesis of isoquinoline alkaloids [138]. Subsequently, the key central ring closure is the conversion of the *N*-CH_3 of (*S*)-reticuline to the berberine bridge carbon, C8 of (*S*)-scoulerine, thereby forming the protoberberine carbon skeleton. This step is accomplished by berberine bridge enzyme (BBE), which is also an important step in the biosynthesis of other isoquinoline alkaloids including protopine, protoberberine, and benzophenanthridine alkaloids [143]. BBE is a key rate-limiting enzyme in the synthesis of (*S*)-scoulerine. More recently, Li et al. [144] achieved high expression of McBBE derived from *Macleaya cordata* in *S. cerevisiae* through codon optimization, *N*-terminal truncation, and CRISPR-Cas9 technology, obtaining a genetically stable *S. cerevisiae* strain with high McBBE expression. Further methylation of (*S*)-scoulerine was performed by *O*-methyltransferase (SAM: scoulerine 9-*O*-methyltransferase (SMT)) [145,146] to yield (*S*)-tetrahydrocolumbamine, which is stereospecifically converted to (*S*)-canadine under formation of the methylenedioxy bridge through (*S*)-canadine synthase [147]. Finally, (*S*)-canadine is oxidatively aromatized to berberine through tetrahydroprotoberberine oxidase [147,148]. The above is the detailed process of berberine biosynthesis from Tyr, and it has been reported that by combining enzymes from the same or different sources, this pathway could successfully synthesize berberine and a series of important intermediates [130,149–151] (Figure 5).

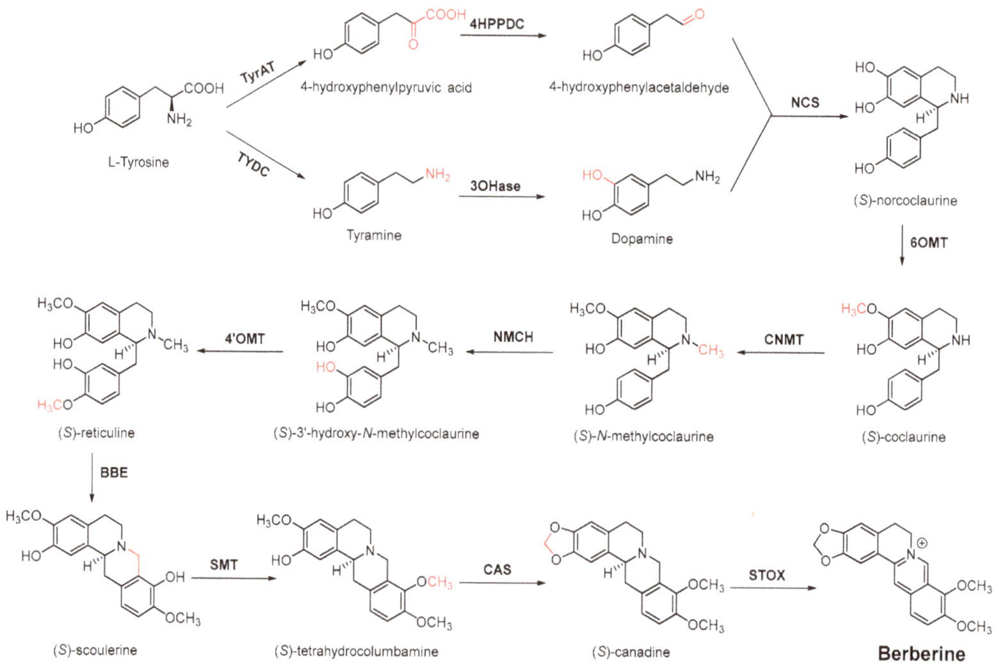

Figure 5. Biosynthesis pathway of the alkaloid antimicrobial agent berberine. TyrAT, L-tyrosine aminotransferase; 4HPPDC, 4-hydroxyphenylpuruvate decarboxylase; TYDC, tyrosine decarboxylase; 3OHase, tyrosine/tyramine 3-hydroxylase; NCS, (*S*)-norcoclaurine synthase; 6OMT, SAM: norcoclaurine 6-*O*-methyltransferase; CNMT, (*S*)-coclaurine *N*-methyltransferase; NMCH, *N*-methylcoclaurine 3′-hydroxylase; 4′OMT, SAM: 3′-hydroxy-*N*-methylcoclaurine 4′-*O*-methyltransferase; BBE, berberine bridge enzyme; SMT, SAM: scoulerine 9-*O*-methyltransferase; CAS, (*S*)-canadine synthase; STOX, (*S*)-tetrahydroprotoberberine oxidase.

3.3. *Biosynthesis Investigations of the Antimicrobial Alkaloid Compound—Colchicine*

Colchicine is an FDA-approved, available, safe, and effective anti-inflammatory drug derived from *Colchicum* and *Gloriosa* species [152–154]. On the basis of its unique efficacy as an anti-inflammatory agent, colchicine has been used in the therapy of cardiovascular diseases. Most recently, there have numerous reports suggesting that colchicine could also be used in the treatment of coronavirus disease 2019 (COVID-19) [155,156]. The antiviral activity of this alkaloid is attributed to its ability to bind tubulin dimers and inhibit microtubule assembly, which not only promotes anti-inflammatory effects but also makes colchicine a potent mitotic poison [154,157]. In addition, colchicine may inhibit inflammasome signaling and reduce proinflammatory cytokines, which is a purported mechanism of COVID-19 pneumonia [158].

For the biosynthesis of colchicine, since Leete conducted the first biosynthetic experiments on colchicine in 1960 [159], the chemical origins of colchicine have been thoroughly studied through an abundance of feeding studies with isotope-labeled substrates in *Colchicum* plants, as well as the structural characterization of colchicine-related alkaloids isolated from species of the Colchicaceae family that helped to define a well-established biosynthetic hypothesis [160–163] (Figure 6). It has been established that colchicine originated from Phe and Tyr [164]. Similar to the former part of the berberine biosynthetic pathway, the phenethylisoquinoline skeleton of colchicine is also formed by the condensation of an aldehyde with an amine [162]. Namely, the initial amino acids Phe and Tyr are processed into 4-hydroxydihydrocinnamaldehyde (4-HDCA) and dopamine, respectively, which are joined through a Pictet–Spengler reaction to form a 1-phenethylisoquinoline scaffold [162,163,165]. The scaffold then undergoes a series of methylations and phenyl ring hydroxylations to yield (*S*)-autumnaline [163], which proceeds to *para–para* phenol coupling to create a bridged tetracycle [166]. An unusual oxidative ring expansion followed, yielding the characteristic tropolone ring of the colchicine carbon scaffold, which is essential for the tubulin-binding activity of colchicine [167]. The biosynthesis of colchicine is further accomplished through final processing and *N*-acetylation of the extruded nitrogen atom [168].

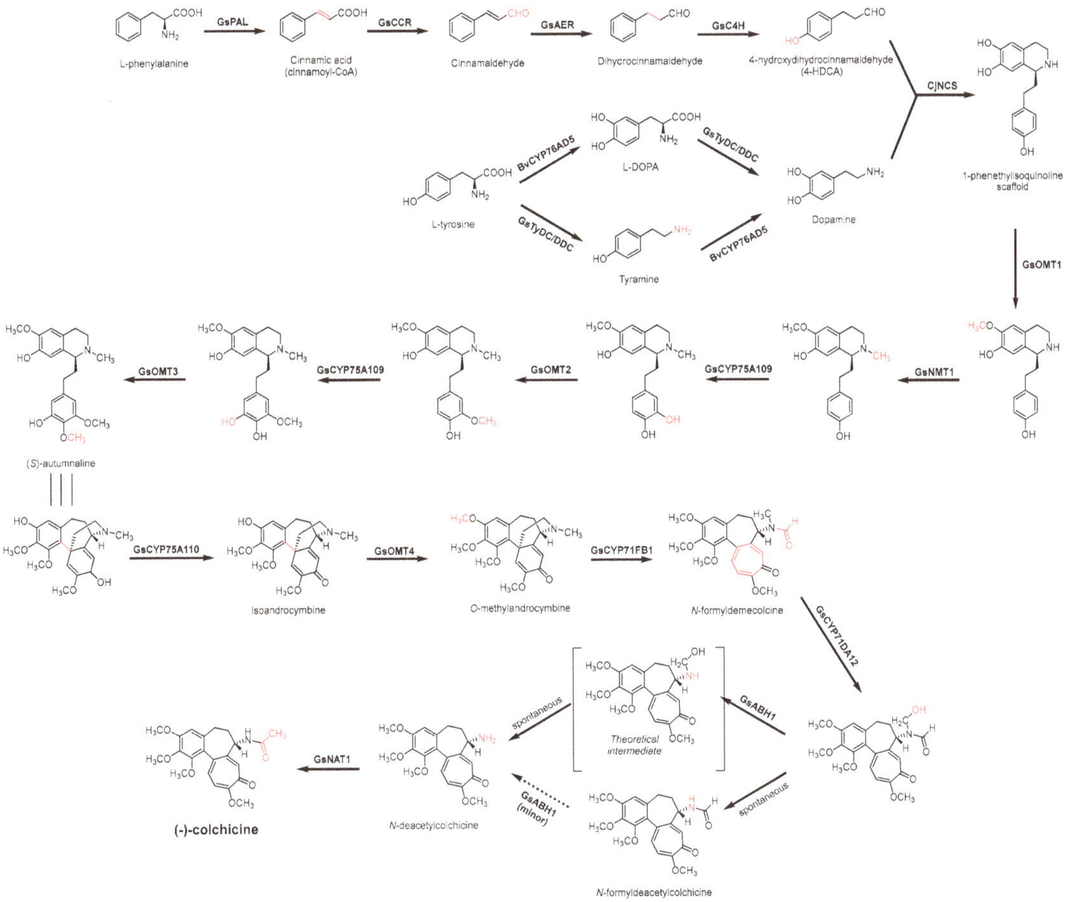

Figure 6. Biosynthesis pathway of alkaloid antimicrobial agent (-)-colchicine. GsPAL, phenylalanine ammonia-lyase; GsCCR, cinnamoyl-CoA reductase; GsAER, alkenal reductase; GsC4H, cinnamate 4-hydroxylase; GsTyDC/DDC, L-tyrosine/L-DOPA decarboxylase; BvCYP76AD5, cytochromes P450 3′-hydroxylase; CjNCS, (*S*)-norcoclaurine synthase; *S*-adenosylmethionine-dependent methyltransferase (MT): GsOMT1, GsOMT2, GsOMT3, GsOMT4, *O*-methyltransferase; GsNMT1, *N*-methyltransferase; GsCYP75A109, GsCYP75A110, GsCYP71FB1, GsCYP71DA12, cytochromes P450; GsABH1, alpha/beta hydrolase; GsNAT1, *N*-acetyltransferase.

3.4. De Novo Biosynthetic Production of Colchicine in *Nicotiana benthamiana*

In view of the above research, Sattely et al. [169,170] established a metabolic pathway of tropolone-containing colchicine alkaloids by using a combination of transcriptomics, metabolic logic, and pathway reconstitution. The first stage is the generation of the key precursor 1-phenethylisoquinoline scaffold, which requires the Pictet–Spengler condensation of 4-HDCA and dopamine derived from the amino acids Phe and Tyr. Labeling studies have shown that 4-HDCA is produced from Phe through a metabolic pathway analogous to the biosynthesis of monolignols [164,171]. Through hierarchical clustering analysis of *Gloriosa superba* transcriptomic data utilizing the other identified colchicine biosynthesis genes, the researchers demonstrated co-clustering of many monolignol biosynthetic gene orthologs (GsPAL, Gs4CL, GsCCR, GsAER, GsC4H, and GsDAHPS), and their heterologous co-expression in *N. benthamiana* resulted in the production of 4-HDCA. For dopamine formation, the incorporation of L-Tyr and tyramine into colchicine demonstrated the activity of L-Tyr/L-DOPA decarboxylase (TyDC/DDC) and 3′-hydroxylase enzymes [161]. The researchers identified a TyDC/DDC homolog (GsTyDC/DDC) highly co-expressed with other identified colchicine biosynthesis genes in the public *G. superba* transcriptome via a similar analysis approach, combining it with 3′-hydroxylase BvCYP76AD5 from *Beta vulgaris* to produce L-DOPA successfully [169,172]. Furthermore, the modified (*S*)-norcoclaurine synthase from *Coptis japonica* (CjNCS) was utilized to catalyze the condensation of 4-HDCA with dopamine to produce the first alkaloidal precursor 1-phenethylisoquinoline. The NCS is a previously characterized plant Pictet–Spenglerase, which condenses 4HPAA and dopamine within the biosynthesis of benzylisoquinoline alkaloid (BIA) [173]. It can also condense a wide range of aldehyde substrates with dopamine [174,175]. The precursor will yield (*S*)-autumnaline by further modification (hydroxylations, methylations). (*S*)-Autumnaline then undergoes enzyme-catalyzed phenolic coupling together with further modification to produce *O*-methylandrocymbine, which is then converted to colchicine via homoallylic ring expansion [176,177]. On the basis of the above information, the researchers utilized eight genes (GsOMT1, GsNMT1, GsCYP75A109, GsOMT2, GsOMT3, GsCYP75A110, GsOMT4, and GsCYP71FB1) explored from *G. superba* to act on 1-phenethylisoquinoline for the biosynthesis of the colchicine precursor *N*-formyldemecolcine, which contains the characteristic tropolone ring and pharmacophore of colchicine. Combining all the above genes, the authors engineered a biosynthetic pathway (16 enzymes in total) in *N. benthamiana* and realized the de novo biosynthetic production of *N*-formyldemecolcine starting from amino acids Phe and Tyr in the commonly used model plant [169]. Subsequently, enzymes that catalyze the *N*-demethylation, *N*-deformylation, and *N*-acetylation (GsCYP71FB1, GsABH1, GsNAT1) of *N*-formyldemecolcine were further excavated and transferred into *N. benthamiana*. Ultimately, through the heterologous system of 20 genes from *G. superba* (17 genes) and other plants (3 genes), total biosynthesis of enantiopure (-)-colchicine was successfully achieved from primary metabolites [170] (Figure 6).

3.5. Biosynthesis Investigations of Other Antimicrobial Alkaloids

As a major source of bioactive natural products, in addition to the above-discussed berberine and colchicine, there are still many alkaloids that showed desirable antimicrobial activities whose biosynthesis pathways have also been clarified. Quinoline alkaloids such as 8-hydroxyquinoline are important kinds of nitrogen-containing heterocyclic aromatic compounds with a broad range of antimalarial, antibacterial, antifungal, and antiviral activities. Quinoline alkaloids mainly exist in the Rutaceae family, and their biosynthesis is derived from 3-hydroxyanthranilic acid, a metabolite formed through a series of enzymatic reactions of tryptophan. Specifically, 3-hydroxyanthranilic acid and malonyl-SCoA are condensed and then cyclized to yield quinoline alkaloids [178]. Monoterpenoid indole alkaloids (MIAs)—a large group of natural products derived from plants, such as camptothecin, quinine, and vinblastine—exhibited anticancer, antimalarial, and antibacterial effects [179,180]. Secologanin is the terminus of the monoterpenoid biosynthesis branch and is coupled to tryptamine by strictosidine synthase (STR) to form strictosidine, which is the universal MIAs precursor in plants. The tomato plant, *S. lycopersicum* L., produces the cholesterol-derived steroidal alkaloids tomatine and tomatidine. Tomatidine selectively and potently inhibits small-colony variants of *S. aureus* that cause opportunistic infections in patients with cystic fibrosis [181], and also has potent fungistatic activity against *Candida* spp. with low toxicity to human cells [182]. Their biosynthesis begins from the precursor dehydrotomatidine via enzymatic dehydrogenation, isomerization, and sequential reductions [183]. Scopolamine is a kind of TA that is present in many different plants of the Solanaceae family and is classified as essential medicine by the WHO. Scopolamine showed considerable antifungal activity [184,185]. Smolke et al. [186] realized the construction of a modular biosynthetic pathway by engineered baker's yeast for the production of medicinal TA scopolamine, starting from simple sugars and amino acids. Genetic-level manipulations they performed included functional genomics to identify missing pathway enzymes, protein engineering to enable expression of functional acyltransferases through trafficking to the vacuole, heterologous transporters to facilitate intracellular routing, and strain optimization to increase titers.

The enormous potential of alkaloids as drug precursors is far from exhausted, and various pharmacological effects continue to be reported and reviewed [187]. In addition, emerging biotechnologies have been optimized for plants, including metabolomics, CRISPR-based gene editing, and heterologous yeast platforms, enabling the production of diverse and complex plant compounds. It is reasonable to expect that with an increased understanding of the biosynthesis of other antimicrobial alkaloids, increasingly more alkaloid antimicrobial agents could be explored and mass produced in the near future.

4. Flavonoids

Flavonoids widely exist in plants, being the general name of a series of compounds derived from 2-phenyl chromogenic ketones. According to the chemical properties, positions, and types of substituents on the ring, flavonoids can be divided into several subclasses, such as flavones, flavonols, dihydroflavones, dihydroflavonols, isoflavones, chalcone, aurone, and anthocyanidin, among others [188]. The abundance and diversity of chemical structures of flavonoids determined their wide-spectrum biological activities. In addition to the traditional antioxidant, anti-radiation, radicals scavenging, anti-inflammatory, and anti-tumor activities, flavonoids are also reported to possess remarkable antimicrobial bioactivities [189]. They could effectively inhibit bacteria, viruses, and fungi, having good therapeutic effects on infections caused by various pathogenic microorganisms, including *S. aureus*, *Bacillus subtilis*, *P. aeruginosa*, *E. coli*, *S. typhimurium*, *C. albicans*, and *Aspergillus flavus*. These compounds are not easy to produce drug resistance and have high clinical therapeutic values. For example, oral candidiasis is one of the most common types of oral mucosal infection caused by the yeast-like fungus *Candida*. The elderly and children with low immunity are very susceptible to infection. Phloretin (Table 1) can inhibit the pathogenicity and virulence factors of *C. albicans* both in vivo and in vitro, and is considered to be an effective candidate for the treatment of oral candidiasis [64]. Other flavonoids such as apigenin (Table 1) and quercetin (Table 1) have been proven to have significant antibacterial and antiviral activities [190,191]. Quercetin, when taken together with vitamin C, is helpful to prevent and treat patients with early respiratory tract infections. According to the report, when quercetin is used for phytotherapy, patients with mild COVID-19 symptoms have a shorter time to clear the virus [192]. Thus, the plant-originated flavonoids can be used as an ideal natural source to explore novel antimicrobial agents [193].

4.1. Structure–Activity Relationship Study on Antimicrobial Activity of Flavonoids

The antibacterial activity of flavonoids has attracted extensive attention from researchers. Correspondingly, the relationship between the chemical structure and biological activity has been discussed in depth. It was found that the antibacterial activity of flavonoids was mainly related to the existence of hydroxyl groups on the aromatic skeletons of flavonoids and the types of substituents. In particular, flavonoids substituted by hydrophobic groups, such as propenyl, acyl, alkyl amino chain, alkyl chain, and nitrogen-containing or oxygen-containing heterocyclic groups, have been proven to have better antibacterial potential [194]. Smejkal et al. [195] tested the antibacterial activities of eight flavonoids isolated from *Paulownia tomentosa* towards *S. aureus*. The results showed that hydroxylation at the C-5 position of ring A was very important to enhance the antibacterial activity of flavonoids. The in vitro antibacterial activities of a variety of chalcone derivatives towards MSSA and MRSA were tested. The results showed that chalcones with hydroxyl substitution at the 2 or 4 positions of the B ring had antibacterial activity. However, the methylation of the active hydroxyl groups generally eliminated or weakened its antibacterial activity [196]. Celiz et al. [52] found that acylated flavonoid derivatives usually showed a high inhibitory effect on Gram-positive bacteria. For example, the activity of hesperidin (Table 1) against *S. aureus* and *L. monocytogenes* can be greatly increased by connecting the saturated fatty chain containing 10-12 carbon atoms to the ring of hesperidin. Similar results were also obtained by Babu et al. [54]. They found that the introduction of the acyl group at the C-7 position of oroxylin A (Table 1) can significantly enhance the antibacterial activity. When the acyl group contains long-chain alkyl or phenyl, derivatives of oroxylin A showed even stronger antibacterial activity. Moreover, the antibacterial potential of nitrogen-containing flavone derivatives was also investigated. It was found that the antibacterial activity of nitrogen-substituted apigenin derivatives was much higher than that of apigenin [55].

4.2. Antibacterial Effects and Action Mechanisms of Flavonoid Antimicrobial Agents

The antimicrobial mechanism of flavonoids mainly includes the following aspects: inhibiting the energy metabolism of bacteria, interfering with the cell wall of bacteria, destroying the integrity of the cell membrane and increasing its permeability, inhibiting bacterial efflux pumps, inhibiting the metabolism of bacterial nuclear acid, inhibiting bacterial mobility, and reducing the expression of virulence factors to weaken the pathogenicity [194]. Chinnam et al. [56] reported that flavonoids such as hesperidin, morin (Table 1), and silymarin (Table 1) can inhibit the F_1F_o ATPase activity of *E. coli* by binding to the polyphenol binding bag of ATP synthase, thus inhibiting the energy metabolism of bacteria and further exerting the bacteriostatic effect. Navarro-Martinez found for the first time that the epigallocatechin gallate (Table 1) in green tea has strong antibacterial activity against *Stenotrophomonas maltophilia*, mainly by inhibiting the dihydrofolate reductase of *S. maltophilia* [57]. The flavonol compound quercetin could inhibit the growth of various drug-resistant microorganisms. It can suppress the herpesvirus and poliovirus by inhibiting viral polymerase and viral nucleic acid. The flavonol compound galangin (Table 1) could directly destroy the plasma membrane or weaken the cell wall of *S. aureus*, which will lead to osmotic lysis, thus resulting in a bacteriostatic effect [59]. Catechin (Table 1), a flavonoid in green tea, could inhibit the bacterial DNA gyrase by binding to the ATP binding site of the B subunit of the bacterial gyrase, the inactivation of which will cause the death of the bacteria [60]. Studies have shown that bacteria can migrate on semi-solid surfaces in a flagella-driven manner, and this coordinated movement form is considered to be related to the antibiotic resistance of various human pathogens [197]. Pejin et al. [58] discussed the antibacterial mechanism of catechin, caffeic acid, quercetin, and morin against *P. aeruginosa PAO1*. The results showed that quercetin could inhibit the formation of its biofilm at 0.5 MIC concentration. *P. aeruginosa* biofilm formation also depends on the flagellum (swimming motility) and type IV pili (twitching motility). Moreover, among the four compounds tested, quercetin was the only one found to effectively reduce the convulsive movement of *P. aeruginosa*. Fathima et al. [61] used the Gram-positive bacteria *B. subtilis* and Gram-negative bacteria *E. coli* as model organisms to prove that catechins play an antibacterial role mainly by producing active oxygen to cause bacterial cell membrane rupture. Wang et al. [65] reported that silybin (Table 1), a flavonoid compound, combined with ciprofloxacin can improve the antibacterial efficiency by inhibiting the efflux pump of MRSA. Liu et al. [64] established a mouse oral candidiasis model to explore the inhibitory effect of phloretin on the pathogenicity of *Candida albicans*. The results show that phloretin can eliminate virulence factors in vitro, such as inhibition of biofilm formation, yeast-to-hyphae transition, and secretion of protease and phospholipase, in order to play an inhibitory role.

4.3. *Plant Type III Polyketide Synthase*

The key enzyme involved in the biosynthesis of flavonoids in plants is type III polyketide synthase (PKS), which not only is the key rate-limiting enzyme in the biosynthesis and metabolism pathway of flavonoids but also determines the formation of the basic molecular skeleton of these compounds. Plant type III PKS can repeatedly catalyze the initiation, extension, and cyclization reactions to form polyketone products. At present, nearly 30 plant PKSs genes with different functions have been successively explored and verified in the biosynthetic pathway of various polyketides, such as chalcone syntheses (CHS), benzophenone synthase (BPS), 2-pyrone syntheses (2-PS), pentaketide chromone synthase (PCS), and benzalacetone synthase (BAS). Among them, CHS is involved in the synthesis of all flavonoids of plant origin and has been deeply studied [198]. This enzyme was first cloned from parsley. It could catalyze the three acetyl groups of malonyl CoA to be connected to the molecule of 4-coumaroyl CoA through a continuous condensation reaction, and then generate naringenin chalcone through Claisen-type cyclization reaction, which is the critical intermediate for the biosynthesis of many flavonoid compounds [199], and will be then converted to various flavonoid compounds by downstream tailoring enzymes such as chalcone isomerase (CHI), flavanone 3-hydroxylase (F3H), flavone synthase (FNS), flavonol synthase (FLS), isoflavone synthase (IFS), and polyketide reductase [200] (Figure 7). Since then, researchers have isolated hundreds of CHS genes from lily, rice, corn, alfalfa, and other plants. The protein structure and catalysis mechanism of CHS have also been elucidated [200]. In 1999, Ferrer and his colleagues reported the X-ray crystal structure of *Medicago sativa* CHS2 at 1.56 Å resolution [201]. Through crystallography and site-directed mutagenesis, it was clarified that the key amino acid residues that determined CHS catalytic activities were Cys164, His303, and Asn336 [198]. During the formation of flavonoids, Cys164 acts as a nucleophilic active site and an attachment site for intermediates, while His303 and Asn336 play an important role in the decarboxylation of malonyl CoA. In addition to the key ternary amino acid residues, its internal active site also includes a coenzyme, a binding tunnel, a promoter-substrate-binding pocket, and a cyclization pocket. In recent years, with the elucidation of flavonoids' biosynthetic metabolic pathway and the development of synthetic biology, it is possible to obtain large-scale flavonoids by building microbial cell factories. *E. coli* and *S. cerevisiae* are common microbial hosts. Genetic engineering strategies such as optimization of culture conditions, modular co-culture technology, and iterative high-throughput screening methods have been used in the construction and improvement of the engineering strains to obtain high-yield target compounds [202,203].

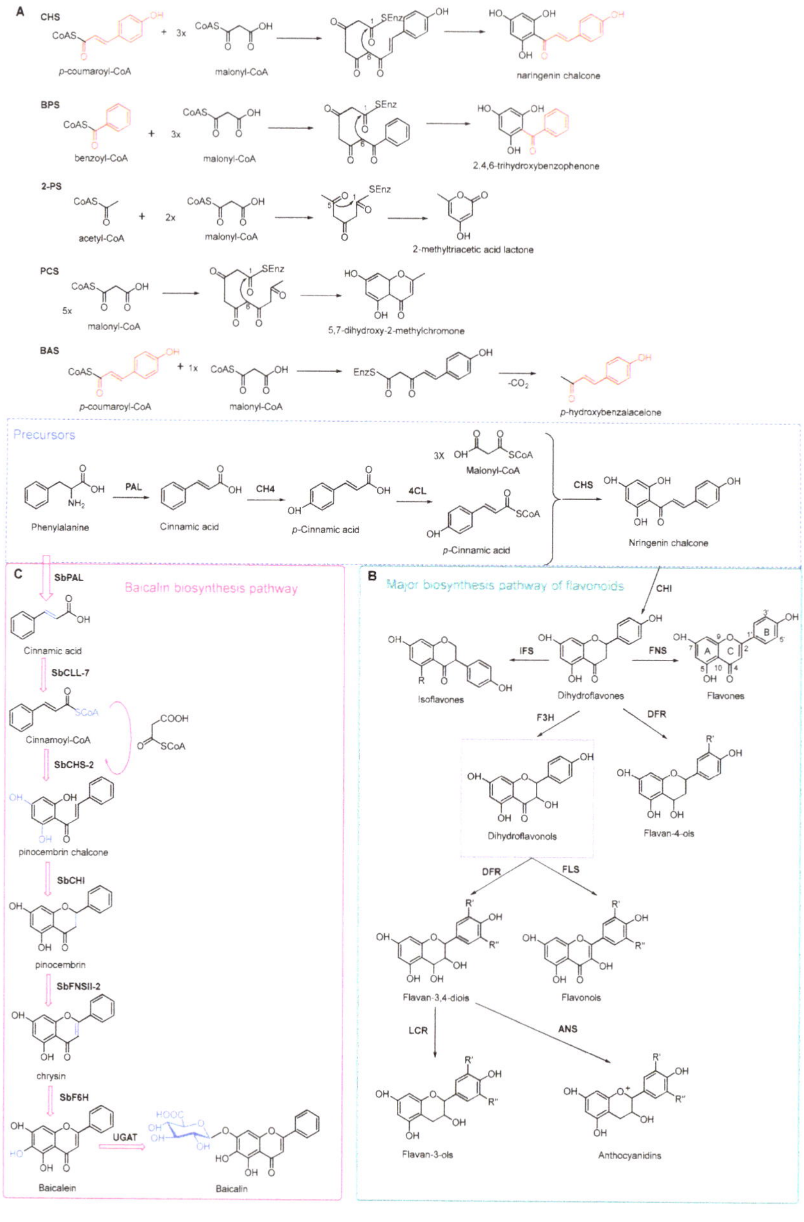

Figure 7. (**A**) Typical reactions catalyzed by plant type III PKSs. (**B**) Main skeleton types of plant-derived flavonoid compounds. (**C**) Biosynthetic pathway of flavonoid antimicrobial agent baicalin. CHS, chalcone syntheses; BPS, benzophenone synthase; 2-PS, 2-pyrone syntheses; PCS, pentaketide chromone synthase; BAS, benzalacetone synthase; PAL, phenylalanine ammonia lyase; CH4, cinnamic acid 4-hydroxylase; 4CL, 4-coumarin CoA ligase; CHS, chalcone synthase; CHI, chalcone isomerase; IFS, isoflavone synthase; FNS, flavone synthase; F3H, flavanone 3-hydroxylase; DFR, dihydroflavonol 4-reductase; FLS, flavonol synthase; ANS, anthocyanidin synthase; LCR, leucoanthocyanidin reductase; SbCLL-7, cinnamate CoA ligase; SbCHS-2, pinocembrin chalcone synthase; SbFNS II-2, flavone synthase; SbF6H, flavone 6 hydroxylase; UGAT, UDP-glucuronic acid transferase [188,204].

4.4. Biosynthesis Investigations of the Antimicrobial Flavonoid Compound—Baicalin

Baicalin (Table 1) is one of the representative flavonoid antimicrobial agents isolated from *Scutellaria baicalensis*. It has been applied as a natural antibacterial agent against foodborne pathogens such as *Salmonella* and *Staphylococcus* spp. [62]. Moreover, this compound also showed significant anti-HIV-1 activity as a nonnucleoside reverse transcriptase inhibitor [205]. Meanwhile, it can prevent the entry of HIV-1 into animal cells by perturbing the interaction between HIV-1 Env protein and HIV-1 co-receptors on the cell surface [206]. As one of the popular lead natural products from medicinal plants for preventing HIV infection, the biosynthesis pathway of baicalin has been fully analyzed, as shown in Figure 7 [188,204]. There are two different biosynthetic metabolic pathways existing in *S. baicalensis*, namely, the aerial flavone part pathway and the root-specific flavone pathway. Baicalein, the precursor of baicalin, is synthesized through the root-specific flavone pathway. The reaction process is as follows: amino acid Phe was used as a biosynthesis precursor that could generate cinnamic acid under the action of SbPAL. Subsequently, cinnamic acid forms cinnamoyl CoA under the catalysis of cinnamate CoA ligase (SbCLL-7). Then, pinocembrin chalcone synthase SbCHS-2 will catalyze cinnamoyl CoA to generate pinocembrin chalcone, which will be further converted to dihydroflavone pinocembrin by SbChI. Next, pinocembrin could be catalyzed by flavone synthase (SbFNSII-2) to form chrysin, finally leading to the production of baicalein under the catalysis of flavone 6-hydroxylase (SbF6H) [207]. As the 7-*O*-glucuronic acid product of baicalein, baicalin could be biosynthesized by UDP glycosyltransferase (UGT) to transfer glucuronic acid to the 7-hydroxyl group of baicalein. Pei et al. [208] identified the baicalin metabolic accumulation pattern and tissue-specific expression patterns of a total of 124 UGTs in *S. baicalensis*. Combined with phylogenetic analysis, four SbUGAT genes were screened out to be able to use UDP-glucuronic acid as a sugar donor to catalyze the conversion of baicalein to baicalin. On the basis of the illumination of the biosynthesis pathway, heterologous production of baicalin has been successfully realized in both *E. coli* [209] and model plant *Lycopersicon esculentum* [210].

5. Conclusions and Perspective

Currently, plant-derived antimicrobial agents are still in the early stage of research. The developed plants only account for a very small number of global plants. In addition, the potential synergy or antagonism between plant compounds and antibiotics is still uncertain. Moreover, there is no research showing the resistance of plant antimicrobial agents, so whether there is resistance is unknown. Moreover, some plant compounds have not been tested to prove their effectiveness and safety. Furthermore, studies on the mechanisms of action, exploring the potential synergistic or antagonistic effects and improving the bioavailability, stability, and physicochemical property of the candidate compounds, were also very important before their clinic uses.

Terpenoids, alkaloids, and flavonoids made up the predominant part of the currently reported phytochemicals with antimicrobial activities. Synthetic biology research around these compounds is one of the hotspot fields in recent years. It can be seen that even for artemisinin—one of the most famous antimalaria drugs with in-depth biosynthesis investigation—its whole biosynthesis pathway still has several key enzymes to be discovered, let alone many plant-originated compounds whose biosynthesis pathway is obscure. Thus, a lot of challenges have remained in the investigations of plant-derived compounds.

In our opinion, one of the greatest challenges may be the discovery of genes because normally functional genes in plants are not clustered. Although recently there are reports that found that the co-expression of physically linked genes occurs frequently, it is still very difficult to explore a new gene through gene cluster searching, especially considering the limited number of plant species with the known whole-genome sequence. In addition, plants usually have different organs and tissues, and the gene expression level could be tissue specific, which makes the selection of the gene extraction material more complex. Moreover, genes in plants often exist in homologs, and their expression could be different and affected by multiple factors such as the environment, living position, temperature, and developmental stages. Currently, the frequently used method for gene discovery in plants is homology-based cloning. However, it is hard to realize in those enzymes with new functions or that do not have enough known templates. With the rapid development of sequencing and transcriptomic advances, enzymes can be discovered by the 'omics' tools, such as genomics, transcriptomics, proteomics, and metabolomics. By comparing transcriptome, proteome, and metabolome data from different conditions, candidate genes could be selected and subsequently tested for their putative activity—this approach has been successfully used in the discovery of a large number of unknown genes in phytochemical substances' biosynthesis pathways.

The other important field that synthetic biology focuses on is the biosynthetic pathways reconstructing and optimizing the production of secondary metabolites. Recent plant genome editing/engineering methods such as transcription activator-like effector nucleases (TALENS), zinc-finger nucleases (ZFNs), and CRISPR-Cas open new avenues for ration design of the biosynthesis pathway. Using these approaches, genes of interest could be constructed in a highly effective way; meanwhile, the side pathways could be eliminated to a large extent. The targeted antimicrobial or resistance–reversal agents could be produced in the transgenic microorganisms or plants, which have had great success in the production of artemisinin [91], as well as other valuable compounds such as vinblastine [211,212], etoposide aglycone [213], vindoline, and catharanthine [212]. Lastly, with the continuous discovery of new phytochemicals, deep clarification of pharmacological mechanisms, and comprehensive understanding of specific biosynthesis pathways, plant-derived natural products will become increasingly more useful therapeutic antimicrobial candidates in the future.

Author Contributions: W.H., Y.W., W.T., X.C. and X.L.: writing—review and editing; P.T., J.L. and S.S.: supervision; X.L.: funding acquisition. All authors have read and agreed to the published version of the manuscript.

Funding: This work was financially supported by the National Natural Science Foundation of China (grant nos. 82173922, 81402809); the Beijing Natural Science Foundation (grant no. 7192112); the Young Elite Scientists Sponsorship Program by CAST (grant no. CACM−2018−QNRC1−02); the Foundation supported by the State Key Laboratory of Natural and Biomimetic Drugs (grant no. K202119); and the Young Scientist Program by Beijing University of Chinese Medicine.

Institutional Review Board Statement: Not applicable.

Informed Consent Statement: Not applicable.

Data Availability Statement: Not applicable.

Conflicts of Interest: The authors declare no conflicts of interest.

References

1. Reinhardt, U.; Cheng, T. The world health report 2000–Health systems: Improving performance. *Bull. World Health Organ.* **2000**, *78*, 1064.
2. Mohr, K.I. History of Antibiotics Research. *Curr. Top Microbiol. Immunol.* **2016**, *398*, 237–272.
3. Ali, S.M.; Siddiqui, R.; Khan, N.A. Antimicrobial discovery from natural and unusual sources. *J. Pharm. Pharmacol.* **2018**, *70*, 1287–1300. [CrossRef] [PubMed]
4. Fauci, A.S. Infectious diseases: Considerations for the 21st century. *Clin. Infect. Dis.* **2001**, *32*, 675–685. [CrossRef]
5. Keita, K.; Darkoh, C.; Okafor, F. Secondary plant metabolites as potent drug candidates against antimicrobial-resistant pathogens. *SN Appl. Sci.* **2022**, *4*, 209. [CrossRef]
6. Davies, S.C.; Fowler, T.; Watson, J.; Livermore, D.M.; Walker, D. Annual Report of the Chief Medical Officer: Infection and the rise of antimicrobial resistance. *Lancet* **2013**, *381*, 1606–1609. [CrossRef]
7. Jia, H.; Li, L.; Li, W.; Hou, T.; Ma, H.; Yang, Y.; Wu, A.; Liu, Y.; Wen, J.; Yang, H.; et al. Impact of Healthcare-Associated Infections on Length of Stay: A Study in 68 Hospitals in China. *Biomed. Res. Int.* **2019**, *2019*, 2590563. [CrossRef]
8. Porras, G.; Chassagne, F.; Lyles, J.T.; Marquez, L.; Dettweiler, M.; Salam, A.M.; Samarakoon, T.; Shabih, S.; Farrokhi, D.R.; Quave, C.L. Ethnobotany and the Role of Plant Natural Products in Antibiotic Drug Discovery. *Chem. Rev.* **2021**, *121*, 3495–3560. [CrossRef]

9. Tiwari, P.; Khare, T.; Shriram, V.; Bae, H.; Kumar, V. Plant synthetic biology for producing potent phyto-antimicrobials to combat antimicrobial resistance. *Biotechnol. Adv.* **2021**, *48*, 107729. [CrossRef] [PubMed]
10. Vaou, N.; Stavropoulou, E.; Voidarou, C.; Tsigalou, C.; Bezirtzoglou, E. Towards Advances in Medicinal Plant Antimicrobial Activity: A Review Study on Challenges and Future Perspectives. *Microorganisms* **2021**, *9*, 2041. [CrossRef]
11. Gonzalez-Lamothe, R.; Mitchell, G.; Gattuso, M.; Diarra, M.S.; Malouin, F.; Bouarab, K. Plant antimicrobial agents and their effects on plant and human pathogens. *Int. J. Mol. Sci.* **2009**, *10*, 3400–3419. [CrossRef] [PubMed]
12. Cowan, M.M. Plant products as antimicrobial agents. *Clin. Microbiol. Rev.* **1999**, *12*, 564–582. [CrossRef]
13. Savoia, D. Plant-derived antimicrobial compounds: Alternatives to antibiotics. *Future Microbiol.* **2012**, *7*, 979–990. [CrossRef]
14. Shankar, S.R.; Rangarajan, R.; Sarada, D.V.L.; Kumar, C.S. Evaluation of antibacterial activity and phytochemical screening of *Wrightia tinctoria* L. *Pharmacogn. J.* **2010**, *2*, 19–22. [CrossRef]
15. Li, J.M.; Feng, S.S.; Liu, X.; Jia, X.; Qiao, F.L.; Guo, J.L.; Deng, S.S. Effects of traditional chinese medicine and its active ingredients on drug-resistant bacteria. *Front. Pharmacol.* **2022**, *13*, 837907. [CrossRef]
16. Kannappan, A.; Sivaranjani, M.; Srinivasan, R.; Rathna, J.; Pandian, S.K.; Ravi, A.V. Inhibitory efficacy of geraniol on biofilm formation and development of adaptive resistance in *Staphylococcus epidermidis* RP62A. *J. Med. Microbiol.* **2017**, *66*, 1506–1515. [CrossRef]
17. Upadhyay, H.C.; Dwivedi, G.R.; Roy, S.; Sharma, A.; Darokar, M.P.; Srivastava, S.K. Phytol derivatives as drug resistance reversal agents. *ChemMedChem* **2014**, *9*, 1860–1868. [CrossRef]
18. Nhiri, M.; Mrid, R.B.; Omari, R.E.; Bouargalne, Y. New insights into the therapeutic effects of phenolic acids from sorghum seeds. *J. Rep. Pharm. Sci.* **2019**, *8*, 91–101. [CrossRef]
19. Sharma, A.; Biharee, A.; Kumar, A.; Jaitak, V. Antimicrobial terpenoids as a potential substitute in overcoming antimicrobial resistance. *Curr. Drug. Targets* **2020**, *21*, 1476–1494. [CrossRef]
20. Nazzaro, F.; Fratianni, F.; De Martino, L.; Coppola, R.; De Feo, V. Effect of essential oils on pathogenic bacteria. *Pharmaceuticals* **2013**, *6*, 1451–1474. [CrossRef]
21. Burt, S.A.; Reinders, R.D. Antibacterial activity of selected plant essential oils against *Escherichia coli* O157:H7. *Lett. Appl. Microbiol.* **2003**, *36*, 162–167. [CrossRef]
22. Mulyaningsih, S.; Sporer, F.; Zimmermann, S.; Reichling, J.; Wink, M. Synergistic properties of the terpenoids aromadendrene and 1,8-cineole from the essential oil of *Eucalyptus globulus* against antibiotic-susceptible and antibiotic-resistant pathogens. *Phytomedicine* **2010**, *17*, 1061–1066. [CrossRef]
23. Di Pasqua, R.; Betts, G.; Hoskins, N.; Edwards, M.; Ercolini, D.; Mauriello, G. Membrane toxicity of antimicrobial compounds from Essential oils. *J. Agric. Food Chem.* **2007**, *55*, 4863–4870. [CrossRef]
24. Mulat, M.; Pandita, A.; Khan, F. Medicinal plant compounds for combating the multi-drug resistant pathogenic bacteria: A Review. *Curr. Pharm. Biotechnol.* **2019**, *20*, 183–196. [CrossRef]
25. Niu, C.; Afre, S.; Gilbert, E.S. Subinhibitory concentrations of cinnamaldehyde interfere with quorum sensing. *Lett. Appl. Microbiol.* **2006**, *43*, 489–494. [CrossRef]
26. Myszka, K.; Schmidt, M.T.; Majcher, M.; Juzwa, W.; Olkowicz, M.; Czaczyk, K. Inhibition of quorum sensing-related biofilm of *Pseudomonas fluorescens* KM121 by *Thymus vulgare* essential oil and its major bioactive compounds. *Int. Biodeter. Biodegr.* **2016**, *114*, 252–259. [CrossRef]
27. Mittal, R.P.; Rana, A.; Jaitak, V. Essential Oils: An impending substitute of synthetic antimicrobial agents to overcome antimicrobial resistance. *Curr. Drug Targets* **2019**, *20*, 605–624. [CrossRef]
28. Ahmad, A.; Khan, A.; Yousuf, S.; Khan, L.A.; Manzoor, N. Proton translocating ATPase mediated fungicidal activity of eugenol and thymol. *Fitoterapia* **2010**, *81*, 1157–1162. [CrossRef]
29. Khan, I.; Bahuguna, A.; Shukla, S.; Aziz, F.; Chauhan, A.K.; Ansari, M.B.; Bajpai, V.K.; Huh, Y.S.; Kang, S.C. Antimicrobial potential of the food-grade additive carvacrol against uropathogenic *E. coli* based on membrane depolarization, reactive oxygen species generation, and molecular docking analysis. *Microb. Pathog.* **2020**, *142*, 104046. [CrossRef]
30. Domadia, P.; Swarup, S.; Bhunia, A.; Sivaraman, J.; Dasgupta, D. Inhibition of bacterial cell division protein FtsZ by cinnamaldehyde. *Biochem. Pharmacol.* **2007**, *74*, 831–840. [CrossRef]
31. Naz, F.; Kumar, M.; Koley, T.; Sharma, P.; Haque, M.A.; Kapil, A.; Kumar, M.; Kaur, P.; Ethayathulla, A.S. Screening of plant-based natural compounds as an inhibitor of FtsZ from *Salmonella typhi* using the computational, biochemical and in vitro cell-based studies. *Int. J. Biol. Macromol.* **2022**, *219*, 428–437. [CrossRef]
32. Zhou, F.; Ji, B.; Zhang, H.; Jiang, H.; Yang, Z.; Li, J.; Li, J.; Ren, Y.; Yan, W. Synergistic effect of thymol and carvacrol combined with chelators and organic acids against *Salmonella Typhimurium*. *J. Food Prot.* **2007**, *70*, 1704–1709. [CrossRef]
33. Fontanay, S.; Grare, M.; Mayer, J.; Finance, C.; Duval, R.E. Ursolic, oleanolic and betulinic acids: Antibacterial spectra and selectivity indexes. *J. Ethnopharmacol.* **2008**, *120*, 272–276. [CrossRef]
34. Hikal, D.M. Antibacterial Activity of Piperine and Black Pepper Oil. *Biosci. Biotechnol. Res. Asia* **2018**, *15*, 877–880. [CrossRef]
35. Khan, I.A.; Mirza, Z.M.; Kumar, A.; Verma, V.; Qazi, G.N. Piperine, a phytochemical potentiator of ciprofloxacin against *Staphylococcus aureus*. *Antimicrob. Agents Chemother.* **2006**, *50*, 810–812. [CrossRef]
36. Mittal, R.P.; Jaitak, V. Plant-derived natural alkaloids as new antimicrobial and adjuvant agents in existing antimicrobial therapy. *Curr. Drug Targets* **2019**, *20*, 1409–1433. [CrossRef]

37. Su, F.; Wang, J. Berberine inhibits the MexXY-OprM efflux pump to reverse imipenem resistance in a clinical carbapenem-resistant *Pseudomonas aeruginosa* isolate in a planktonic state. *Exp. Ther. Med.* **2018**, *15*, 467–472. [CrossRef] [PubMed]
38. Laudadio, E.; Cedraro, N.; Mangiaterra, G.; Citterio, B.; Mobbili, G.; Minnelli, C.; Bizzaro, D.; Biavasco, F.; Galeazzi, R. Natural alkaloid berberine activity against *Pseudomonas aeruginosa* MexXY-mediated aminoglycoside resistance: In Silico and in vitro Studies. *J. Nat. Prod.* **2019**, *82*, 1935–1944. [CrossRef]
39. Mujtaba, M.A.; Akhter, M.H.; Alam, M.S.; Ali, M.D.; Hussain, A. An updated review on therapeutic potential and recent advances in drug delivery of berberine: Current status and future prospect. *Curr. Pharm. Biotechnol.* **2022**, *23*, 60–71. [CrossRef]
40. Wei, W.Y.; Du, H.X.; Shao, C.Y.; Zhou, H.F.; Lu, Y.Y.; Yu, L.; Wan, H.T.; He, Y. Screening of antiviral components of ma huang tang and investigation on the ephedra alkaloids efficacy on influenza virus type A. *Front. Pharmacol.* **2019**, *10*, 961. [CrossRef]
41. He, N.; Wang, P.; Wang, P.; Ma, C.; Kang, W. Antibacterial mechanism of chelerythrine isolated from root of *Toddalia asiatica* (Linn) Lam. *BMC Complement. Altern. Med.* **2018**, *18*, 261. [CrossRef] [PubMed]
42. Prachayasittikul, V.; Prachayasittikul, S.; Ruchirawat, S.; Prachayasittikul, V. 8-Hydroxyquinolines: A review of their metal chelating properties and medicinal applications. *Drug Des. Dev. Ther.* **2013**, *7*, 1157–1178. [CrossRef]
43. Houdkova, M.; Rondevaldova, J.; Doskocil, I.; Kokoska, L. Evaluation of antibacterial potential and toxicity of plant volatile compounds using new broth microdilution volatilization method and modified MTT assay. *Fitoterapia* **2017**, *118*, 56–62. [CrossRef] [PubMed]
44. McMahon, J.B.; Currens, M.J.; Gulakowski, R.J.; Buckheit, R.W.; Lackman-Smith, C.; Hallock, Y.F.; Boyd, M.R. Michellamine B, a novel plant alkaloid, inhibits human immunodeficiency virus-induced cell killing by at least two distinct mechanisms. *Antimicrob. Agents Chemother.* **1995**, *39*, 484–488. [CrossRef] [PubMed]
45. Hamoud, R.; Reichling, J.; Wink, M. Synergistic antibacterial activity of the combination of the alkaloid sanguinarine with EDTA and the antibiotic streptomycin against multidrug resistant bacteria. *J. Pharm. Pharmacol.* **2015**, *67*, 264–273. [CrossRef] [PubMed]
46. Beuria, T.K.; Santra, M.K.; Panda, D. Sanguinarine blocks cytokinesis in bacteria by inhibiting FtsZ assembly and bundling. *Biochemistry* **2005**, *44*, 16584–16593. [CrossRef] [PubMed]
47. Yin, S.J.; Rao, G.X.; Wang, J.; Luo, L.Y.; He, G.H.; Wang, C.Y.; Ma, C.Y.; Luo, X.X.; Hou, Z.; Xu, G.L. Roemerine improves the survival rate of septicemic BALB/c mice by increasing the cell membrane permeability of *Staphylococcus aureus*. *PLoS ONE* **2015**, *10*, e0143863. [CrossRef] [PubMed]
48. Avci, F.G.; Atas, B.; Aksoy, C.S.; Kurpejovic, E.; Toplan, G.G.; Gurer, C.; Guillerminet, M.; Orelle, C.; Jault, J.M.; Akbulut, B.S. Repurposing bioactive aporphine alkaloids as efflux pump inhibitors. *Fitoterapia* **2019**, *139*, 104371. [CrossRef] [PubMed]
49. Costa, R.S.; Lins, M.O.; Le Hyaric, M.; Barros, T.F.; Velozo, E.S. In vitro antibacterial effects of Zanthoxylum tingoassuiba root bark extracts and two of its alkaloids against multiresistant *Staphylococcus aureus*. *Rev. Bras. Farmacogn.* **2017**, *27*, 195–198. [CrossRef]
50. Guzman, J.D.; Wube, A.; Evangelopoulos, D.; Gupta, A.; Hufner, A.; Basavannacharya, C.; Rahman, M.M.; Thomaschitz, C.; Bauer, R.; McHugh, T.D.; et al. Interaction of N-methyl-2-alkenyl-4-quinolones with ATP-dependent MurE ligase of *Mycobacterium tuberculosis*: Antibacterial activity, molecular docking and inhibition kinetics. *J. Antimicrob. Chemother.* **2011**, *66*, 1766–1772. [CrossRef] [PubMed]
51. Hochfellner, C.; Evangelopoulos, D.; Zloh, M.; Wube, A.; Guzman, J.D.; McHugh, T.D.; Kunert, O.; Bhakta, S.; Bucar, F. Antagonistic effects of indoloquinazoline alkaloids on antimycobacterial activity of evocarpine. *J. Appl. Microbiol.* **2015**, *118*, 864–872. [CrossRef] [PubMed]
52. Celiz, G.; Daz, M.; Audisio, M.C. Antibacterial activity of naringin derivatives against pathogenic strains. *J. Appl. Microbiol.* **2011**, *111*, 731–738. [CrossRef]
53. Pyrzynska, K. Hesperidin: A Review on Extraction Methods, Stability and Biological Activities. *Nutrients* **2022**, *14*, 2387. [CrossRef] [PubMed]
54. Babu, K.S.; Babu, T.H.; Srinivas, P.V.; Sastry, B.S.; Kishore, K.H.; Murty, U.S.; Rao, J.M. Synthesis and in vitro study of novel 7-*O*-acyl derivatives of Oroxylin A as antibacterial agents. *Bioorg. Med. Chem. Lett.* **2005**, *15*, 3953–3956. [CrossRef] [PubMed]
55. Liu, R.; Zhao, B.; Wang, D.E.; Yao, T.Y.; Pang, L.; Tu, Q.; Ahmed, S.M.; Liu, J.J.; Wang, J.Y. Nitrogen-containing apigenin analogs: Preparation and biological activity. *Molecules* **2012**, *17*, 14748–14764. [CrossRef] [PubMed]
56. Chinnam, N.; Dadi, P.K.; Sabri, S.A.; Ahmad, M.; Kabir, M.A.; Ahmad, Z. Dietary bioflavonoids inhibit *Escherichia coli* ATP synthase in a differential manner. *Int. J. Biol. Macromol.* **2010**, *46*, 478–486. [CrossRef] [PubMed]
57. Navarro-Martinez, M.D.; Navarro-Peran, E.; Cabezas-Herrera, J.; Ruiz-Gomez, J.; Garcia-Canovas, F.; Rodriguez-Lopez, J.N. Antifolate activity of epigallocatechin gallate against *Stenotrophomonas maltophilia*. *Antimicrob. Agents Chemother.* **2005**, *49*, 2914–2920. [CrossRef] [PubMed]
58. Pejin, B.; Ciric, A.; Markovic, J.D.; Glamoclija, J.; Nikolic, M.; Stanimirovic, B.; Sokovic, M. Quercetin potently reduces biofilm formation of the strain *Pseudomonas aeruginosa* PAO1 in vitro. *Curr. Pharm. Biotechnol.* **2015**, *16*, 733–737. [CrossRef] [PubMed]
59. Cushnie, T.P.; Lamb, A.J. Detection of galangin-induced cytoplasmic membrane damage in *Staphylococcus aureus* by measuring potassium loss. *J. Ethnopharmacol.* **2005**, *101*, 243–248. [CrossRef]
60. Gradisar, H.; Pristovsek, P.; Plaper, A.; Jerala, R. Green tea catechins inhibit bacterial DNA gyrase by interaction with its ATP binding site. *J. Med. Chem.* **2007**, *50*, 264–271. [CrossRef]
61. Fathima, A.; Rao, J.R. Selective toxicity of Catechin-a natural flavonoid towards bacteria. *Appl. Microbiol. Biotechnol.* **2016**, *100*, 6395–6402. [CrossRef] [PubMed]

62. Zhao, Q.; Chen, X.Y.; Martin, C. *Scutellaria baicalensis*, the golden herb from the garden of Chinese medicinal plants. *Sci. Bull.* **2016**, *61*, 1391–1398. [CrossRef]
63. Wang, J.; Zhu, J.; Meng, J.; Qiu, T.; Wang, W.; Wang, R.; Liu, J. Baicalin inhibits biofilm formation by influencing primary adhesion and aggregation phases in *Staphylococcus saprophyticus*. *Vet. Microbiol.* **2021**, *262*, 109242. [CrossRef] [PubMed]
64. Liu, N.; Zhang, N.; Zhang, S.; Zhang, L.; Liu, Q. Phloretin inhibited the pathogenicity and virulence factors against *Candida albicans*. *Bioengineered* **2021**, *12*, 2420–2431. [CrossRef]
65. Wang, D.; Xie, K.; Zou, D.; Meng, M.; Xie, M. Inhibitory effects of silybin on the efflux pump of methicillin-resistant *Staphylococcus aureus*. *Mol. Med. Rep.* **2018**, *18*, 827–833.
66. Yin, Z.; Dickschat, J.S. Engineering fungal terpene biosynthesis. *Nat. Prod. Rep.* **2022**. *Advance Article*. [CrossRef] [PubMed]
67. Bergman, M.E.; Davis, B.; Phillips, M.A. Medically useful plant terpenoids: Biosynthesis, occurrence, and mechanism of action. *Molecules* **2019**, *24*, 3961. [CrossRef] [PubMed]
68. Saha, P.; Rahman, F.I.; Hussain, F.; Rahman, S.M.A.; Rahman, M.M. Antimicrobial diterpenes: Recent development from natural sources. *Front. Pharmacol.* **2021**, *12*, 820312. [CrossRef]
69. Tu, Y. The discovery of artemisinin (qinghaosu) and gifts from Chinese medicine. *Nat. Med.* **2011**, *17*, 1217–1220. [CrossRef]
70. Tu, Y. Artemisinin-a gift from traditional Chinese medicine to the world (Nobel Lecture). *Angew. Chem. Int. Ed. Engl.* **2016**, *55*, 10210–10226. [CrossRef]
71. O'Neill, P.M.; Barton, V.E.; Ward, S.A. The molecular mechanism of action of artemisinin—The debate continues. *Molecules* **2010**, *15*, 1705–1721. [CrossRef] [PubMed]
72. Meshnick, S.R.; Taylor, T.E.; Kamchonwongpaisan, S. Artemisinin and the antimalarial endoperoxides: From herbal remedy to targeted chemotherapy. *Microbiol. Rev.* **1996**, *60*, 301–315. [CrossRef] [PubMed]
73. Klonis, N.; Crespo-Ortiz, M.P.; Bottova, I.; Abu-Bakar, N.; Kenny, S.; Rosenthal, P.J.; Tilley, L. Artemisinin activity against *Plasmodium falciparum* requires hemoglobin uptake and digestion. *Proc. Natl. Acad. Sci. USA* **2011**, *108*, 11405–11410. [CrossRef] [PubMed]
74. Wen, W.; Yu, R. Artemisinin biosynthesis and its regulatory enzymes: Progress and perspective. *Pharmacogn. Rev.* **2011**, *5*, 189–194. [PubMed]
75. Schramek, N.; Wang, H.; Römisch-Margl, W.; Keil, B.; Radykewicz, T.; Winzenhörlein, B.; Beerhues, L.; Bacher, A.; Rohdich, F.; Gershenzon, J.; et al. Artemisinin biosynthesis in growing plants of *Artemisia annua*. A $^{13}CO_2$ study. *Phytochemistry* **2010**, *71*, 179–187. [CrossRef] [PubMed]
76. Bouwmeester, H.J.; Wallaart, T.E.; Janssen, M.H.; van Loo, B.; Jansen, B.J.; Posthumus, M.A.; Schmidt, C.O.; De Kraker, J.W.; König, W.A.; Franssen, M.C. Amorpha-4,11-diene synthase catalyses the first probable step in artemisinin biosynthesis. *Phytochemistry* **1999**, *52*, 843–854. [CrossRef]
77. Mercke, P.; Bengtsson, M.; Bouwmeester, H.J.; Posthumus, M.A.; Brodelius, P.E. Molecular cloning, expression, and characterization of amorpha-4,11-diene synthase, a key enzyme of artemisinin biosynthesis in *Artemisia annua* L. *Arch. Biochem. Biophys.* **2000**, *381*, 173–180. [CrossRef]
78. Picaud, S.; Mercke, P.; He, X.; Sterner, O.; Brodelius, M.; Cane, D.E.; Brodelius, P.E. Amorpha-4,11-diene synthase: Mechanism and stereochemistry of the enzymatic cyclization of farnesyl diphosphate. *Arch. Biochem. Biophys.* **2006**, *448*, 150–155. [CrossRef]
79. Kim, S.H.; Heo, K.; Chang, Y.J.; Park, S.H.; Rhee, S.K.; Kim, S.U. Cyclization mechanism of amorpha-4,11-diene synthase, a key enzyme in artemisinin biosynthesis. *J. Nat. Prod.* **2006**, *69*, 758–762. [CrossRef]
80. Teoh, K.H.; Polichuk, D.R.; Reed, D.W.; Nowak, G.; Covello, P.S. *Artemisia annua* L. (Asteraceae) trichome-specific cDNAs reveal CYP71AV1, a cytochrome P450 with a key role in the biosynthesis of the antimalarial sesquiterpene lactone artemisinin. *FEBS Lett.* **2006**, *580*, 1411–1416. [CrossRef]
81. Paddon, C.J.; Westfall, P.J.; Pitera, D.J.; Benjamin, K.; Fisher, K.; McPhee, D.; Leavell, M.D.; Tai, A.; Main, A.; Eng, D.; et al. High-level semi-synthetic production of the potent antimalarial artemisinin. *Nature* **2013**, *496*, 528–532. [CrossRef] [PubMed]
82. Teoh, K.; Polichuk, D.; Reed, D.; Covello, P. Molecular cloning of an aldehyde dehydrogenase implicated in artemisinin biosynthesis in *Artemisia annua*. *Botany* **2009**, *87*, 635–642. [CrossRef]
83. Zhang, Y.; Teoh, K.H.; Reed, D.W.; Maes, L.; Goossens, A.; Olson, D.J.; Ross, A.R.; Covello, P.S. The molecular cloning of artemisinic aldehyde Δ11(13) reductase and its role in glandular trichome-dependent biosynthesis of artemisinin in *Artemisia annua*. *J. Biol. Chem.* **2008**, *283*, 21501–21508. [CrossRef] [PubMed]
84. Bertea, C.M.; Freije, J.R.; van der Woude, H.; Verstappen, F.W.; Perk, L.; Marquez, V.; De Kraker, J.W.; Posthumus, M.A.; Jansen, B.J.; de Groot, A.; et al. Identification of intermediates and enzymes involved in the early steps of artemisinin biosynthesis in *Artemisia annua*. *Planta Med.* **2005**, *71*, 40–47. [CrossRef]
85. Wallaart, T.E.; Pras, N.; Quax, W.J. Isolation and identification of dihydroartemisinic acid hydroperoxide from *Artemisia annua*: A novel biosynthetic precursor of artemisinin. *J. Nat. Prod.* **1999**, *62*, 1160–1162. [CrossRef]
86. Sy, L.-K.; Brown, G.D. The mechanism of the spontaneous autoxidation of dihydroartemisinic acid. *Tetrahedron* **2002**, *58*, 897–908. [CrossRef]
87. Arsenault, P.R.; Wobbe, K.K.; Weathers, P.J. Recent advances in artemisinin production through heterologous expression. *Curr. Med. Chem.* **2008**, *15*, 2886–2896. [CrossRef]
88. Zhang, Y.S.; Liu, B.Y.; Li, Z.Q.; Ye, H.C.; Wang, H.; Li, G.F.; Han, J.L. Molecular cloning of a classical plant peroxidase from *Artemisia annua* and its effect on the biosynthesis of artemisinin in vitro. *Acta Bot. Sin.* **2004**, *46*, 1338–1346.

89. Brown, G.D.; Sy, L.-K. In vivo transformations of artemisinic acid in *Artemisia annua* plants. *Tetrahedron* **2007**, *63*, 9548–9566. [CrossRef]
90. Brown, G.D.; Sy, L.-K. In vivo transformations of dihydroartemisinic acid in *Artemisia annua* plants. *Tetrahedron* **2004**, *60*, 1139–1159. [CrossRef]
91. Paddon, C.J.; Keasling, J.D. Semi-synthetic artemisinin: A model for the use of synthetic biology in pharmaceutical development. *Nat. Rev. Microbiol.* **2014**, *12*, 355–367. [CrossRef] [PubMed]
92. Martin, V.J.J.; Pitera, D.J.; Withers, S.T.; Newman, J.D.; Keasling, J.D. Engineering a mevalonate pathway in *Escherichia coli* for production of terpenoids. *Nat. Biotechnol.* **2003**, *21*, 796–802. [CrossRef] [PubMed]
93. Martin, V.J.; Yoshikuni, Y.; Keasling, J.D. The in vivo synthesis of plant sesquiterpenes by *Escherichia coli*. *Biotechnol. Bioeng.* **2001**, *75*, 497–503. [CrossRef] [PubMed]
94. Newman, J.D.; Marshall, J.; Chang, M.; Nowroozi, F.; Paradise, E.; Pitera, D.; Newman, K.L.; Keasling, J.D. High-level production of amorpha-4,11-diene in a two-phase partitioning bioreactor of metabolically engineered *Escherichia coli*. *Biotechnol. Bioeng.* **2006**, *95*, 684–691. [CrossRef]
95. Chang, M.C.; Keasling, J.D. Production of isoprenoid pharmaceuticals by engineered microbes. *Nat. Chem. Biol.* **2006**, *2*, 674–681. [CrossRef]
96. Ro, D.K.; Paradise, E.M.; Ouellet, M.; Fisher, K.J.; Newman, K.L.; Ndungu, J.M.; Ho, K.A.; Eachus, R.A.; Ham, T.S.; Kirby, J.; et al. Production of the antimalarial drug precursor artemisinic acid in engineered yeast. *Nature* **2006**, *440*, 940–943. [CrossRef] [PubMed]
97. Jimenez-Arellanes, A.; Meckes, M.; Torres, J.; Luna-Herrera, J. Antimycobacterial triterpenoids from *Lantana hispida* (Verbenaceae). *J. Ethnopharmacol.* **2007**, *111*, 202–205. [CrossRef]
98. Kozai, K.; Suzuki, J.; Okada, M.; Nagasaka, N. Effect of oleanolic acid-cyclodextrin inclusion compounds on dental caries by in vitro experiment and rat-caries model. *Microbios* **1999**, *97*, 179–188.
99. Horiuchi, K.; Shiota, S.; Hatano, T.; Yoshida, T.; Kuroda, T.; Tsuchiya, T. Antimicrobial activity of oleanolic acid from *Salvia officinalis* and related compounds on vancomycin-resistant enterococci (VRE). *Biol. Pharm. Bull.* **2007**, *30*, 1147–1149. [CrossRef]
100. Castellano, J.M.; Ramos-Romero, S.; Perona, J.S. Oleanolic Acid: Extraction, Characterization and Biological Activity. *Nutrients* **2022**, *14*, 623. [CrossRef]
101. Wolska, K.; Grudniak, A.M.; Fiecek, B.; Kraczkiewicz-Dowjat, A.; Kurek, A. Antibacterial activity of oleanolic and ursolic acids and their derivatives. *Open Life Sci.* **2010**, *5*, 543–553. [CrossRef]
102. Pollier, J.; Goossens, A. Oleanolic acid. *Phytochemistry* **2012**, *77*, 10–15. [CrossRef] [PubMed]
103. Phillips, D.R.; Rasbery, J.M.; Bartel, B.; Matsuda, S.P. Biosynthetic diversity in plant triterpene cyclization. *Curr. Opin. Plant Biol.* **2006**, *9*, 305–314. [CrossRef]
104. Abe, I. Enzymatic synthesis of cyclic triterpenes. *Nat. Prod. Rep.* **2007**, *24*, 1311–1331. [CrossRef] [PubMed]
105. Kushiro, T.; Shibuya, M.; Ebizuka, Y. β-amyrin synthase—Cloning of oxidosqualene cyclase that catalyzes the formation of the most popular triterpene among higher plants. *Eur. J. Biochem.* **1998**, *256*, 238–244. [CrossRef] [PubMed]
106. Abe, I.; Rohmer, M.; Prestwich, G.D. Enzymatic Cyclization of Squalene and Oxidosqualene to Sterols and Triterpenes. *Chem. Rev.* **1993**, *93*, 2189–2206. [CrossRef]
107. Seo, S.; Yoshimura, Y.; Uomori, A.; Takeda, K.; Seto, H.; Ebizuka, Y.; Sankawa, U. Biosynthesis of triterpenes, ursolic acid and oleanolic acid in tissue cultures of *Rabdosia japonica* Hara fed [5-$^{13}C^2H_2$] mevalonolactone and [2-$^{13}C^2H_3$] acetate. *J. Am. Chem. Soc.* **1988**, *110*, 1740–1745. [CrossRef]
108. Fukushima, E.O.; Seki, H.; Ohyama, K.; Ono, E.; Umemoto, N.; Mizutani, M.; Saito, K.; Muranaka, T. CYP716A subfamily members are multifunctional oxidases in triterpenoid biosynthesis. *Plant Cell Physiol.* **2011**, *52*, 2050–2061. [CrossRef] [PubMed]
109. Carelli, M.; Biazzi, E.; Panara, F.; Tava, A.; Scaramelli, L.; Porceddu, A.; Graham, N.; Odoardi, M.; Piano, E.; Arcioni, S.; et al. *Medicago truncatula* CYP716A12 is a multifunctional oxidase involved in the biosynthesis of hemolytic saponins. *Plant Cell* **2011**, *23*, 3070–3081. [CrossRef] [PubMed]
110. Han, Y.; Sun, Z.; Chen, W. Antimicrobial Susceptibility and Antibacterial Mechanism of Limonene against Listeria monocytogenes. *Molecules* **2019**, *25*, 33. [CrossRef] [PubMed]
111. Cheng, S.; Liu, X.; Jiang, G.; Wu, J.; Zhang, J.L.; Lei, D.; Yuan, Y.J.; Qiao, J.; Zhao, G.R. Orthogonal Engineering of Biosynthetic Pathway for Efficient Production of Limonene in *Saccharomyces cerevisiae*. *ACS Synth. Biol.* **2019**, *8*, 968–975. [CrossRef] [PubMed]
112. Parthasarathy, A.; Borrego, E.J.; Savka, M.A.; Dobson, R.C.J.; Hudson, A.O. Amino acid-derived defense metabolites from plants: A potential source to facilitate novel antimicrobial development. *J. Biol. Chem.* **2021**, *296*, 100438. [CrossRef] [PubMed]
113. Othman, L.; Sleiman, A.; Abdel-Massih, R.M. Antimicrobial activity of polyphenols and alkaloids in middle eastern plants. *Front. Microbiol.* **2019**, *10*, 911. [CrossRef] [PubMed]
114. Cushnie, T.P.T.; Cushnie, B.; Lamb, A.J. Alkaloids: An overview of their antibacterial, antibiotic-enhancing and antivirulence activities. *Int. J. Antimicrob. Agents* **2014**, *44*, 377–386. [CrossRef]
115. Thawabteh, A.; Juma, S.; Bader, M.; Karaman, D.; Scrano, L.; Bufo, S.A.; Karaman, R. The biological activity of natural alkaloids against herbivores, cancerous cells and pathogens. *Toxins* **2019**, *11*, 656. [CrossRef]
116. Khameneh, B.; Iranshahy, M.; Ghandadi, M.; Atashbeyk, D.G.; Bazzaz, B.S.F.; Iranshahi, M. Investigation of the antibacterial activity and efflux pump inhibitory effect of co-loaded piperine and gentamicin nanoliposomes in methicillin-resistant *Staphylococcus aureus*. *Drug Dev. Ind. Pharm.* **2015**, *41*, 989–994. [CrossRef] [PubMed]

117. Mabhiza, D.; Chitemerere, T.; Mukanganyama, S. Antibacterial Properties of Alkaloid Extracts from *Callistemon citrinus* and *Vernonia adoensis* against *Staphylococcus aureus* and *Pseudomonas aeruginosa*. *Int. J. Med. Chem.* **2016**, *2016*, 6304163. [CrossRef] [PubMed]
118. Wink, M.; Ashour, M.L.; El-Readi, M.Z. Secondary metabolites from plants inhibiting ABC transporters and reversing resistance of cancer cells and microbes to cytotoxic and antimicrobial agents. *Front. Microbiol.* **2012**, *3*, 130. [CrossRef]
119. Gibbons, S.; Udo, E.E. The effect of reserpine, a modulator of multidrug efflux pumps, on the *in vitro* activity of tetracycline against clinical isolates of methicillin resistant *Staphylococcus aureus* (MRSA) possessing the tet (K) determinant. *Phytother. Res.* **2000**, *14*, 139–140. [CrossRef]
120. Stermitz, F.R.; Lorenz, P.; Tawara, J.N.; Zenewicz, L.A.; Lewis, K. Synergy in a medicinal plant: Antimicrobial action of berberine potentiated by 5′-methoxyhydnocarpin, a multidrug pump inhibitor. *Proc. Natl. Acad. Sci. USA* **2000**, *97*, 1433–1437. [CrossRef] [PubMed]
121. Ahmed, M.; Borsch, C.M.; Neyfakh, A.A.; Schuldiner, S. Mutants of the Bacillus subtilis multidrug transporter Bmr with altered sensitivity to the antihypertensive alkaloid reserpine. *J. Biol. Chem.* **1993**, *268*, 11086–11089. [CrossRef]
122. Xie, Q.; Johnson, B.R.; Wenckus, C.S.; Fayad, M.I.; Wu, C.D. Efficacy of berberine, an antimicrobial plant alkaloid, as an endodontic Irrigant against a mixed-culture biofilm in an in vitro tooth model. *J. Endodont.* **2012**, *38*, 1114–1117. [CrossRef] [PubMed]
123. Yi, Z.B.; Yan, Y.; Liang, Y.Z.; Bao, Z. Evaluation of the antimicrobial mode of berberine by LC/ESI-MS combined with principal component analysis. *J. Pharm. Biomed. Anal.* **2007**, *44*, 301–304. [CrossRef]
124. Albert, A.; Magrath, D. The choice of a chelating agent for inactivating trace metals: II. Derivatives of oxine (8-hydroxyquinoline). *Biochem. J.* **1947**, *41*, 534–545. [CrossRef] [PubMed]
125. Morita, Y.; Nakashima, K.; Nishino, K.; Kotani, K.; Tomida, J.; Inoue, M.; Kawamura, Y. Berberine is a novel type efflux inhibitor which attenuates the MexXY-mediated aminoglycoside resistance in *Pseudomonas aeruginosa*. *Front. Microbiol.* **2016**, *7*, 1223. [CrossRef] [PubMed]
126. Tegos, G.; Stermitz, F.R.; Lomovskaya, O.; Lewis, K. Multidrug pump inhibitors uncover remarkable activity of plant antimicrobials. *Antimicrob. Agents Chemother.* **2002**, *46*, 3133–3141. [CrossRef]
127. Das, S.; Kumar, G.S.; Ray, A.; Maiti, M. Spectroscopic and thermodynamic studies on the binding of sanguinarine and berberine to triple and double helical DNA and RNA structures. *J. Biomol. Struct. Dyn.* **2003**, *20*, 703–714. [CrossRef] [PubMed]
128. Bhadra, K.; Maiti, M.; Kumar, G.S. Berberine-DNA complexation: New insights into the cooperative binding and energetic aspects. *Biochim. Biophys. Acta.* **2008**, *1780*, 1054–1061. [CrossRef] [PubMed]
129. Eckhardt, K.; Zeller, K.P.; Siehl, H.U.; Berger, S.; Sicker, D. Look, the Yellow is so near: Berberine chloride from berberine bark. *Chem. Unserer. Zeit* **2017**, *51*, 344–356. [CrossRef]
130. He, S.M.; Liang, Y.L.; Cong, K.; Chen, G.; Zhao, X.; Zhao, Q.M.; Zhang, J.J.; Wang, X.; Dong, Y.; Yang, J.L.; et al. Identification and characterization of genes involved in benzylisoquinoline alkaloid biosynthesis in *Coptis* species. *Front. Plant Sci.* **2018**, *9*, 731. [CrossRef] [PubMed]
131. Samanani, N.; Facchini, P.J. Purification and characterization of norcoclaurine synthase. The first committed enzyme in benzylisoquinoline alkaloid biosynthesis in plants. *J. Biol. Chem.* **2002**, *277*, 33878–33883. [CrossRef]
132. Samanani, N.; Liscombe, D.K.; Facchini, P.J. Molecular cloning and characterization of norcoclaurine synthase, an enzyme catalyzing the first committed step in benzylisoquinoline alkaloid biosynthesis. *Plant J.* **2004**, *40*, 302–313. [CrossRef]
133. Minami, H.; Dubouzet, E.; Iwasa, K.; Sato, F. Functional analysis of norcoclaurine synthase in *Coptis japonica*. *J. Biol. Chem.* **2007**, *282*, 6274–6282. [CrossRef]
134. Rueffer, M.; Nagakura, N.; Zenk, M.H. Partial purification and properties of *S*-adenosylmethionine: (*R*), (*S*)-Norlaudanosoline-6-*O*-methyltransferase from *Argemone platyceras* cell cultures. *Planta Med.* **1983**, *49*, 131–137. [CrossRef]
135. Sato, F.; Tsujita, T.; Katagiri, Y.; Yoshida, S.; Yamada, Y. Purification and characterization of *S*-adenosyl-L-methionine: Norcoclaurine 6-*O*-methyltransferase from cultured *Coptis japonica* cells. *Eur. J. Biochem.* **1994**, *225*, 125–131. [CrossRef]
136. Loeffler, S.; Deusneumann, B.; Zenk, M.H. *S*-adenosyl-L-methionine-(*S*)-coclaurine-*N*-methyltransferase from *Tinospora-Cordifolia*. *Phytochemistry* **1995**, *38*, 1387–1395. [CrossRef]
137. Choi, K.B.; Morishige, T.; Shitan, N.; Yazaki, K.; Sato, F. Molecular cloning and characterization of coclaurine *N*-methyltransferase from cultured cells of *Coptis japonica*. *J. Biol. Chem.* **2002**, *277*, 830–835. [CrossRef]
138. Morishige, T.; Tsujita, T.; Yamada, Y.; Sato, F. Molecular characterization of the *S*-adenosyl-L-methionine: 3′-hydroxy-*N*-methylcoclaurine 4′-*O*-methyltransferase involved in isoquinoline alkaloid biosynthesis in *Coptis japonica*. *J. Biol. Chem.* **2000**, *275*, 23398–23405. [CrossRef]
139. Frenzel, T.; Zenk, M.H. *S*-Adenosyl-L-Methionine—3′-Hydroxy-*N*-Methyl-(*S*)-Coclaurine-4′-*O*-Methyl Transferase, a Regioselective and Stereoselective Enzyme of the (*S*)-Reticuline Pathway. *Phytochemistry* **1990**, *29*, 3505–3511. [CrossRef]
140. Chapple, C. Molecular-genetic analysis of plant cytochrome P450-dependent monooxygenases. *Annu. Rev. Plant Physiol. Plant Mol. Biol.* **1998**, *49*, 311–343. [CrossRef]
141. Mizutani, M.; Sato, F. Unusual P450 reactions in plant secondary metabolism. *Arch. Biochem. Biophys.* **2011**, *507*, 194–203. [CrossRef] [PubMed]
142. Ikezawa, N.; Tanaka, M.; Nagayoshi, M.; Shinkyo, R.; Sakaki, T.; Inouye, K.; Sato, F. Molecular cloning and characterization of CYP719, a methylenedioxy bridge-forming enzyme that belongs to a novel P450 family, from cultured *Coptis japonica* cells. *J. Biol. Chem.* **2003**, *278*, 38557–38565. [CrossRef] [PubMed]

143. Kutchan, T.M.; Dittrich, H. Characterization and mechanism of the berberine bridge enzyme, a covalently flavinylated oxidase of benzophenanthridine alkaloid biosynthesis in plants. *J. Biol. Chem.* **1995**, *270*, 24475–24481. [CrossRef] [PubMed]
144. Xu, Z.X.; Xia, L.Q.; Sun, M.S.; Huang, P.; Zeng, J.G. Effects of codon optimization, *N*-terminal truncation and gene dose on the heterologous expression of berberine bridge enzyme. *World J. Microb. Biot.* **2022**, *38*, 77. [CrossRef]
145. Muemmler, S.; Rueffer, M.; Nagakura, N.; Zenk, M.H. *S*-adenosyl-L-methionine: (*S*)-scoulerine 9-*O*-methyltransferase, a highly stereo- and regio-specific enzyme in tetrahydroprotoberberine biosynthesis. *Plant Cell Rep.* **1985**, *4*, 36–39. [CrossRef]
146. Takeshita, N.; Fujiwara, H.; Mimura, H.; Fitchen, J.H.; Yamada, Y.; Sato, F. Molecular cloning and characterization of *S*-adenosyl-L-methionine: Scoulerine-9-*O*-methyltransferase from cultured cells of *Coptis japonica*. *Plant Cell Physiol.* **1995**, *36*, 29–36.
147. Galneder, E.; Rueffer, M.; Wanner, G.; Tabata, M.; Zenk, M.H. Alternative final steps in berberine biosynthesis in *Coptis japonica* cell cultures. *Plant Cell Rep.* **1988**, *7*, 1–4. [CrossRef]
148. Okada, N.; Koizumi, N.; Tanaka, T.; Ohkubo, H.; Nakanishi, S.; Yamada, Y. Isolation, sequence, and bacterial expression of a cDNA for (*S*)-tetrahydroberberine oxidase from cultured berberine-producing *Coptis japonica* cells. *Proc. Natl. Acad. Sci. USA* **1989**, *86*, 534–538. [CrossRef]
149. Payne, J.T.; Valentic, T.R.; Smolke, C.D. Complete biosynthesis of the bisbenzylisoquinoline alkaloids guattegaumerine and berbamunine in yeast. *Proc. Natl. Acad. Sci. USA* **2021**, *118*, e2112520118. [CrossRef]
150. Hawkins, K.M.; Smolke, C.D. Production of benzylisoquinoline alkaloids in *Saccharomyces cerevisiae*. *Nat. Chem. Biol.* **2008**, *4*, 564–573. [CrossRef]
151. Park, S.U.; Facchini, P.J. *Agrobacterium rhizogenes*-mediated transformation of opium poppy, *Papaver somniferum* L., and California poppy, *Eschscholzia californica* cham., root cultures. *J. Exp. Bot.* **2000**, *51*, 1005–1016. [CrossRef]
152. Sivakumar, G. Upstream biomanufacturing of pharmaceutical colchicine. *Crit. Rev. Biotechnol.* **2018**, *38*, 83–92. [CrossRef]
153. Jana, S.; Shekhawat, G.S. Critical review on medicinally potent plant species: *Gloriosa superba*. *Fitoterapia* **2010**, *82*, 293–301. [CrossRef]
154. Slobodnick, A.; Shah, B.; Pillinger, M.H.; Krasnokutsky, S. Colchicine: Old and New. *Am. J. Med.* **2015**, *128*, 461–470. [CrossRef]
155. Karamanou, M.; Tsoucalas, G.; Pantos, K.; Androutsos, G. Isolating Colchicine in 19th Century: An old drug revisited. *Curr. Pharm. Des.* **2018**, *24*, 654–658. [CrossRef]
156. Zhang, F.S.; He, Q.Z.; Qin, C.H.; Little, P.J.; Weng, J.P.; Xu, S.W. Therapeutic potential of colchicine in cardiovascular medicine: A pharmacological review. *Acta. Pharmacol. Sin.* **2022**, *43*, 2173–2190. [CrossRef]
157. Bhattacharyya, B.; Panda, D.; Gupta, S.; Banerjee, M. ChemInform Abstract: Anti-mitotic activity of colchicine and the structural basis for its interaction with tubulin. *ChemInform* **2008**, *28*, 155–183. [CrossRef]
158. Chiu, L.; Lo, C.H.; Shen, M.; Chiu, N.; Aggarwal, R.; Lee, J.; Choi, Y.G.; Lam, H.; Prsic, E.H.; Chow, R.; et al. Colchicine use in patients with COVID-19: A systematic review and meta-analysis. *PLoS ONE* **2021**, e0261358. [CrossRef]
159. Leete, E.; Nemeth, P.E. The Biogenesis of the Alkaloids of *Colchicum*. I. The incorporation of phenylalanine into colchicine. *J. Am. Chem. Soc.* **1960**, *82*, 6055–6057. [CrossRef]
160. Herbert, R.B. The biosynthesis of plant alkaloids and nitrogenous microbial metabolites. *Nat. Prod. Rep.* **2003**, *20*, 494–508. [CrossRef]
161. Battersby, A.R.; Herbert, R.B.; McDonald, E.; Ramage, R.; Clements, J.H. Alkaloid biosynthesis. 18. Biosynthesis of colchicine from the 1-phenethylisoquinoline system. *J. Chem. Soc. Perkin. 1* **1972**, *14*, 1741–1746. [CrossRef]
162. Herbert, R.B.; Kattah, A.E.; Knagg, E. The Biosynthesis of the phenethylisoquinoline alkaloid colchicine. Early and intermediate stages. *Tetrahedron* **1990**, *46*, 7119–7138. [CrossRef]
163. Nasreen, A.; Gundlach, H.; Zenk, M.H. Incorporation of phenethylisoquinolines into colchicine in isolated seeds of *Colchicum autumnale*. *Phytochemistry* **1997**, *46*, 107–115. [CrossRef]
164. Leete, E. Biosynthesis of the tropolone ring of colchicine. *Tetrahedron Lett.* **1965**, *6*, 333–336. [CrossRef]
165. Herbert, R.B.; Knagg, E. The biosynthesis of the phenethylisoquinoline alkaloid, colchicine, from cinnamaldehyde and dihydrocinnamaldehyde. *Tetrahedron Lett.* **1986**, *27*, 1099–1102. [CrossRef]
166. Maier, U.H.; Zenk, M.H. Colchicine is formed by *para-para* phenol coupling from autumnaline. *Tetrahedron Lett.* **1997**, *38*, 7357–7360. [CrossRef]
167. Rueffer, M.; Zenk, M.H. Microsome-mediated transformation of *O*-methylandrocymbine to demecolcine and colchicine. *FEBS Lett.* **1998**, *438*, 111–113. [CrossRef]
168. Sheldrake, P.W.; Suckling, K.E.; Woodhouse, R.N.; Murtagh, A.J. Biosynthesis Part 30.1 Colchicine: Studies on the ring expansion step focusing on the fate of the hydrogens at C-4 of autumnaline. *J. Chem. Soc. Perkin Trans. 1* **1998**, *18*, 3003–3010. [CrossRef]
169. Nett, R.S.; Lau, W.; Sattely, E.S. Discovery and engineering of colchicine alkaloid biosynthesis. *Nature* **2020**, *584*, 148–153. [CrossRef]
170. Nett, R.S.; Sattely, E.S. Total biosynthesis of the tubulin-binding alkaloid colchicine. *J. Am. Chem. Soc.* **2021**, *143*, 19454–19465. [CrossRef]
171. Maeda, H.; Dudareva, N. The shikimate pathway and aromatic amino acid biosynthesis in plants. *Annu. Rev. Plant Biol.* **2012**, *63*, 73–105. [CrossRef]
172. Sunnadeniya, R.; Bean, A.; Brown, M.; Akhavan, N.; Hatlestad, G.; Gonzalez, A.; Symonds, V.V.; Lloyd, A. Tyrosine Hydroxylation in betalain pigment biosynthesis is performed by cytochrome P450 enzymes in beets (*Beta vulgaris*). *PLoS ONE* **2016**, *11*, e0149417. [CrossRef] [PubMed]

173. Luk, L.Y.P.; Bunn, S.; Liscombe, D.K.; Facchini, P.J.; Tanner, M.E. Mechanistic studies on norcoclaurine synthase of benzylisoquinoline alkaloid biosynthesis: An enzymatic Pictet-Spengler reaction. *Biochemistry* **2007**, *46*, 10153–10161. [CrossRef] [PubMed]
174. Ruff, B.M.; Brase, S.; O'Connor, S.E. Biocatalytic production of tetrahydroisoquinolines. *Tetrahedron Lett.* **2012**, *53*, 1071–1074. [CrossRef]
175. Nishihachijo, M.; Hirai, Y.; Kawano, S.; Nishiyama, A.; Minami, H.; Katayama, T.; Yasohara, Y.; Sato, F.; Kumagai, H. Asymmetric synthesis of tetrahydroisoquinolines by enzymatic Pictet-Spengler reaction. *Biosci. Biotech. Bioch.* **2014**, *78*, 701–707. [CrossRef]
176. Barker, A.C.; Julian, D.R.; Ramage, R.; Woodhouse, R.N.; Hardy, G.; McDonald, E.; Battersby, A.R. Biosynthesis. Part 28.[1,2] Colchicine: Definition of intermediates between *O*-mehylandrocymbine and colchicine and studies on speciosine. *J. Chem. Soc. Perk Trans.* **1998**, *30*, 2989–2994. [CrossRef]
177. McDonald, E.; Ramage, R.; Woodhouse, R.N.; Underhill, E.W.; Wetter, L.R.; Battersby, A.R. Biosynthesis. Part 27.[1,2] Colchicine: Studies of the phenolic oxidative coupling and ring-expansion processes based on incorporation of multiply labelled 1-phenethylisoquinolines. *J. Chem. Soc. Perk Trans.* **1998**, *18*, 2979–2987. [CrossRef]
178. Shang, X.F.; Morris-Natschke, S.L.; Liu, Y.Q.; Guo, X.; Xu, X.S.; Goto, M.; Li, J.C.; Yang, G.Z.; Lee, K.H. Biologically active quinoline and quinazoline alkaloids part I. *Med. Res. Rev.* **2018**, *38*, 775–828. [CrossRef] [PubMed]
179. Subramani, R.; Narayanasamy, M.; Feussner, K.D. Plant-derived antimicrobials to fight against multi-drug-resistant human pathogens. *3 Biotech* **2017**, *7*, 172. [CrossRef]
180. Casciaro, B.; Mangiardi, L.; Cappiello, F.; Romeo, I.; Loffredo, M.R.; Iazzetti, A.; Calcaterra, A.; Goggiamani, A.; Ghirga, F.; Mangoni, M.L.; et al. Naturally-Occurring alkaloids of plant origin as potential antimicrobials against antibiotic-resistant infections. *Molecules* **2020**, *25*, 3619. [CrossRef]
181. Mitchell, G.; Gattuso, M.; Grondin, G.; Marsault, E.; Bouarab, K.; Malouin, F. Tomatidine Inhibits Replication of *Staphylococcus aureus* Small-Colony Variants in Cystic Fibrosis Airway Epithelial Cells. *Antimicrob. Agents Chemother.* **2011**, *55*, 1937–1945. [CrossRef]
182. Dorsaz, S.; Snaka, T.; Favre-Godal, Q.; Maudens, P.; Boulens, N.; Furrer, P.; Ebrahimi, S.N.; Hamburger, M.; Allemann, E.; Gindro, K.; et al. Identification and mode of action of a plant natural product targeting human fungal pathogens. *Antimicrob. Agents Chemother.* **2017**, *61*, e00829-17. [CrossRef] [PubMed]
183. Akiyama, R.; Lee, J.H.; Nakayasu, M.; Osakabe, K.; Osakabe, Y.; Umemoto, N.; Saito, K.; Muranaka, T.; Sugimoto, Y.; Mizutani, M. Characterization of steroid 5α-reductase involved in α-tomatine biosynthesis in tomatoes. *Plant Biotechnol.* **2019**, *36*, 253–263. [CrossRef] [PubMed]
184. Abdel-Motaal, F.F.; Nassar, M.S.M.; El-Zayat, S.A.; El-Sayed, M.A.; Ito, S. Responses of fungi to tropane alkaloids produced by a medicinal plant *Hyoscyamus muticus* (Egyptian Henbane). *Folia Microbiol.* **2009**, *54*, 207–212. [CrossRef] [PubMed]
185. Abdel-Motaal, F.F.; El-Zayat, S.A.; Kosaka, Y.; El-Sayed, M.A.; Kashima, R.; Maeda, Y.; Nassar, M.S.M.; Ito, S. Antifungal activities of hyoscyamine and scopolamine against two major rice pathogens: *Magnaporthe oryzae* and *Rhizoctonia solani*. *J. Gen. Plant Pathol.* **2010**, *76*, 102–111. [CrossRef]
186. Srinivasan, P.; Smolke, C.D. Biosynthesis of medicinal tropane alkaloids in yeast. *Nature* **2020**, *585*, 614–619. [CrossRef] [PubMed]
187. Davison, E.K.; Brimble, M.A. Natural product derived privileged scaffolds in drug discovery. *Curr. Opin. Chem. Biol.* **2019**, *52*, 1–8. [CrossRef]
188. Shen, N.; Wang, T.F.; Gan, Q.; Liu, S.; Wang, L.; Jin, B. Plant flavonoids: Classification, distribution, biosynthesis, and antioxidant activity. *Food Chem.* **2022**, *383*, 132531. [CrossRef] [PubMed]
189. Chiu, H.H.; Hsieh, Y.C.; Chen, Y.H.; Wang, H.Y.; Lu, C.Y.; Chen, C.J.; Li, Y.K. Three important amino acids control the regioselectivity of flavonoid glucosidation in glycosyltransferase-1 from *Bacillus cereus*. *Appl. Microbiol. Biotechnol.* **2016**, *100*, 8411–8424. [CrossRef] [PubMed]
190. Roy, A.; Khan, A.; Ahmad, I.; Alghamdi, S.; Rajab, B.S.; Babalghith, A.O.; Alshahrani, M.Y.; Islam, S.; Islam, M.R. Flavonoids a bioactive compound from medicinal plants and its therapeutic applications. *Biomed. Res. Int.* **2022**, *2022*, 5445291. [CrossRef] [PubMed]
191. Li, Y.; Kumar, P.S.; Tan, S.; Huang, C.; Xiang, Z.; Qiu, J.; Tan, X.; Luo, J.; He, M. Anticancer and antibacterial flavonoids from the callus of *Ampelopsis grossedentata*; a new weapon to mitigate the proliferation of cancer cells and bacteria. *RSC Adv.* **2022**, *12*, 24130–24138. [CrossRef] [PubMed]
192. Di Pierro, F.; Iqtadar, S.; Khan, A.; Mumtaz, S.U.; Chaudhry, M.M.; Bertuccioli, A.; Derosa, G.; Maffioli, P.; Togni, S.; Riva, A.; et al. Potential clinical benefits of quercetin in the early stage of COVID-19: Results of a second, pilot, randomized, controlled and open-label clinical trial. *Int. J. Gen. Med.* **2021**, *14*, 2807–2816. [CrossRef] [PubMed]
193. Biharee, A.; Sharma, A.; Kumar, A.; Jaitak, V. Antimicrobial flavonoids as a potential substitute for overcoming antimicrobial resistance. *Fitoterapia* **2020**, *146*, 104720. [CrossRef] [PubMed]
194. Xie, Y.X.; Yang, W.J.; Tang, F.; Chen, X.Q.; Ren, L.C. Antibacterial activities of flavonoids: Structure-activity relationship and mechanism. *Curr. Med. Chem.* **2015**, *22*, 132–149. [CrossRef] [PubMed]
195. Smejkal, K.; Chudík, S.; Kloucek, P.; Marek, R.; Cvacka, J.; Urbanová, M.; Julínek, O.; Kokoska, L.; Slapetová, T.; Holubová, P.; et al. Antibacterial *C*-geranylflavonoids from *Paulownia tomentosa* Fruits. *J. Nat. Prod.* **2008**, *71*, 706–709. [CrossRef]
196. Tran, T.D.; Do, T.H.; Tran, N.C.; Ngo, T.D.; Huynh, T.N.; Tran, C.D.; Thai, K.M. Synthesis and anti Methicillin resistant *Staphylococcus aureus* activity of substituted chalcones alone and in combination with non-beta-lactam antibiotics. *Bioorg. Med. Chem. Lett.* **2012**, *22*, 4555–4560. [CrossRef] [PubMed]

197. Rütschlin, S.; Böttcher, T. Inhibitors of Bacterial Swarming Behavior. *Chemistry* **2020**, *26*, 964–979. [CrossRef]
198. Austin, M.B.; Noel, J.P. The chalcone synthase superfamily of type III polyketide synthases. *Nat. Prod. Rep.* **2003**, *20*, 79–110. [CrossRef] [PubMed]
199. Reimold, U.; Kröger, M.; Kreuzaler, F.; Hahlbrock, K. Coding and 3′ non-coding nucleotide sequence of chalcone synthase mRNA and assignment of amino acid sequence of the enzyme. *Embo. J.* **1983**, *2*, 1801–1805. [CrossRef]
200. Pandith, S.A.; Ramazan, S.; Khan, M.I.; Reshi, Z.A.; Shah, M.A. Chalcone synthases (CHSs): The symbolic type III polyketide synthases. *Planta* **2019**, *251*, 15. [CrossRef]
201. Ferrer, J.L.; Jez, J.M.; Bowman, M.E.; Dixon, R.A.; Noel, J.P. Structure of chalcone synthase and the molecular basis of plant polyketide biosynthesis. *Nat. Struct. Biol.* **1999**, *6*, 775–784. [PubMed]
202. Jones, J.A.; Vernacchio, V.R.; Sinkoe, A.L.; Collins, S.M.; Ibrahim, M.H.A.; Lachance, D.M.; Hahn, J.; Koffas, M.A.G. Experimental and computational optimization of an *Escherichia coli* co-culture for the efficient production of flavonoids. *Metab. Eng.* **2016**, *35*, 55–63. [CrossRef] [PubMed]
203. Zhou, S.; Lyu, Y.; Li, H.; Koffas, M.A.G.; Zhou, J. Fine-tuning the (2S)-naringenin synthetic pathway using an iterative high-throughput balancing strategy. *Biotechnol. Bioeng.* **2019**, *116*, 1392–1404. [CrossRef]
204. Zhao, Q.; Zhang, Y.; Wang, G.; Hill, L.; Weng, J.K.; Chen, X.Y.; Xue, H.; Martin, C. A specialized flavone biosynthetic pathway has evolved in the medicinal plant, *Scutellaria baicalensis*. *Sci. Adv.* **2016**, *2*, e1501780. [CrossRef]
205. Kitamura, K.; Honda, M.; Yoshizaki, H.; Yamamoto, S.; Nakane, H.; Fukushima, M.; Ono, K.; Tokunaga, T. Baicalin, an inhibitor of HIV-1 production in vitro. *Antiviral Res.* **1998**, *37*, 131–140. [CrossRef]
206. Li, B.Q.; Fu, T.; Dongyan, Y.; Mikovits, J.A.; Ruscetti, F.W.; Wang, J.M. Flavonoid baicalin inhibits HIV-1 infection at the level of viral entry. *Biochem. Biophys. Res. Commun.* **2000**, *276*, 534–538. [CrossRef]
207. Liu, W.X.; Feng, Y.; Yu, S.H.; Fan, Z.Q.; Li, X.L.; Li, J.Y.; Yin, H.F. The flavonoid biosynthesis network in plants. *Int. J. Mol. Sci.* **2021**, *22*, 12824. [CrossRef]
208. Pei, T.L.; Yan, M.X.; Li, T.; Li, X.Q.; Yin, Y.J.; Cui, M.Y.; Fang, Y.M.; Liu, J.; Kong, Y.; Xu, P.; et al. Characterization of UDP-glycosyltransferase family members reveals how major flavonoid glycoside accumulates in the roots of *Scutellaria baicalensis*. *BMC Genom.* **2022**, *23*, 169. [CrossRef] [PubMed]
209. Han, H.D.; Lee, Y.S.; Ahn, J.H. Biological synthesis of baicalein derivatives using *Escherichia coli*. *J. Microbiol. Biotechnol.* **2016**, *26*, 1918–1923. [CrossRef]
210. Liao, J.J.; Xie, L.; Liu, T.Y.; Mo, C.M.; Cui, S.R.; Jia, X.L.; Huang, X.Y.; Luo, Z.L.; Ma, X.J. Heterologous biosynthesis of health-promoting baicalein in *Lycopersicon esculentum*. *Molecules* **2022**, *27*, 3086. [CrossRef]
211. Caputi, L.; Franke, J.; Farrow, S.C.; Chung, K.; Payne, R.M.E.; Nguyen, T.D.; Dang, T.T.; Soares Teto Carqueijeiro, I.; Koudounas, K.; Dugé de Bernonville, T.; et al. Missing enzymes in the biosynthesis of the anticancer drug vinblastine in *Madagascar periwinkle*. *Science* **2018**, *360*, 1235–1239. [CrossRef] [PubMed]
212. Zhang, J.; Hansen, L.G.; Gudich, O.; Viehrig, K.; Lassen, L.M.M.; Schrübbers, L.; Adhikari, K.B.; Rubaszka, P.; Carrasquer-Alvarez, E.; Chen, L.; et al. A microbial supply chain for production of the anti-cancer drug vinblastine. *Nature* **2022**, *609*, 341–347. [CrossRef] [PubMed]
213. Lau, W.; Sattely, E.S. Six enzymes from mayapple that complete the biosynthetic pathway to the etoposide aglycone. *Science* **2015**, *349*, 1224–1228. [CrossRef] [PubMed]

antibiotics

Review

Diverse Metabolites and Pharmacological Effects from the Basidiomycetes *Inonotus hispidus*

Zhen-xin Wang [1,†], Xi-long Feng [1,†], Chengwei Liu [2], Jin-ming Gao [1] and Jianzhao Qi [1,*]

1 Shaanxi Key Laboratory of Natural Products Chemistry and Biology, College of Chemistry and Pharmacy, Northwest Agriculture and Forestry University, Xiangyang 712100, China
2 College of Life Sciences, Northeast Forestry University, Harbin 150040, China
* Correspondence: qjz@nwafu.edu.cn; Tel.: +86-29-87092381
† These authors contributed equally to this work.

Abstract: *Inonotus hispidus* mushroom is a popular edible and medicinal mushroom with a long history of use. It is well known as a medicinal fungus with various health benefits for its significant anticancer and immunomodulatory activities. Over the last 60 years, secondary metabolites derived from *I. hispidus* and their biological activities have been discovered and investigated. Structurally, these compounds are mainly polyphenols and triterpenoids, which have anticancer, anti-inflammatory, antioxidant, antimicrobial, and enzyme inhibitor activities. Here, the secondary metabolites derived from *I. hispidus* and their activities were systematically and comprehensively classified and summarized, and the biosynthetic pathway of stylylpyrones was deduced and analyzed further. This review contributes to our understanding of *I. hispidus* and will help with research into natural product chemistry, pharmacology, and the biosynthesis of *I. hispidus* metabolites. According to this review, *I. hispidus* could be a promising source of bioactive compounds for health promotion and the development of functional foods.

Keywords: *Inonotus hispidus*; natural product; biological activity; medicinal fungi

Citation: Wang, Z.-x.; Feng, X.-l.; Liu, C.; Gao, J.-m.; Qi, J. Diverse Metabolites and Pharmacological Effects from the Basidiomycetes *Inonotus hispidus*. *Antibiotics* **2022**, *11*, 1097. https://doi.org/10.3390/antibiotics11081097

Academic Editors: Yaojun Tong, Zixin Deng and Linquan Bai

Received: 14 July 2022
Accepted: 11 August 2022
Published: 12 August 2022
Corrected: 21 November 2022

1. Introduction

The fungus *Inonotus hispidus* (Bull.: Fr.) Karst. is a facultative saprophytic (brown-rot Basidiomycete), which has been found to be parasitic on various broadleaf trees in China and Europe including mulberry (*Morus alba* L.), ash (*Fraxinus mandshurica*), *Populus euphratica*, *Ulmus campestris*, *Sorbus aucuparia*, and *Acer saccharum* [1,2]. According to modern classification systems, *I. hispidus* belongs to the *Inonotus* genus in Hymenochaetaceae (Hymenochaetales, Agaricomycetes, Basidiomycota) [3], whose obsolete synonyms include *Polyporus hispidus* (Bull.) Fr., *Boletus hispidus* (Bull.), *Xanthochrous hispidus* (Bull.) Pat. The fruit bodies of *I. hispidus* are annual, sessile, woody mushroom. The fruiting bodies are bright yellow at the initial growth stage (edible) (Figure 1A), and the color gradually changes from light brown to brown as they grow, becoming dark brown when fully matured with hairy surfaces (Figure 1B), and eventually turns to black [4].

Since *I. hispidus* are frequently parasitic on the mulberry tree (pronounced "sang" in Chinese), their fruit bodies are yellow (pronounced "huang" in Chinese) in color, highly similar in morphology to the modern taxonomic definition of the genus *Sanghuangporus* [5], and relatively close in herbal activity; *I. hispidus* were described as "Sanghuang" in the ancient Chinese *medical classics*. In the Aksu region of Xinjiang and Xiajin County of Shandong in China, *I. hispidus* parasites on mulberry trees has been used as a traditional indigenous medicine "Sanghuang" for the treatment of dyspepsia, cancer, diabetes, and stomach problems [6]. In fact, "Sanghuang" is a traditional collective name including *I. hispidus*, which has a long history of medicinal use and high economic value in the Southeast Asia region ("Meshimakobu" in Japan, and Sanghwang in South Korea) [7]. In China, the earliest medicinal records on "Sanghuang" can be traced back to *Shen Nong's Materia*

Medica in the Qin and Han Dynasties. The "*Shen Nong's Materia Medica*" and "*Compendium of Materia Medica*" have detailed descriptions of the efficacy of "Sanghuang". After the Second World War, the number of cancer cases in Nagasaki, Japan, increased dramatically. Danjo-gunto in Nagasaki is rich in "Sanghuang" due to the cultivation of mulberries and silkworms, and the residents were more healthy and had fewer cancer cases due to taking "Sanghuang". It was later found that it was the anti-tumor properties of mulberry that played a key role [8]. Recent studies in natural product chemistry and pharmacology have shown that *I. hispidus* contains a diverse spectrum of monomeric compounds and exhibits a diversity of physiological actions in in vitro and in vivo assays.

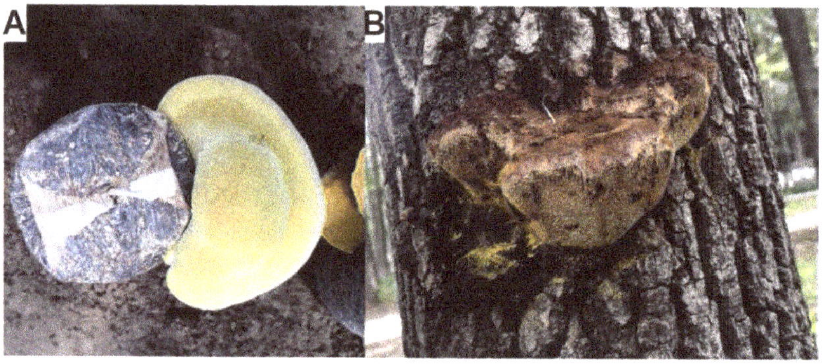

Figure 1. Morphological photographs of fruiting bodies of *I. hispidus* at the initial growth stage in artificial cultivation (**A**) and at the mature stage in the wild (**B**). The photo of mature fruiting body taken on ash on the campus of Northeast Forestry University in August 2021.

This review summarizes English and Chinese articles published in international academic journals after 1961, based on retrieving electronic databases PubMed, Wiley Online Library, Science Direct, and Web of Science. Totals of 64 compounds, including polyphenols, terpenoids, fatty acids and other types, were classified and sorted. These metabolites with outstanding medicinal value, and crude extracts of the fruit bodies were then inventoried and summarized according to their antitumor, anti-inflammatory, antioxidant, antineurodegenerative, and antimicrobial biological activities. The biosynthetic pathways of styrylpyrones derived from *I. hispidus* and related fungi were deduced and summarized. Related studies on phenols and sesquiterpenes, and food properties of *I. hispidus* are discussed. This work will provide a reference for subsequent studies on the natural product chemistry, pharmacological activities, and biosynthesis of secondary metabolites of *I. hispidus*.

2. Compounds with Diverse Structures

2.1. Polyphenol Compounds

Polyphenols (Figure 2) are the main pigment components present in the fruiting bodies of *I. hispidus*. Zopf first isolated a series of yellow substances from the fruiting bodies of *I. hispidus* in 1889, and Edwards et al. isolated the yellow substances to obtain a yellow crystalline hispidin (**1**) in 1961 [9]. Hispidin was the first reported monomeric compound from *I. hispidus*, and its structure was identified as 6-(3,4-dihydroxystryl)-4-hydroxy-2-pyrone by infrared techniques and derivatization reactions. Bis-noryangonin (**2**) was found in the fermentation broth of *P. hispidus* (synonym of *I. hispidus*) in liquid medium with glucose as the main carbon source. Its structure was determined as 4-bydroxy-6-(4-hydroxystylyl)-2-pyrone by isotopic label feeding, spectral characterization, and thin-layer chromatography (TLC) [10]. Fiasson and Jean-Louis isolated hispidin (**1**) from the fruiting bodies of *I. hispidus* in 1982, along with its two dimer derivatives, hypholomin B (**3**) and 3,14'-bishispidinyl (**4**) [11]. In 1996, Nasser Ali et al. obtained two pigment polyphenols, hispolon (**5**) and known hispidin (**1**), from the crude ethanolic extract of fruiting bodies under the guidance of immunomodulatory and antiviral activity. The structure of compound **5** was identified as 6-(9,10-dihydroxyphenyl)-3,5(*E*)-hexadien-4-ol-2-one by combined spectroscopic, mass spectrometry (MS) and nuclear magnetic resonance (NMR) techniques [12]. In 2009, Yousfi et al. extracted and isolated two other known polyphenols (**1** and **5**) and a new compound methyl 5-(3,4-dihydroxyphenyl)-3-hydroxypenta-2,4-dienoate (**6**) from the antioxidant components of the fruiting bodies parasitic on *Pistacia atlantica*, and its structure was determined by NMR and EI-MS [13]. In the experiment to determine the trace element composition of anti-influenza virus mushrooms, compounds **1** and **5** were once more isolated and extracted from the fruiting bodies of *I. hispidus* in 2011 [14]. In 2011 and 2012, Zan et al. isolated ten phenolic compounds from methanol extracts of fruiting bodies living on ash, among which inonotusins A (**7**) and B (**8**) were new natural products. The other known compounds were hispidin (**1**), hispolon (**5**), osmundacetone (**9**), protocatechualdehyde (**10**), inoscavin C (**11**), and inoscavin D (**12**), as well as protocatechuic acid (**13**) [15]. Compound **7** contains a highly oxidized functional group with a 2,3,4,5-tetrahydrooxepine backbone, which is the only phenolic compound containing a heptameric ring from *I. hispidus*. A combined approach of MS, spectroscopy and NMR methods clarified the structures of **7** and **8** are 6-(11,12-dihydroxystyryl)-2-hydroxy-2-methyl-5,6-dihydro-oxepine-1,4-dione and 5-(10,11-dihydroxystyryl)-3,4-dihydroxy7-(2-oxopropyl) pyrano-isochromen-1-one [16]. Compounds **9–13** were isolated from this strain for the first time [15]. A content comparison revealed that the fruit bodies contained higher concentrations of hispidin (**1**), hispolon (**5**), and osmundacetone (**9**), which were isolated for the first time from this strain [15]. Gruendemann et al. isolated known compounds hispidin (**1**) and hispolon (**5**) from methanol extracts of their fruiting bodies with the help of bioactivity-guided fractionation and analyzed their immunomodulatory effects in 2016 [17]. In 2017, Ren et al. isolated a new compound hispinine (**14**) and five known polyphenols (**1, 4, 5, 9,** and **15**) from the methanol extract of fruiting bodies of Xinjiang (China) indigenous medicinal fungus *I. hispidus* [18]. Despite the fact that the structure of compound **14** was determined by spectroscopic analysis to be 2-(3,4-dihydroxyphenyl)-6-methyl-2H-pyran-4(3H)-one [18], its absolute conformation has not yet been verified. Compound **15**, pinidine, isolated from *I. hispidus* in 2017, was first isolated from *Phellinus pini* [19]. *Phellinus pini* is a fungus that is morphologically similar to the fungi of the genus *Sanghuangporus* defined by the latest taxonomy [5]. Five polyphenols (**16–20**) were isolated and identified by Li et al. from the ethanolic extract of mature yellow fruiting bodies in 2017. Two compounds 3′4′-dihydroxy-5-[11-hydroxyphenyl]-6,7-vinyl]-3,5-dioxafluoren-5-one (**16**) and phelligridin C (**17**) were undescribed previously, while phellibaumin A (**18**), phelligridin C′ (**19**) and phelligridin D (**20**) were isolated for the first time from *I. hispidus* [2]. In 2019, Yang et al. isolated the compound 3,3′-methylene-bis [6-(3,4-dihydroxystyryl)-4-hydroxy-2H-pyran-2-one (MBP, **21**) from a methanol extract of the fruiting body, together with four reported polyphenols (**9, 10, 13,** and **18**) derived from *I. hispidus* [20]. From a methanol extract of the fruiting body, Kou et al. isolated six polyphenols in 2021, including three compounds inonophenols A-C (**22–24**), as well as **5, 9,** and 4-(3′,4′-dihydroxyphenyl)-2-butanone (**25**) [21]. The structures of compounds **22–24** were identified by NMR, HRMS, and/or computational circular dichroism data [21]. In the work on the analysis of antitumor metabolites in the mycelial fermentation broth of *I. hispidus*, hispolon (**5**) was isolated again, which was the first reported isolation from the liquid fermentation [22]. Bis-noryangonin (**2**), isolated from the mycelial broth of *I. hispidus*, is the only phenolic metabolite derived solely from the mycelial broth [10]. Structurally, compound **21**, MBP, is the symmetrical isomer of compound **1**, while compounds **3, 4,** and **15** are its non-symmetrical isomers.

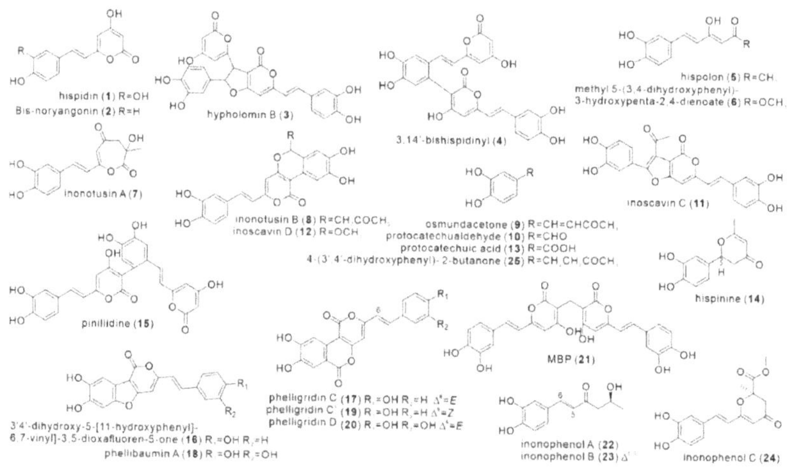

Figure 2. Structures of polyphenol compounds (**1–25**).

2.2. Triterpenoids

Triterpenoids are another important group of secondary metabolites found in *I. hispidus*, with a total of 15 triterpene compounds isolated (Figure 3). Yang et al. first reported the isolation and identification of four triterpenoid monomers from the ethanolic extract of fruiting bodies in 2008, which included 3,22-dihydroxy-lanosta-8,24-diene (**26**), ergosterol (**27**), ergosterol-5,8-peroxide (**28**), and eburicoic acid (**29**), with ergosterol (**27**) being the main chemical constituent isolated from 95% ethanolic extract of *I. hispidus* [23]. Zan et al. isolated five triterpenoids from the methanolic extract of fruiting bodies in 2012, including 7(8),22(23)-dien-3-one-ergostane (**30**), 4,6,8(14),22(23)-tetraen-3-one-ergostane (**31**) for the first time from this fungus, along with known compounds **27–29** [15]. Compound **28** is a high-content component in the fruiting bodies of *I. hispidus* [15].

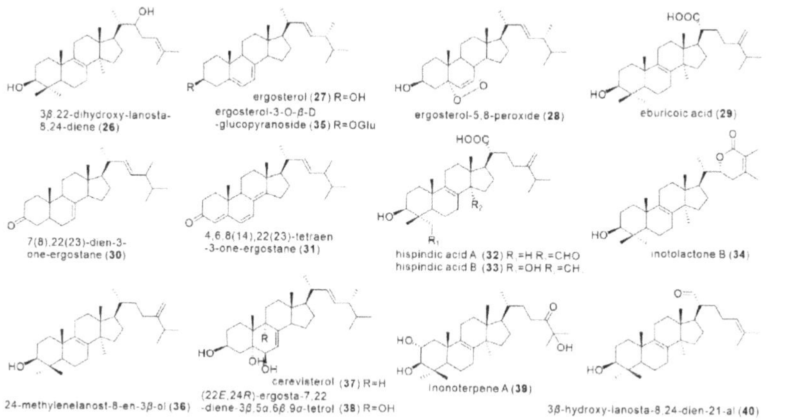

Figure 3. Structures of triterpenoids (**26–40**).

Ren et al. isolated six triterpenoids, including hispindic acids A and B (**32–33**), as well as the known compounds **27**, **29**, inotolactone B (**34**), and ergosterol-3-*O*-*β*-D-glucopyranoside (**35**) from the methanolic extract of fruiting bodies in 2017. The two new triterpenoids **32** and **33** were structurally identified as 24-exomethylene-3*β*-hydroxy-30-oxo-lanost-8-en-21-oic acid and 24-exomethylene-3*β*, 28-dihydroxy-lanost-8-en-21 oic acid by a combination of mass spectrometry and NMR techniques [18]. Seven triterpenoids were isolated from the methanolic extract of the fruiting bodies of *I. hispidus* by Kou et al. in 2021, including compound **28**, **35**, 24-methylenelanost-8-en-3*β*-ol (**36**), cerevisterol (**37**), (22*E*, 24*R*)-ergosta-7,22-diene-3*β*,5*α*,6*β*,9*α*-tetrol (**38**), and Inonoterpene A (**39**), and 3*β*-hydroxy-lanosta-8,24-dien-21-al (**40**). Among them, Inonoterpene A (**39**) is a new Lanostane triterpenoid, and compound **40** was isolated from the fungus for the first time. Interestingly, the above 15 triterpenoids are all lanosterol-type triterpenoids, all derived from the fruiting bodies of *I. hispidus*. Only three triterpenes **32**, **33**, and **39**, have been isolated as new compounds for the first time from *I. hispidus*. Compounds **32** and **40** contain formyl groups at C30 and C21, respectively, while compound **35** possesses glycosylated modifications. It was discovered that specific combinations of four compounds, methyl jasmonate, salicylic acid, oleic acid, and Cu^{2+}, and their combinations, can increase the content of total triterpenoids in an orthogonal experiment for liquid fermentation of *I. hispidus* [24].

2.3. Fatty Acid Compounds

The fatty acid analogues of *I. hispidus* were first reported by Yang et al. in 2008, and two monomer compounds hexadecanoic acid (**41**) and octadecanoic acid (**42**) were obtained in the ethanolic extract of its fruit bodies [23] (Figure 4). The structure of the former was identified by EI-MS and TLC comparison of the standard, and the identification of the latter was done with additional infrared techniques [23]. Zan et al. published 2012 the detection results of gas mass spectrometry of the yellow oily substance obtained from the petroleum ether extraction part of the methanol extract of fruit bodies. A total of 12 fatty acid compounds were identified by consulting the Wiley Online Library. These compounds are hexadecanoic acid (**41**), (*E*)-9-hexadecenoic acid methyl ester (**43**), hexadecanoic acid methyl ester (**44**), hexadecanoic acid ethyl ester (**45**), 10,13-octadecadienoic acid methyl ester (**46**), ethyl oleate (**47**), linoleic acid ethyl ester (**48**), 9,12-octadecadienoic acid (Z, Z)-methyl ester (**49**), Isopropyl linoleate (**50**), *cis*-erucic acid (**51**), and *cis*-9,17 octadecadienoic acid methyl ester (**52**) [15]. Due to their low polarity and insignificant drug properties, fatty acid compounds are not preferred subjects in natural product chemistry studies of fungi. Only two cases of *I. hispidus* have been reported in this regard.

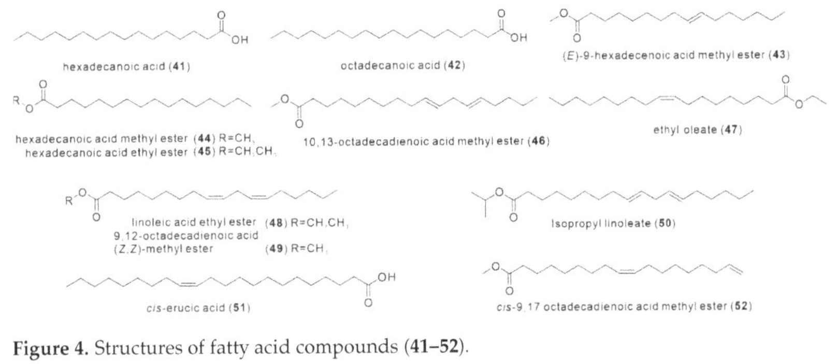

Figure 4. Structures of fatty acid compounds (**41–52**).

2.4. Miscellaneous Compounds

During the constituent analysis of dried fruiting bodies of *I. hispidus* in 2007, Politi et al. identified a known compound, phenylalaninopine (**53**), by spectral and mass spectrometric characterization collected by LC-DAD-MS and comparing with the literature, while attempting to develop a phytochemical extraction method for mushrooms using hot water as a solvent [25] (Figure 5). The compound was isolated initially from Clitocybe acromelalga, a poisonous mushroom [26]. Yang et al. obtained two monosaccharides, D-arabitol (**54**) and D-glucose (**55**), along with the isolation of four triterpenoids [23]. In their experiments to examine the fatty acid composition of fruiting bodies, Zan et al. discovered two hydrocarbons, squalene (**56**) and docosane (**57**), and one phthalate, 1,2-benzenedicarboxylic acid mono(2-ethylhexyl) ester (**58**) [15]. In the separation of anticancer active components, Yang et al. isolated a new compound from the methanol extract of fruiting bodies in 2019 and confirmed its structure as (4S,5S)-4-Hydroxy-3,5-dimethoxycyclohex-2-enone (HDE, **59**) by LC-MS and NMR [27]. When analyzing the methanol extract composition of *I. hispidus* fruiting bodies, Yang et al. identified a nucleoside compound, inosine (**60**), based on a comparison of UV absorption characteristics and retention time with a standard [20]. Kou et al. isolated an unsaturated sesquiterpene, xylaritriol (**61**) [21], from the methanol extract of fruiting bodies, which had previously been from the metabolites of endophytic fungus *Xylaria cubensis* (Ascomycete) [28]. Tang et al. isolated compounds, cinnamic acid (**62**), uridine (**63**), and Cyclo (*L*-Leu-*L*-Phe) (**64**) from the fermentation broth of *I. hispidus* mycelium in 2021 [22]. Among these 12 compounds with no shared structural features, compound **59** is the only new compound isolated for the first time from *I. hispidus*, and compound **61** was the only sesquiterpene isolated from *I. hispidus*. Cyclic dipeptide compounds are uncommon in mushrooms (Basidomycota), but compound **64** was discovered in *I. hispidus*. This could be related to the fact that compound **64** is derived from mycelial fermentation broth.

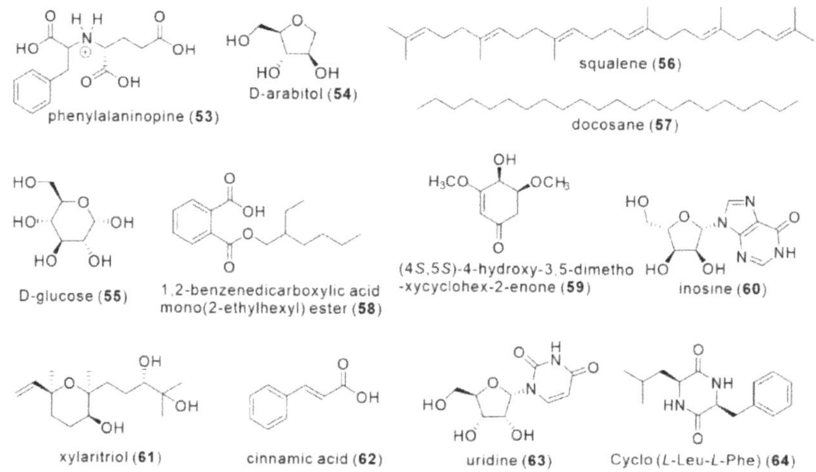

Figure 5. Structures of miscellaneous compounds (**53–64**).

3. Biological and Pharmacological Activities

Inonotus hispidus has long been used to treat dyspepsia, cancer, diabetes, and stomach problems in Xinjiang residents [6]. Several studies have been published on the pharmacological activities and mechanisms of major compounds from *I. hispidus* (Table 1). These studies have been summarized here, with a special emphasis on polyphenols with medicinal potential (Figure 6).

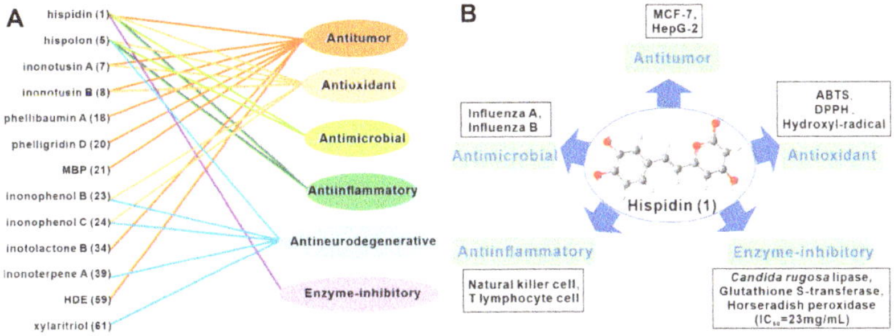

Figure 6. The bioactivities of representative compounds (**A**) and hispidin (**1**) (**B**) from *I. hispidus* compounds.

Table 1. Occurrence of compounds from *Inonotus hispidus*.

No.	Name	Resource	Method	Activity	Reference
1	hispidin	fruiting body, mycelium	EtOH, MeOH	Antitumor, ABTS, DPPH, influenza A and B viruses, Enzyme-inhibitory, antineurodegenerative	Edwards et al. [9]. Zan et al. [15]. Ali, N.A.A. et al. [29]. Zan et al. [16]. Zan et al. [30]. Benarous, K. et al. [31]. Benarous, K. et al. [32]. Gruendemann, C. et al. [17].
2	Bis-noryangonin	mycelium	NA	NA	Perrin, P.W.et al. [10].
3	hypholomin B	fruiting body	NA	NA	Fiasson et al. [11].
4	3,14′-bishispidinyl	fruiting body	MeOH	NA	Fiasson et al. [11]. Ren et al. [18].
5	hispolon	fruiting body mycelium	EtOH, MeOH EAC	DPPH, influenza A and B viruses, PC-12 cell, BV-2 microglia, ntineurodegenerative	Nasser Ali et al. [12]. Zan et al. [15]. Tang et al. [22]. Kou et al. [21]. Ali, N.A.A. et al. [29]. Gruendemann, C. et al. [17].
6	methyl 5-(3,4-dihydroxyphenyl)-3-hydroxypenta-2,4-dienoate	fruiting body	MeOH	NA	Yousfi et al. [13].
7	inonotusin A	fruiting body	MeOH	MCF-7, ABTS	Zan et al. [15]. Zan et al. [16].
8	inonotusin B	fruiting body	MeOH	Antitumor, ABTS	Zan et al. [15]. Zan et al. [16].
9	osmundacetone	fruiting body	MeOH	DPPH, PC-12 cell, BV-2 microglia	Zan et al. [15]. Kou et al. [21].
10	protocatechualdehyde	fruiting body	MeOH	NA	Zan et al. [15].
11	inoscavin C	fruiting body	MeOH	NA	Zan et al. [15].
12	inoscavin D	fruiting body	MeOH	NA	Zan et al. [15].
13	protocatechuic acid	fruiting body	MeOH	NA	Zan et al. [15].
14	hispinine	fruiting body	MeOH	NA	Ren et al. [18].
15	pinillidine	fruiting body	MeOH	NA	Ren et al. [18].
16	3′4′-dihydroxy-5-[11-hydroxyphenyl]-6,7-vinyl]-3,5-dioxafluoren-5-one	fruiting body	EtOH	mouse macrophage cell	Li et al. [2].
17	phelligridin C	fruiting body	EtOH	NA	Li et al. [2].

Table 1. *Cont.*

No.	Name	Resource	Method	Activity	Reference
18	phellibaumin A	fruiting body	EtOH MeOH	mouse macrophage cell	Li et al. [2]. Yang et al. [20].
19	phelligridin C′	fruiting body	EtOH	NA	Li et al. [2].
20	phelligridin D	fruiting body	EtOH	mouse macrophage cell	Li et al. [2].
21	MBP	fruiting body	MeOH	HepG2, MCF-7, Hela and A549 cells	Yang et al. [20].
22	inonophenol A	fruiting body	MeOH	DPPH, PC-12 cell, BV-2 microglia	Kou et al. [21].
23	inonophenol B	fruiting body	MeOH	DPPH, PC-12 cell, BV-2 microglia	Kou et al. [21].
24	inonophenol C	fruiting body	MeOH	DPPH, PC-12 cell, BV-2 microglia	Kou et al. [21].
25	4-(3′,4′-dihydroxyphenyl)-2-butanone	fruiting body	MeOH	DPPH, PC-12 cell, BV-2 microglia	Kou et al. [21].
26	3*β*, 22-dihydroxy-lanosta-8,24-diene	fruiting body	EtOH	U937, Hela, QRH-7701	Yang et al. [23].
27	ergosterol	fruiting body mycelium	EtOH, MeOH EAC	NA	Yang et al. [23]. Zan et al. [15]. Tang et al. [22].
28	ergosterol-5,8-peroxide	fruiting body	EtOH, MeOH	BV-2 microglia	Yang et al. [23]. Zan et al. [15]. Kou et al. [21].
29	eburicoic acid	fruiting body	EtOH, MeOH	NA	Yang et al. [23]. Zan et al. [15].
30	7(8),22(23)-dien-3-one-ergostane	fruiting body	MeOH	NA	Zan et al. [15].
31	4,6,8(14),22(23)-tetraen-3-one-ergostane	fruiting body	MeOH	NA	Zan et al. [15].
32	hispindic acid A	fruiting body	MeOH	NA	Ren et al. [18].
33	hispindic acid B	fruiting body	MeOH	NA	Ren et al. [18].
34	inotolactone B	fruiting body	MeOH	B16 melanoma cell	Ren et al. [18].
35	ergosterol-3-*O*-*β*-D-glucopyranoside	fruiting body	MeOH	BV-2 microglia	Ren et al. [18]. Kou et al. [21].
36	24-methylenelanost-8-en-3*β*-ol	fruiting body	MeOH	BV-2 microglia	Kou et al. [21].
37	cerevisterol	fruiting body	MeOH	NA	Kou et al. [21].
38	(22*E*, 24*R*)-ergosta-7,22-diene-3*β*,5*α*,6*β*,9*α*-tetrol	fruiting body	MeOH	NA	Kou et al. [21].
39	Inonoterpene A	fruiting body	MeOH	BV-2 microglia	Kou et al. [21].
40	3*β*-hydroxy-lanosta-8,24-dien-21-al	fruiting body	MeOH	BV-2 microglia	Kou et al. [21].
41	hexadecanoic acid	fruiting body	EtOH, MeOH	NA	Yang et al. [23]. Zan et al. [15].
42	octadecanoic acid	fruiting body	EtOH	NA	Yang et al. [23].
43	(*E*)-9-hexadecenoic acid methyl ester	fruiting body	MeOH	NA	Zan et al. [15].

Table 1. *Cont.*

No.	Name	Resource	Method	Activity	Reference
44	hexadecanoic acid methyl ester	fruiting body	MeOH	NA	Zan et al. [15].
45	hexadecanoic acid ethyl ester	fruiting body	MeOH	NA	Zan et al. [15].
46	10,13-octadecadienoic acid methyl ester	fruiting body	MeOH	NA	Zan et al. [15].
47	ethyl oleate	fruiting body	MeOH	NA	Zan et al. [15].
48	linoleic acid ethyl ester	fruiting body	MeOH	NA	Zan et al. [15].
49	9,12-octadecadienoic acid (*Z*, *Z*)-methyl ester	fruiting body	MeOH	NA	Zan et al. [15].
50	Isopropyl linoleate	fruiting body	MeOH	NA	Zan et al. [15].
51	*cis*-erucic acid	fruiting body	MeOH	NA	Zan et al. [15].
52	*cis*-9,17 octadecadienoic acid methyl ester	fruiting body	MeOH	NA	Zan et al. [15].
53	phenylalaninopine	fruiting body	Water	NA	Politi et al. [25].
54	D-arabitol	fruiting body	EtOH	NA	Yang et al. [23].
55	D-glucose	fruiting body	EtOH	NA	Yang et al. [23].
56	squalene	fruiting body	MeOH	NA	Zan et al. [15].
57	docosane	fruiting body	MeOH	NA	Zan et al. [15].
58	1,2-benzenedicarboxylic acid mono(2-ethylhexyl) ester	fruiting body	MeOH	NA	Zan et al. [15].
59	HDE	fruiting body	MeOH	HepG2, McF-7, Hela, A549 and H22 cells	Yang et al. [27].
60	inosine	fruiting body	MeOH	NA	Yang et al. [20].
61	xylaritriol	fruiting body	MeOH	BV-2 microglia	Kou et al. [21].
62	cinnamic acid	mycelium	NA	NA	Tang et al. [22].
63	uridine	mycelium	NA	NA	Tang et al. [22].
64	Cyclo (*L*-Leu-*L*-Phe)	mycelium	NA	NA	Tang et al. [22].

NA indicated the data is not available.

3.1. Antitumor Activity

As a traditional medicinal fungus with a long-applied history, anti-cancer activity is the principal medicinal value of *I. hispidus*. Multiple monomers with anticancer activity from *I. hispidus* have been discovered and preliminarily resolved in modern chemical chemistry and pharmacological investigations. In vitro screening of five compounds including hispidin (**1**), inonotusin A (**7**) and inonotusin B (**8**) against breast cancer cells MCF-7 using the SRB method revealed that inonotusin A (**7**) had moderate activity against MCF-7 with an IC_{50} value of 19.6 μM [16]. The anti-tumor screening study of MBP (**21**) was carried out by MTT method. The results showed that MBP (**21**) had a dosage-dependent inhibitory effect on the proliferation of HepG2, MCF-7, Hela and A549 cells. Among them, HepG2 showed the best inhibitory effect, with an IC_{50} of 2.3 μg/mL.

According to the results of the apoptosis analysis, it is hypothesized that the compound (**21**) achieves its anti-tumor activity by inducing apoptosis [20]. A study of compound 26's antitumor activity discovered that it inhibited the growth of lymphoma cells U937, cervical tumor cells Hela, and liver cancer cells QRH-7701 [23]. In vitro screening showed that HDE (**59**) inhibited the proliferation of HepG2, McF-7, Hela, A549 and H_{22} cells, with a greater effect on HepG2 cells (IC_{50} = 7.9μg/mL) [27]. Not only did HDE (**59**) increase the relative activities of Caspase-3 and Caspase-8 and induce HepG2 apoptosis, but it also promoted HepG2 apoptosis by up-regulating *Fas* expression and down-regulating *FasL* expression [27] (Figure 7A). In vivo mouse experiments revealed that HDE (**59**) does not cross the blood–brain barrier and is rapidly metabolized with no adverse effects on organs. Overall, HDE (**59**) has the potential to be an effective antineoplastic agent [27]. At 10 mmol/L, the compound inotolactone B (**34**) still showed activity in activating melanogenesis and tyrosinase in B16 melanoma cells, outperforming 8-MOP (a drug used clinically in the treatment of vitiligo). This indicates that inotolactone B (**34**) has the potential to be developed as an anti-vitiligo agent [18]. The evaluation of the cellular activity of mouse macrophage cells RAW264.7 revealed that compounds **16**, **18** and **20** all had anti-inflammatory activity, with compound **16** being the most active. Cytotoxicity studies showed that all three compounds have a low impact on cell proliferation [2]. This result suggests that compound **16** has the potential to be an anti-leukemic agent. In the supernatant of fermented mycelium of a suspected new subspecies of *I. hispidus*, IH3, a single fraction with high purity of anti-tumor activity, WIH3, was obtained. WIH3 had strong inhibitory against melanoma cells B16 (IC_{50} 29.32 μg/mL), liver cancer cells Hep-3B (37.39 μg/mL), human cervical cancer cells Hela (47.03 μg/mL), and human breast cancer cells MCF-7 (58.01 μg/mL) [33].

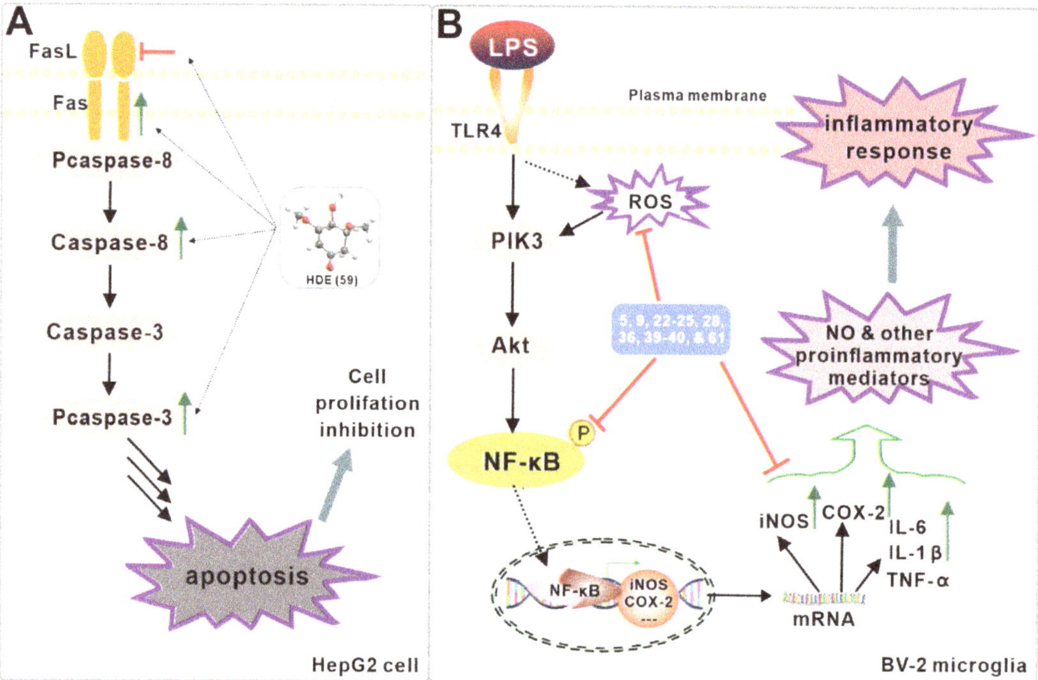

Figure 7. Schematic diagram of the proposed inhibitory mechanism of compound **59** on HepG2 cell (**A**) and compounds including **24** on inflammatory BV-2 microglia induced by LPS (**B**).

In vivo antitumor activity experiments with extracts of fruit bodies from different growth stages of *I. hispidus* revealed that both petroleum ether extracts (IPE) and aqueous extracts at the mature stage had the best tumor suppressive effect on H_{22}-bearing mice [15]. Using non-targeted metabolomics techniques, it was found that the antitumor effects of IPE on H_{22} tumor-bearing mice were mainly mediated by energy modulation and regulation of biosynthetic pathways including amino acids and corticosteroids [34]. A series of key regulators of the antitumor activity of IPE were identified in H_{22} tumor suppressor model mice through transcriptomic and proteomic approaches, and the results provide a useful reference for the pharmacological study of the antitumor activity of IPE [35]. The feeding experiment of solid fermentation powder of *I. hispidus* on H_{22}-bearing mice demonstrated that it has antitumor effect, whose mechanism may be related to antioxidation, improving immunity and inhibiting tumor tissue angiogenesis [36].

3.2. Antioxidant Activity

One of the important properties of mushrooms is their antioxidant function, and antioxidant active molecules derived from *I. hispidus* have been investigated and studied. Antioxidant assays revealed that compounds **1**, **7**, and **8** have significant oxidative protective activity against 2,2′-azino-bis (3-ethylbenzthiazoline)-6-sulfonic acid (ABTS) [16]. Compounds **5**, **9**, **22–25** have remarkable inhibitory activity against 1,1-Diphenyl-2-picrylhydrazyl radical (DPPH) with IC_{50} values of 9.82−21.43 μM, with compound **24** being more effective than the positive control [21]. Preliminary structure–activity relationship analysis revealed that catechol unit, double bond on the side-chain, and pyran ring may be important pharmacophores for increasing activity [21]. A study of the antioxidant activity of the compound hispidin (**1**) to scavenge free radicals discovered that at a low concentration (12.5 mg/L), the scavenging rate of **1** to free radical DPPH was 84.70%. At low concentrations, the DPPH scavenging rate was higher than that of BHA, a commercial antioxidant, and compound **1** has natural antioxidant value [30]. The antioxidant activity of compound **1** for scavenging hydroxyl radicals was found to be superior to that of BHA at concentrations as low as 25 mg/kg [30]. DPPH, ABTS and phosphomolybdenum were used as indicators to determine the antioxidant activity of the ethyl acetate extract of *I. hispidus*. The extract was found to be more powerful than the commonly tested antioxidant compounds gallic acid and quercetin [31]. The scavenging effect of hydroxyl radical, DPPH radical and superoxide anion of the volatile oil fragments (H-2) obtained by the diethyl ether reflux extraction of the fruiting bodies of *I. hispidus* was significantly better than that of the volatile oil fragments (H-1) obtained by the steam distillation extraction of the fruiting bodies of *I. hispidus* [37].

3.3. Antimicrobial Activity

The antibacterial activity tests performed on hispidin (**1**) revealed that compound **1** lacked adequate inhibitory activity against *Escherichia coli*, *Staphylococcus aureus*, and *Bacillus subtilis* [15]. The inhibition activity of a 70% ethanolic extract of fruit bodies was tested, and it was discovered that the ethanolic extract inhibited only *E. coli*, with the high dosage extract (40.96 mg/mL) achieving a significant level of inhibition. *S. aureus* and *B. subtilis* were not inhibited significantly [15]. At a concentration of 1×10^{-5} g/mL, both the H-1 (water extract) and H-2 (ether extract) fractions of fruit bodies inhibited *S. aureus*, *Candida albicans*, *Aspergillus niger*, and *Pseudomonas aeruginosa*, with a positive correlation with concentration, but neither inhibited *E. coli* [15]. The disc diffusion assay of ethanolic extracts of *I. hispidus* fruit bodies revealed a minor antimicrobial effect against *S. aureus*, *E. coli*, and *P. aeruginosa* [38]. Hispolon (**5**) and hispidin (**1**) were discovered to have antiviral activity, inhibiting the growth of influenza A and B viruses [29]. After feeding mice *I. hispidus* mushroom containing compounds **1** and **5**, the anti-influenza virus ability of mice was improved, as was the content of trace elements zinc (194 ± 16.9 mg/kg), selenium, and iron in mouse serum. These metal elements are thought to enhance the antiviral ability of the *I. hispidus* mushroom [14].

3.4. Neurotrophic and Neuroprotective Activity

Compounds **5**, **9**, and **22–25** exhibited a growth-promoting effect on PC-12 cells. Of these, compounds **23** (33.21 ± 0.8%) and **24** (33.34 ± 1.0%) were the most significant in promoting growth at a concentration of 10 μM [21]. In screening assays for anti-neuroinflammatory activity, compounds **5**, **9**, **22–25**, **28**, **36**, **39–40**, and **61** all inhibited of LPS-induced NO production in BV-2 microglia, with compound **24** showing significant activity with an IC_{50} value of 11.56 μM. None of the compounds tested toxic [21]. Additional molecular immunological experiments revealed that compounds **5**, **9**, and **22–25** inhibited the expression of TLR-4 to down-regulate the NF-κB signaling pathway, and then inhibited the expression of COX-2 and iNOS to reduce inflammation (Figure 7B) [21]. These findings show that these polyphenols and triterpenoids have anti-neurodegenerative activity.

3.5. Enzyme Inhibitory Activity

Yousfi et al. discovered in 2013 that the polyphenol-rich ethyl acetate extract of *I. hispidus* fruit bodies was very effective in inhibiting *Candida rugosa* lipase, but the specific inhibitory compounds were unknown [39]. In 2015, they isolated hispidin (**1**) from the fruit bodies and demonstrated that hispidin (**1**) strongly inhibited *C. rugosa* lipase, proposing the use of hispidin (**1**) in the treatment of candidiasis [31]. Hispidin (**1**) was discovered to have anti-peroxidase activity, with an IC_{50} of 23 mg/mL showing a strong competitive inhibition of horseradish peroxidase (HRP) activity. The inhibition mechanism of compound **1** against peroxidases (horseradish and thyroid) was studied using molecular docking simulations in terms of hispidin interaction with amino acid residues, and its drug-forming potential was predicted using ADEMT and Lipinski filtering analyses [40]. According to the combined evaluation, hispidin (**1**) is a more potent irreversible thyroperoxidase inhibitor than the anti-thyroid drug 6-hropylthiouracil [40]. The discovery of this result, to some extent, revealed the mechanism of the antioxidant activity of hispidin (**1**). Furthermore, low doses of ethanolic extracts of *I. hispidus* fruiting bodies increased GST enzyme activity [38]. Compound **34** (IC_{50} = 0.24 mM) has excellent α-glycosidase inhibitory activity and is more potent than the clinically used acarbose (IC_{50} = 0.46 mM) [40]. At the same concentration, the ethyl acetate extract of fruiting bodies had higher α-glucosidase inhibitory activity than the *n*-butanol, methanol, and petroleum ether extracts [41].

3.6. Immunomodulatory Effects and Other Biological activities

Cellular activity assays on NK cells showed that compounds **1** and **5** both inhibited the activation and proliferation of activated lymphocytes, but the combination of the two did not enhance this effect [17]. This result demonstrates the immunomodulatory activity of these phenolic compounds. In addition, the fermentation broth exopolysaccharide of *I. hispidus* was shown to attenuate inflammatory responses in mice with acute alcoholic liver disease [42].

4. Biosynthetic Progresses on Styrylpyrones

Styrylpyrone compounds are a specific class of polyphenols with a wide range of structures and high yields. The majority of *I. hispidus*-derived yellow polyphenol pigments have a styrylpyrone backbone, and 10 styrylpyrone compounds (**1**, **3–4**, **8**, **12**, **15–16**, **20–21**, **24**) have been identified from the secondary metabolites of *I. hispidus*. Nearly hundreds of styrylpyrone pigments have been discovered in a variety of fungi dominated by the Hymenochaetaceae family since the discovery of hispidin (**1**) as the first naturally occurring styrylpyrone metabolite [43–45]. Experiments on light-regulated enzymatic activity revealed that the cinnamic acid (**62**) pathway involved in phenylalanine metabolism is linked to hispidin synthesis in *P. hispidus*. Cinnamic acid (**62**) is thought to be a key intermediate in the biosynthesis of hispidin and other phenylpropanoids [46]. Isotope labeling traces revealed that the styrene unit was incorporated with phenylalanine, tyrosine, cinnamic acid (**62**), p-coumarate, and caffeic acids, while the pyrone ring was incorporated with acetate and malonate [10]. The isolation of five phenylpropanoids from the fruiting bodies of *Inonotus* sp., including iso-hispidin (**65**), inonotic acid methyl ester, and inotilone, provided clues to the biosynthesis of styrylpyrones. The phenylpropanoid polyketide is an important intermediate that is thought to be a side chain elongation of caffeic acid catalyzed by PKS [47]. Integrating metabolomic and proteomic findings that the up-regulation of phenylpropanoid biosynthesis leads to increased levels of hispidin and other polyphenols in *P. baumii*, genes involved in the up-regulation of expression include PAL, C4H, and 4- coumarate-CoA ligase (4CL) (Figure 8) [48]. PAL, 4CL, C4H, polyketide synthase (PKS), palmitoyl protein thioesterase (PPT), decarboxylase, and dehydrogenase genes were identified in a combined investigation of molecular docking simulations and RNA-seq-based analysis of differentially expressed genes in *P. gilvus* (Figure 8) [49]. Based on biosynthetic research and the analysis of existing styrylpyrone compounds derived from the *Phellinus* and *Inonotus* fungi, the comprehensive biosynthetic pathways of styrylpyrone compounds, including hispidin (**1**), hispolon (**5**) and their dimers, as well as inoscavin A (**66**), have been deduced in this review (Figure 8) [47–49]. The natural structural diversity of styrylpyrones suggests that their biosynthetic gene clusters are active and contain multiple post-modification genes. It is speculated that the study of its biosynthetic pathway and related synthetic gene clusters may aid in understanding the biosynthetic network of styrylpyrones and their derivatives in *I. hispidus*.

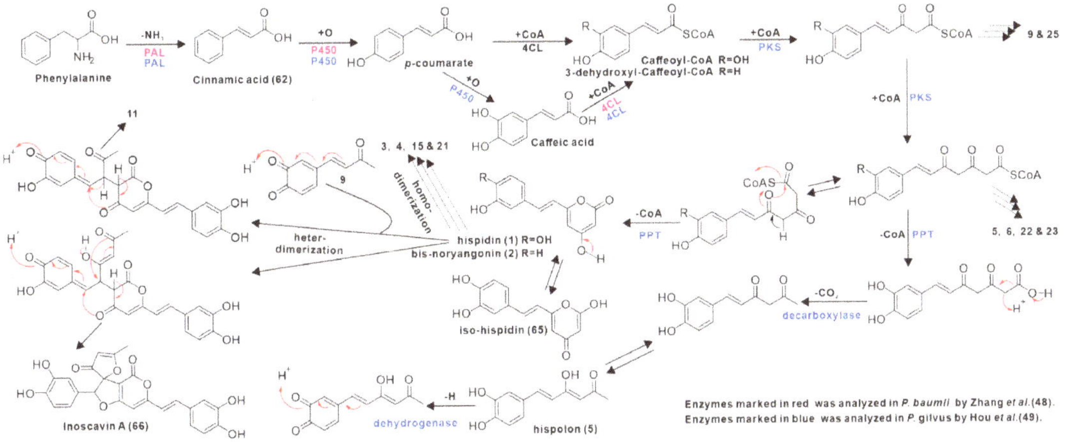

Figure 8. Proposed biosynthesis pathways for styrylpyrones in *Inonotus* and *Phellinus* fungi.

5. Discussions and Perspectives

Triterpenes are characteristic pharmacological components of lignified traditional medicinal fungi, such as Ganoderic acid from *Ganoderma* species [50], and Eburicoic acid and its derivatives from polypore species [51]; however, they are not monolithic. Phenols, a class of high-value natural products with medicinal potential [44,52], are prominent bioactive components of the highly lignified and well-known medicinal mushrooms of the genus *Phellinus* and *Inonotus* belonging to the order of Hymenochaetales [43]. The diversely modified phenolic derivatives yielded by these mushrooms have promising applications in drug discovery and development [43,44]. Hispidin (**1**) is the first compound isolated and identified from *I. hispidus*, and these phenolic compounds with styryl and pyrone moieties are gathered here with diverse activities such as antitumor, antioxidant, antimicrobial, and anti-inflammatory activities (Figure 6). These diverse activities of the natural product reflect the biological activities of its producer. Despite the fact that **5** was initially isolated from *I. hispidus* in 1996 [12], it is more of a characteristic component of fungi belonging to the genus *Phellinus*, as it has also been isolated from various species of the genus *Phellinus*, including *P. linteus*, *P. igniarius*, *P. lonicerinus* and *P. merrillii* [43].

A summary of the anticancer, antidiabetic, antioxidant, antiviral, and anti-inflammatory properties of compound **5** revealed its potential development value as a complementary and alternative medicine [43].

As one of the most abundant secondary metabolites found in mushrooms, numerous mushroom-derived sesquiterpenes have been isolated and identified [53]. Sesquiterpene synthases are one of the core gene types involved in the biosynthesis of secondary metabolites. Mushroom genomes contain typically at least ten sesquiterpene synthases. *I. obliquus* is a congener species of *I. hispidus*. Although the genome of *I. obliquus* contains at least 4 types of 20 sesquiterpene synthases [54], only 8 sesquiterpenoids of the drimane-type have been reported [55]. So far, only one sesquiterpenoid (**61**) has been reported for *I. obliquus* [21], and the type and number of sesquiterpene synthases are also unknown. The identification of sesquiterpene cyclases derived from lots of mushrooms including *Coprinus cinereus* [56], *Omphalotus olearius* [57], *Stereum hirsutum* [58,59], *Flammulina velutipes* [60], and *Agrocybe aegerita* [61] have led to the discovery of numerous novel sesquiterpenoids, which are not found in their natural producer instead. As a result, we speculate that in most mushrooms, only one or a few types of sesquiterpene synthases can be expressed to produce specific sesquiterpenoids, while the majority of sesquiterpene synthases are silent and have no corresponding products.

Although *I. hispidus* has been widely considered a traditional medicinal fungus due to its secondary metabolites with remarkable significant pharmacological value, the food properties of *I. hispidus* that are rich in health and nutritional functional factors should not be overlooked. The edible fruiting body of *I. hispidus* is a scarce ingredient. *I. hispidus* possess a good selenium-enriching function [62], and the antiviral ability of *I. hispidus* rich in selenium has also been enhanced. The dihydroxy phenylalanine-melanin contained in *I. hispidus* could effectively remove DPPH radicals and can therefore be used in health food or as a food additive [63]. Furthermore, the extracellular exopolysaccharide of *I. hispidus* was investigated and found to have antioxidant activity [64] and hepatoprotective functions [42].

Mushrooms are often perceived more for their anticancer or immunomodulatory activity than for their antibacterial activity. In fact, mushrooms with antimicrobial activity are uncommon [65,66], with *Pleurotus mutilus* and *I. hispidus* being the few examples. Pleuromutilin was found to have good antifungal inhibitory activity and was used as the lead compound in the development of the human antibiotic retapamulin, which was approved by the FDA in 2007 [67]. Both aqueous and ether extracts of *I. hispidus* mushrooms inhibit pathogenic microorganisms such as *S. aureus* [15]; however, monomeric compounds with antibacterial activity have yet to be identified. As a result, the discovery and development of antimicrobial compounds derived from *I. hispidus* is worth pursuing.

Herein, a total of 64 compounds derived from *I. hispidus* were collected and classified according to their structural features. Biological and pharmacological activities including antitumor, antioxidant, as well as anti-inflammatory and neurotrophic and neuroprotective activities were summarized. These studies, however, were limited to the isolation, identification, and activity evaluation of compounds. Current research has not paid enough attention to *I. hispidus*, despite its significant medicinal efficacy and functional food value. Although the genomes of multiple rare medicinal fungi including *I. obliquus* [54] have been sequenced, the genome of *I. hispidus* has not been sequenced. The lack of genomic information of *I. hispidus* severely hampered the biosynthetic pathway elucidation of its active compounds. As a result, it is critical to complete the genome sequencing of *I. hispidus* [54] using rapidly developing sequencing technology, which will lay a firm foundation for biosynthesis research and synthetic biology production of valuable compounds to meet the needs of the drug development and functional food industries.

Author Contributions: Z.-x.W. and X.-l.F. collected data and prepared figures, J.Q. wrote the manuscript, and C.L. and J.-m.G. reviewed the manuscript. All authors have read and agreed to the published version of the manuscript.

Funding: This work was supported by the National Natural Science Foundation of China (31800031), the China Postdoctoral Science Foundation (2019M653760), Natural Science Basic Research Plan in Shaanxi Province of China (2019JQ-046).

Institutional Review Board Statement: Not applicable.

Informed Consent Statement: Not applicable.

Conflicts of Interest: The authors declare no conflict of interest.

References

1. Piątek, M. *Inonotus hispidus* (Bull.: Fr.) Karst. In *Atlas of the Geographical Distribution of Fungi in Poland, Fascicle 1*; Polish Academy of Sciences: Kraków, Poland, 2000; pp. 35–40.
2. Qing-Jie, L. Study on the Active Substances and Quality Standards of Sanghuang. Ph.D. Thesis, Jilin Agricultural University, Changchun, China, 2017.
3. Kirk, P.; Cannon, P.; Stalpers, J.; Minter, D.W. *Dictionary of the Fungi*, 10th ed.; CABI Publishing: Oxfordshire, UK, 2008.
4. Zhang, F.; Xue, F.; Xu, H.; Yuan, Y.; Wu, X.; Zhang, J.; Fu, J. Optimization of Solid-State Fermentation Extraction of *Inonotus hispidus* Fruiting Body Melanin. *Foods* **2021**, *10*, 2893. [CrossRef]
5. Wu, S.H.; Chang, C.C.; Wei, C.L.; Jiang, G.Z.; Cui, B.K. *Sanghuangporus toxicodendri* sp. nov. (Hymenochaetales, Basidiomycota) from China. *MycoKeys* **2019**, *57*, 101–111. [CrossRef] [PubMed]
6. Zan, L.-F.; Bao, H.-Y. Progress in *Inonotus hispidus* research. *Acta Edulis. Fungi.* **2011**, *18*, 78–82. [CrossRef]
7. Wu, S.-H.; Dai, Y.-C.; Hattori, T.; Yu, T.W.; Wang, D.-M.; Parmasto, É.K.; Chang, H.-Y.; Shih, S.-Y. Species clarification for the medicinally valuable 'sanghuang' mushroom. *Bota. Stu.* **2012**, *53*, 135–149.
8. Ikekawa, T.; Nakanishi, M.; Uehara, N.; Chihara, G.; Fukuoka, F. Antitumor action of some Basidiomycetes, especially Phllinus linteus. *GaN* **1968**, *59*, 155–157. [CrossRef] [PubMed]
9. Edwards, R.L.; Lewis, D.G.; Wilson, D.V. 983. Constituents of the higher fungi. Part I. Hispidin, a new 4-hydroxy-6-styryl-2-pyrone from polyporus hispidus (Bull.). *Fr. J. Chem. Soc.* **1961**, 4995–5002. [CrossRef]
10. Perrin, P.W.; Towers, G.H.N. Hispidin biosynthesis in cultures of *Polyporus hispidus*. *Phytochemistry* **1973**, *12*, 589–592. [CrossRef]
11. Fiasson, J.-L. Distribution of styrylpyrones in the basidiocarps of various Hymenochaetaceae. *Biochem. Syst. Ecol.* **1982**, *10*, 289–296. [CrossRef]
12. Ali, N.A.A.; Jansen, R.; Pilgrim, H.; Liberra, K.; Lindequist, U. Hispolon, a yellow pigment from *Inonotus hispidus*. *Phytochemistry* **1996**, *41*, 927–929. [CrossRef]
13. Yousfi, M.; Djeridane, A.; Bombarda, I.; Chahrazed, H.; Duhem, B.; Gaydou, E.M. Isolation and Characterization of a New Hispolone Derivative from Antioxidant Extracts of *Pistacia atlantica*. *Phytother. Res.* **2009**, *23*, 1237–1242. [CrossRef] [PubMed]
14. Wang, L.; Hou, Y. Determination of trace elements in anti-influenza virus mushrooms. *Biol. Trace Elem. Res.* **2011**, *143*, 1799–1807. [CrossRef] [PubMed]
15. Lifeng, Z. Studies on the Chemical Constituents and Pharmacological Activities of Inonotus hispidus and Fomitiporia ellipsoidea. Ph.D. Thesis, Jilin Agricultural University, Changchun, China.
16. Zan, L.-F.; Qin, J.-C.; Zhang, Y.-M.; Yao, Y.-H.; Bao, H.-Y.; Li, X. Antioxidant Hispidin Derivatives from Medicinal Mushroom *Inonotus hispidus*. *Chem. Pharm. Bull.* **2011**, *59*, 770–772. [CrossRef] [PubMed]
17. Gruendemann, C.; Arnhold, M.; Meier, S.; Baecker, C.; Garcia-Kaeufer, M.; Grunewald, F.; Steinborn, C.; Klemd, A.M.; Wille, R.; Huber, R.; et al. Effects of *Inonotus hispidus* Extracts and Compounds on Human Immunocompetent Cells. *Planta Med.* **2016**, *82*, 1359–1367. [CrossRef]
18. Ren, Q.; Lu, X.-Y.; Han, J.-X.; Aisa, H.A.; Yuan, T. Triterpenoids and phenolics from the fruiting bodies of *Inonotus hispidus* and their activations of melanogenesis and tyrosinase. *Chin. Chem. Lett.* **2017**, *28*, 1052–1056. [CrossRef]
19. Kemami Wangun, H.V.; Hertweck, C. Squarrosidine and Pinillidine: 3,3′-Fused Bis(styrylpyrones) from *Pholiota squarrosa* and *Phellinus pini*. *Eur. J. Org. Chem.* **2007**, *2007*, 3292–3295. [CrossRef]
20. Yang, S.; Bao, H.; Wang, H. Chemical components and anti-tumour compounds from *Inonotus hispidus*. *Mycosystema* **2019**, *38*, 127–133. [CrossRef]
21. Kou, R.W.; Du, S.T.; Xia, B.; Zhang, Q.; Yin, X.; Gao, J.M. Phenolic and Steroidal Metabolites from the Cultivated Edible *Inonotus hispidus* Mushroom and Their Bioactivities. *J. Agric. Food Chem.* **2021**, *69*, 668–675. [CrossRef] [PubMed]
22. Shaojun, T.; Ping, L.; Wenhua, M.; Chenxia, S.; Shenglian, W.; Yi, Y.; Yuelin, H.; Jun, X. Isolation and Identification of Antitumor Metabolites from Fermentation Broth of *Inonotus hispidus*. *Acta Edulis. Fungi* **2021**, *28*, 109–116. [CrossRef]
23. Yang, X. Studies on the Chemical Constituents of *Xanthochrous hispidus*. Master Thesis, Shandong University of Traditional Chinese Medicine, Jinan, China, 2008.
24. Zhou, J.; Lin, X.; Liu, S.; Wang, Z.; Liu, D.; Huo, Y.; Li, D. Effects of Compound Elicitors on the Biosynthesis of Triterpenoids and Activity of Defense Enzymes from *Inonotus hispidus* (Basidiomycetes). *Molecules* **2022**, *27*, 2618. [CrossRef] [PubMed]
25. Politi, M.; Silipo, A.; Siciliano, T.; Tebano, M.; Flamini, G.; Braca, A.; Jiménez-Barbero, J. Current analytical methods to study plant water extracts: The example of two mushrooms species, *Inonotus hispidus* and *Sparassis crispa*. *Phytochem. Anal.* **2007**, *18*, 33–41. [CrossRef] [PubMed]

26. Fushiya, S.; Matsuda, M.; Yamada, S.; Nozoe, S. New opine type amino acids from a poisonous mushroom, *Clitocybe acromelalga*. *Tetrahedron* **1996**, *52*, 877–886. [CrossRef]
27. Yang, S.; Bao, H.; Wang, H.; Li, Q. Anti-tumour Effect and Pharmacokinetics of an Active Ingredient Isolated from *Inonotus hispidus*. *Biol. Pharm. Bull.* **2019**, *42*, 10–17. [CrossRef] [PubMed]
28. Fan, N.-W.; Chang, H.-S.; Cheng, M.-J.; Hsieh, S.-Y.; Liu, T.-W.; Yuan, G.-F.; Chen, I.-S. Secondary Metabolites from the Endophytic Fungus *Xylaria cubensis*. *Helv. Chim. Acta* **2014**, *97*, 1689–1699. [CrossRef]
29. Ali, N.A.A.; Mothana, R.A.A.; Lesnau, A.; Pilgrim, H.; Lindequist, U. Antiviral activity of *Inonotus hispidus*. *Fitoterapia* **2003**, *74*, 483–485. [CrossRef]
30. Zan, L.; Liang, R.; Bao, H. Antioxidant and Antimicrobial Activities of *Inonotus hispidus* Extracts. *North. Hortic.* **2015**, *5*, 151–155. Available online: https://kns.cnki.net/kcms/detail/detail.aspx?FileName=BFYY201505047&DbName=CJFQ2015 (accessed on 10 July 2022).
31. Benarous, K.; Bombarda, I.; Iriepa, I.; Moraleda, I.; Gaetan, H.; Linani, A.; Tahri, D.; Sebaa, M.; Yousfi, M. Harmaline and hispidin from *Peganum harmala* and *Inonotus hispidus* with binding affinity to Candida rugosa lipase: In silico and in vitro studies. *Bioorg. Chem.* **2015**, *62*, 1–7. [CrossRef] [PubMed]
32. Benarous, K.; Benali, F.Z.; Bekhaoua, I.C.; Yousfi, M. Novel potent natural peroxidases inhibitors with in vitro assays, inhibition mechanism and molecular docking of phenolic compounds and alkaloids. *J. Biomol. Struct Dyn.* **2021**, *39*, 7168–7180. [CrossRef] [PubMed]
33. Tang, S.J.; Shen, D.A.; Shao, C.X.; Jun, X.U.; Yang, Y.; Jin, L.; Wu, S.-L. Classification and identification of a wild *Inonotus hispidus* and the antitumor activity of the fermentation supernatant. *J. South. Agric.* **2019**, *50*, 1671–1679. [CrossRef]
34. Li, Z.; Bao, H. Anti-tumor effect of *Inonotus hispidus* petroleum ether extract in H22 tumor-bearing mice and analysis its mechanism by untargeted metabonomic. *J. Ethnopharmacol.* **2022**, *285*, 114898. [CrossRef] [PubMed]
35. Li, Z.; Bao, H. Deciphering key regulators of *Inonotus hispidus* petroleum ether extract involved in anti-tumor through whole transcriptome and proteome analysis in H22 tumor-bearing mice model. *J. Ethnopharmacol.* **2022**, *296*, 115468. [CrossRef] [PubMed]
36. Wang, T.; Bao, H.Y.; Bau, T.; Li, Y. Antitumor Effect of Solid State Fermentation Powder of *Inonotus hispidus* on H22 Bearing Mice. *Zhong Yao Cai* **2016**, *39*, 389–394. [CrossRef]
37. Yang, S.; Bao, H.; Wang, H.; Li, Z.; Han, C. Bacteriostatic and Antioxidative Effects of Volatile Oil from *Inonotus hispidus*. *J. Jilin Agric. Univ.* **2020**, *42*, 148–153. [CrossRef]
38. Shomali, N.; Onar, O.; Alkan, T.; Demirtaş, N.; Akata, I.; Yildirim, Ö. Investigation of the Polyphenol Composition, Biological Activities, and Detoxification Properties of Some Medicinal Mushrooms from Turkey. *Turk. J. Pharm. Sci.* **2019**, *16*, 155–160. [CrossRef]
39. Benarous, K.; Djeridane, A.; Kameli, A.; Yousfi, M. Inhibition of *Candida rugosa* Lipase by Secondary Metabolites Extracts of Three Algerian Plants and their Antioxydant Activities. *Curr. Enzym. Inhib.* **2013**, *9*, 75–82. [CrossRef]
40. Ying, Y.-M.; Zhang, L.-Y.; Zhang, X.; Bai, H.-B.; Liang, D.-E.; Ma, L.-F.; Shan, W.-G.; Zhan, Z.-J. Terpenoids with alpha-glucosidase inhibitory activity from the submerged culture of *Inonotus obliquus*. *Phytochemistry* **2014**, *108*, 171–176. [CrossRef] [PubMed]
41. Tang, M.; Ding, Y.; Yang, Y.; Xie, M.; Wang, S.; Chen, C.; Wang, H.; Li, Y. α-glucosidase Inhibition and Antioxidant Activities of Different Polar Extracts of *Inonotus hispidus*. *Edible Fungi China* **2021**, *40*, 37–41. [CrossRef]
42. Liu, X.; Hou, R.; Yan, J.; Xu, K.; Wu, X.; Lin, W.; Zheng, M.; Fu, J. Purification and characterization of *Inonotus hispidus* exopolysaccharide and its protective effect on acute alcoholic liver injury in mice. *Int. J. Biol. Macromol.* **2019**, *129*, 41–49. [CrossRef] [PubMed]
43. Lee, I.-K.; Yun, B.-S. Styrylpyrone-class compounds from medicinal fungi *Phellinus* and *Inonotus* spp., and their medicinal importance. *J. Antibiot.* **2011**, *64*, 349–359. [CrossRef] [PubMed]
44. Dos Santos, N.C.; Menezes, R.; Stewart, D. Polyphenols as New Leads in Drug Discovery: Biological Activity and Mechanisms. *Curr. Pharm. Des.* **2018**, *24*, 2041–2042. [CrossRef]
45. Sarfraz, A.; Rasul, A.; Sarfraz, I.; Shah, M.A.; Hussain, G.; Shafiq, N.; Masood, M.; Adem, S.; Sarker, S.D.; Li, X. Hispolon: A natural polyphenol and emerging cancer killer by multiple cellular signaling pathways. *Env. Res.* **2020**, *190*, 110017. [CrossRef] [PubMed]
46. Nambudiri, A.M.D.; Vance, C.P.; Towers, G.H.N. Effect of light on enzymes of phenylpropanoid metabolism and hispidin biosynthesis in *Polyporus hispidus*. *Biochem. J.* **1973**, *134*, 891–897. [CrossRef]
47. Kemami Wangun, H.V.; Härtl, A.; Tam Kiet, T.; Hertweck, C. Inotilone and related phenylpropanoid polyketides from *Inonotus* sp. and their identification as potent COX and XO inhibitors. *Org. Biom. Chem.* **2006**, *4*, 2545–2548. [CrossRef] [PubMed]
48. Zhang, H.; Chen, R.; Zhang, J.; Bu, Q.; Wang, W.; Liu, Y.; Li, Q.; Guo, Y.; Zhang, L.; Yang, Y. The integration of metabolome and proteome reveals bioactive polyphenols and hispidin in ARTP mutagenized *Phellinus baumii*. *Sci. Rep.* **2019**, *9*, 16172. [CrossRef] [PubMed]
49. Huo, J.; Zhong, S.; Du, X.; Cao, Y.; Wang, W.; Sun, Y.; Tian, Y.; Zhu, J.; Chen, J.; Xuan, L.; et al. Whole-genome sequence of *Phellinus gilvus* (mulberry Sanghuang) reveals its unique medicinal values. *J. Adv. Res.* **2020**, *24*, 325–335. [CrossRef] [PubMed]
50. Angulo-Sanchez, L.T.; López-Peña, D.; Torres-Moreno, H.; Gutiérrez, A.; Gaitán-Hernández, R.; Esqueda, M. Biosynthesis, Gene Expression, and Pharmacological Properties of Triterpenoids of Ganoderma Species (Agaricomycetes): A Review. *Int. J. Med. Mushrooms* **2022**, *24*, 1–17. [CrossRef] [PubMed]

51. Grienke, U.; Zöll, M.; Peintner, U.; Rollinger, J.M. European medicinal polypores—A modern view on traditional uses. *J. Ethnopharmacol.* **2014**, *154*, 564–583. [CrossRef]
52. Floris, B.; Galloni, P.; Conte, V.; Sabuzi, F. Tailored Functionalization of Natural Phenols to Improve Biological Activity. *Biomolecules* **2021**, *11*, 1325. [CrossRef]
53. Dasgupta, A.; Acharya, K. Chapter 10—Bioactive terpenoids from mushrooms. In *New and Future Developments in Microbial Biotechnology and Bioengineering*; Singh, J., Gehlot, P., Eds.; Elsevier: Amsterdam, The Netherlands, 2021; pp. 145–154.
54. Duan, Y.; Han, H.; Qi, J.; Gao, J.-m.; Xu, Z.; Wang, P.; Zhang, J.; Liu, C. Genome sequencing of *Inonotus obliquus* reveals insights into candidate genes involved in secondary metabolite biosynthesis. *BMC Genom.* **2022**, *23*, 314. [CrossRef]
55. Yingce, D.; Dan, S.; Lin, W.; Xiaoning, Z.; Chunlei, W.; Chengwei, L. Research Progress of Small Molecule Chemical Components and Pharmacological Values of *Inonotus obliquus*. *J. Fungal Res.* **2022**, *20*, 1–14. [CrossRef]
56. Agger, S.; Lopez-Gallego, F.; Schmidt-Dannert, C. Diversity of sesquiterpene synthases in the basidiomycete *Coprinus cinereus*. *Mol. Microbiol.* **2009**, *72*, 1181–1195. [CrossRef]
57. Wawrzyn, G.T.; Quin, M.B.; Choudhary, S.; López-Gallego, F.; Schmidt-Dannert, C. Draft Genome of *Omphalotus olearius* Provides a Predictive Framework for Sesquiterpenoid Natural Product Biosynthesis in Basidiomycota. *Chem. Biol.* **2012**, *19*, 772–783. [CrossRef] [PubMed]
58. Quin, M.B.; Flynn, C.M.; Wawrzyn, G.T.; Choudhary, S.; Schmidt-Dannert, C. Mushroom Hunting by Using Bioinformatics: Application of a Predictive Framework Facilitates the Selective Identification of Sesquiterpene Synthases in Basidiomycota. *ChemBioChem* **2013**, *14*, 2480–2491. [CrossRef] [PubMed]
59. Nagamine, S.; Liu, C.; Nishishita, J.; Kozaki, T.; Sogahata, K.; Sato, Y.; Minami, A.; Ozaki, T.; Schmidt-Dannert, C.; Maruyama, J.-I.; et al. Ascomycete *Aspergillus oryzae* Is an Efficient Expression Host for Production of Basidiomycete Terpenes by Using Genomic DNA Sequences. *Appl. Environ. Microbio.* **2019**, *85*, e00409–e00419. [CrossRef] [PubMed]
60. Tao, Q.; Ma, K.; Yang, Y.; Wang, K.; Chen, B.; Huang, Y.; Han, J.; Bao, L.; Liu, X.-B.; Yang, Z.; et al. Bioactive Sesquiterpenes from the Edible Mushroom *Flammulina velutipes* and Their Biosynthetic Pathway Confirmed by Genome Analysis and Chemical Evidence. *J. Org. Chem.* **2016**, *81*, 9867–9877. [CrossRef] [PubMed]
61. Zhang, C.; Chen, X.; Orban, A.; Shukal, S.; Birk, F.; Too, H.-P.; Rühl, M. *Agrocybe aegerita* Serves as a Gateway for Identifying Sesquiterpene Biosynthetic Enzymes in Higher Fungi. *ACS Chem. Biol.* **2020**, *15*, 1268–1277. [CrossRef] [PubMed]
62. Song, F.; Su, D.; Keyhani, N.O.; Wang, C.; Shen, L.; Qiu, J. Influence of selenium on the mycelia of the shaggy bracket fungus, *Inonotus hispidus*. *J. Sci. Food Agric.* **2022**, *102*, 3762–3770. [CrossRef]
63. Hou, R.; Liu, X.; Xiang, K.; Chen, L.; Wu, X.; Lin, W.; Zheng, M.; Fu, J. Characterization of the physicochemical properties and extraction optimization of natural melanin from *Inonotus hispidus* mushroom. *Food Chem.* **2019**, *277*, 533–542. [CrossRef]
64. Liu, X.; Hou, R.; Xu, K.; Chen, L.; Wu, X.; Lin, W.; Zheng, M.; Fu, J. Extraction, characterization and antioxidant activity analysis of the polysaccharide from the solid-state fermentation substrate of *Inonotus hispidus*. *Int. J. Biol. Macromol.* **2019**, *123*, 468–476. [CrossRef]
65. Alves, M.J.; Ferreira, I.C.F.R.; Dias, J.; Teixeira, V.; Martins, A.; Pintado, M. A Review on Antimicrobial Activity of Mushroom (Basidiomycetes) Extracts and Isolated Compounds. *Planta Med.* **2012**, *78*, 1707–1718. [CrossRef]
66. Patel, D.K.; Dutta, S.D.; Ganguly, K.; Cho, S.-J.; Lim, K.-T. Mushroom-Derived Bioactive Molecules as Immunotherapeutic Agents: A Review. *Molecules* **2021**, *26*, 1359. [CrossRef]
67. Tang, Y.Z.; Liu, Y.H.; Chen, J.X. Pleuromutilin and its derivatives-the lead compounds for novel antibiotics. *Mini Rev. Med. Chem.* **2012**, *12*, 53–61. [CrossRef] [PubMed]

antibiotics

Review

Antifungal and Immunomodulatory Ingredients from Traditional Chinese Medicine

Hua Zhong [1,†], Lei Han [2,†], Ren-Yi Lu [1,†] and Yan Wang [1,*]

1 School of Pharmacy, Second Military Medical University, Shanghai 200433, China
2 School of Pharmacy, Fujian University of Traditional Chinese Medicine, Fuzhou 350122, China
* Correspondence: wangyansmmu@126.com
† These authors contributed equally to this article.

Abstract: Fungal infections have become a growing public health challenge due to the clinical transmission of pathogenic fungi. The currently available antifungal drugs leave very limited choices for clinical physicians to deal with such situation, not to mention the long-standing problems of emerging drug resistance, side effects and heavy economic burdens imposed to patients. Therefore, new antifungal drugs are urgently needed. Screening drugs from natural products and using synthetic biology strategies are very promising for antifungal drug development. Chinese medicine is a vast library of natural products of biologically active molecules. According to traditional Chinese medicine (TCM) theory, preparations used to treat fungal diseases usually have antifungal and immunomodulatory functions. This suggests that if antifungal drugs are used in combination with immunomodulatory drugs, better results may be achieved. Studies have shown that the active components of TCM have strong antifungal or immunomodulatory effects and have broad application prospects. In this paper, the latest research progress of antifungal and immunomodulatory components of TCM is reviewed and discussed, hoping to provide inspiration for the design of novel antifungal compounds and to open up new horizons for antifungal treatment strategies.

Keywords: natural product; traditional Chinese medicine; antifungal drug; immune regulation; combination of drugs; fungal infection

Citation: Zhong, H.; Han, L.; Lu, R.-Y.; Wang, Y. Antifungal and Immunomodulatory Ingredients from Traditional Chinese Medicine. *Antibiotics* **2023**, *12*, 48. https://doi.org/10.3390/antibiotics12010048

Academic Editors: William N. Setzer and Nicholas Dixon

Received: 13 September 2022
Revised: 23 December 2022
Accepted: 23 December 2022
Published: 28 December 2022

1. Introduction

The development of new drugs has a very important role in improving human health and extending human lifespan. There are many strategies for drug development, including screening drugs from natural products and synthetic biology strategies. Traditional Chinese medicine (TCM) is a huge reservoir of natural products, with a long history of practice and great potential for drug development. The discovery of novel drug candidates from TCM and its extracts has become a research hotspot [1].

Fungal infections have been a major public health challenge. Depending on the site of invasion, there are at least three types: superficial, subcutaneous, and deep. Most of these infections are superficial, i.e., non-fatal infections of the skin, nails, and hair, mainly caused by skin fungi, with a global incidence of about 25% [2,3]. Deep mycosis, by contrast, can be life-threatening. The infection can spread to many organ systems and meninges. Under different circumstances, the mortality rate can vary from 35% to 90% [2]. Common pathogens of invasive fungal diseases include *Candida albicans*, *Cryptococcus neoformans*, and *Aspergillus fumigatus* [4,5]. Despite continuous efforts to develop antifungal drugs, the availability of drugs for clinical use remains limited. At present, commonly used antifungal drugs in clinics are azole, polyene, echinomycin, terbinafine, and so on. However, existing drugs have problems such as drug resistance, high price, high toxicity, and serious side effects, such as renal toxicity. It is an urgent and challenging task to find new antifungal drugs with high efficiency and low toxicity.

The TCM treatment of infectious diseases focuses more on the treatment of diseases rather than just killing pathogens. According to the basic theory of TCM, diseases can be divided into *cold, heat, deficiency,* and *excess*. The meaning of the words mentioned here is different from the literal meaning. For example, "*heat*" refers to a range of different kinds of diseases, which include a range of symptoms caused by an excessive inflammatory response, such as vulvovaginal candidiasis. In the case of "*deficiency*", diseases caused by immune deficiency are included, such as invasive pulmonary aspergillosis. Thereby, drugs with anti-inflammatory effects are usually used to treat fungal infections with *heat*. These drugs include *Fructus Forsythiae* (lianqiao in Chinese) and *Lonicera japonica* Flos (jinyinhua in Chinese). Meanw hile, some drugs with immune-modulatory effects are used to tonify "*deficiency*". These drugs include ginseng, *Atractylodes macrocephala* Koidz. (baizhu in Chinese), and *Schisandra chinensis* (Turcz.) Baill. (wuweizi in Chinese). TCM treats the interaction between the fungus and its host as a whole. If we try to explain it in modern medical terms, it could be killing the fungus while modulating the body's immune response.

There are several TCM products that use both antifungal and immunomodulatory strategies to treat fungal infections. For example, Skinguard Lotion, a TCM product for the treatment of vaginitis, contains *Lonicera japonica* Flos, *Taraxacum officinale* L., *Cnidii Fructus*, etc. The pharmacological effects of the lotion include inhibiting inflammation, relieving itching, and killing pathogens. In the lotion, *Cnidii Fructus* is reported to have antifungal effects [6], while the main active molecules in *Lonicera japonica* Flos and *Taraxacum officinale* L. are proven to have anti-inflammatory effects [7,8]. Patients with vaginitis suffer a lot from persistent itch caused by inflammation response. The lotion can reduce the itching and at the same time inhibit pathogens. It can both treat the symptoms and eliminate the causes of the disease. The combination of antifungal drugs and immunomodulators may be a promising antifungal treatment strategy and an effective way to develop new antifungal drugs. Therefore, this paper reviewed two kinds of substances isolated from TCM: those with antifungal effects and those with immunomodulatory effects.

2. Antifungal Ingredients from TCM

Hundreds of TCM have been shown to have antifungal effects, such as *Coptis chinensis* Franch [9], *Scutellaria baicalensis* Georgi [10], *Curcuma longa* Linn [11], and *Allium sativum* L. [12]. We can extract active substances from these TCMs with water, alcohol, chloroform, ethyl acetate, ether, and petroleum ether. The extracts, such as terpenoids, volatile oil, ketones, alkaloids, aldehydes, phenylpropanoids, and saponins, also exhibit potent antifungal activity [13]. We reviewed substances isolated from TCMs with antifungal activities, including dioscin, α-santalol, formyl-phloroglucinol meroterpenoids, asiatic acid, carvacrol, eugenol, thymol, turmeric oil, terpinen-4-ol, silibinin, pinobanksin, tectochrysin, chrysin, licochalcone A, baicalein, baicalin, berberine, sanguinarine, plumbagin, osthole, and ethyl caffeate (Table 1).

Table 1. Natural products from TCMs and their antifungal activities.

Natural Product	Source	Target Fungi	MIC (μg/mL) [a]	FICI [a]	References
dioscin 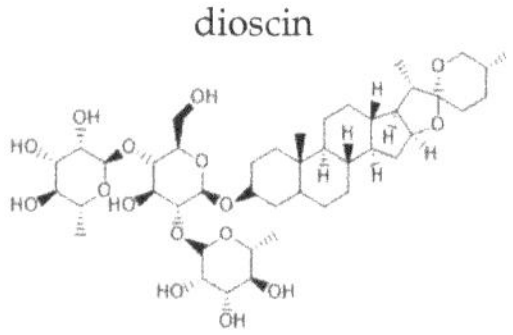	*Dioscoreaceae* family	*C. albicans* (2) [b]	4	-	[14,15]
		C. glabrata	2	-	
		C. parapsilosis	4	-	

Table 1. *Cont.*

Natural Product	Source	Target Fungi	MIC (μg/mL) [a]	FICI [a]	References
α-santalol 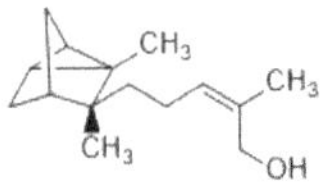	*Santalum* family	*T. rubrum*	12.5 (μg/disc [c])	-	[16,17]
eucarobustol E 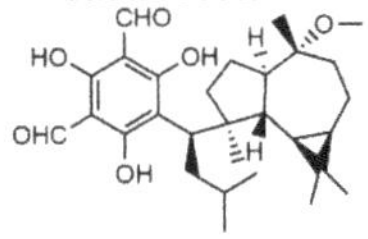	*Eucalyptus robusta*	fluconazole-susceptible *C. albicans* (10)	4–16	-	[18]
		fluconazole-resistant *C. albicans* (10)	32–128	-	[18]
asiatic acid 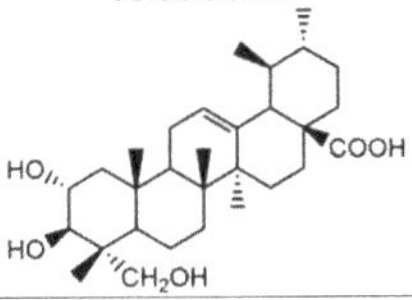	*Centella asiatica*	fluconazole-susceptible *C. albicans* (4)	64	0.75–1.00	[19]
		fluconazole-resistant *C. albicans* (4)	64–128	0.25-	[19]
carvacrol 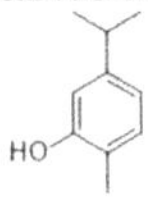	*Oreganum* family	*C. albicans*	250	0.374	[20,21]
thymol 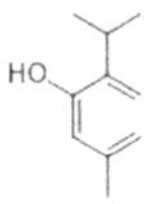	*Oreganum* family	*C. albicans*	500	1.062	[20,22]
		C. neoformans (10)	20–51	-	[23]
eugenol 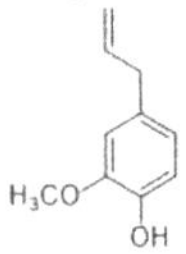	*Oreganum* family	*C. albicans*	1000	0.312	[20,24]
laurel essential oil	*Laurus nobilis*	*C. albicans* (2)	>4 (% *v*/*v*) [d]	-	[25]
anise essential oil	*Pimpinella anisum*	*C. albicans* (6)	4 or >4 (% *v*/*v*)	-	[25]
oregano essential oil	*Thymus capitatus*	*C. albicans* (31)	0.0039–1, or <0.0019 (% *v*/*v*)	-	[25]

Table 1. *Cont.*

Natural Product	Source	Target Fungi	MIC (μg/mL) [a]	FICI [a]	References
basil essential oil	*Ocimum basilicum*	*C. albicans* (11)	4 or >4 (% *v*/*v*)	-	[25]
lavender essential oil	*Lavandula latifolia*	*C. albicans* (15)	0.25–4 (% *v*/*v*)	-	[25]
mint essential oil	*Mentha spicata*	*C. albicans* (25)	2–4 (% *v*/*v*)	-	[25]
rosemary essential oil	*Rosmarinus officinalis*	*C. albicans* (2)	4 (% *v*/*v*)	-	[25]
rosemary extract	*Rosmarinus officinalis* L.	*C. albicans*	50,000	-	[26]
		C. dubliniensis	50,000	-	[26]
		C. glabrata	50,000	-	[26]
		C. krusei	50,000	-	[26]
		C. tropicalis	50,000	-	[26]
tea tree oil	*Melaleuca alternifolia*	*C. albicans* (44)	0.06–4 (% *v*/*v*)	0.25–1.25	[25,27]
terpinen-4-ol 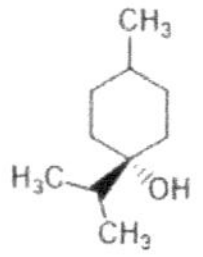	*Melaleuca alternifolia*	*C. albicans* (33)	0.06–0.25	0.250-0.252	[27,28]
grapefruit essential oil	*Citrus paradisi*	*C. albicans* (12)	0.0039–1 (% *v*/*v*)	-	[25]
turmeric essential oil	*Curcuma longa* L.	*M.gypseum* (2)	0.25 (% *v*/*v*), 6.25	-	[29,30]
		T. mentagrophytes (2)	0.25 (% *v*/*v*), 6.25	-	[29,30]
		T. rubrum	1.56	-	[30]
		E. floccosum	1.56	-	[30]

Table 1. *Cont.*

Natural Product	Source	Target Fungi	MIC (μg/mL) [a]	FICI [a]	References
Ar-turmerone 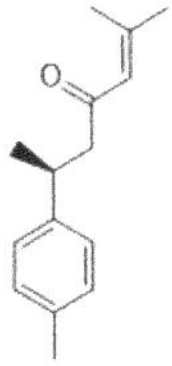	*Curcuma longa* L.	*M.gypseum*	7.81	-	[30,31]
		T. mentagrophytes	7.81	-	[30]
		T. rubrum	3.90	-	[30]
		E. floccosum	3.90	-	[30]
curcumin 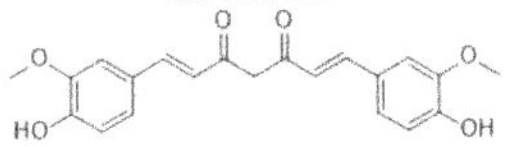	*Curcuma longa* L.	*fluconazole-susceptible Candida* species (27)	250–650	-	[32,33]
		fluconazole-resistant Candida species (11)	250–500	-	[32]
		C. albicans	64	-	[34]
		C. tropicalis	256	-	[34]
		C. krusei	256	-	[34]
		C. parapsilosis	>256	-	[34]
		C. glabrata	>256	-	[34]
		C. dubliniensis (2)	32	-	[34]
		C. neoformans	32	-	[34]
		S. schenckii	32	-	[34]
		P. brasiliensis (7)	0.5–32	-	[34]
		A. fumigatus (2)	>256	-	[34]
		A. nomius	>256	-	[34]
		A. flavus	>256	-	[34]
		A. tamarii	>256	-	[34]

Table 1. *Cont.*

Natural Product	Source	Target Fungi	MIC (μg/mL) [a]	FICI [a]	References
curcumin 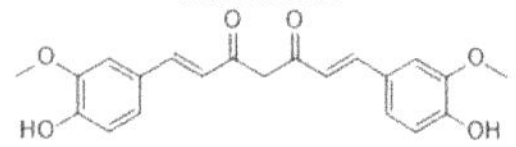	*Curcuma longa* L.	*A. terreus*	>256	-	[34]
		A. clavatus	>256	-	[34]
a blend of essential oils (3.53% of cinnamaldehyde, 3.53% of eugenol, 3.53% of carvacol, 1.04% of carotol, and 88.35% of *Camelina sativa* oil)	*Cinnamomum zeylanicum*, *Syzygium aromaticum*, *Origanum vulgare*, *Daucus carota*, and *Camelina sativa*	*C. albicans* (4)	0.02 (% *v*/*v*)	-	[35]
		C. glabrata	0.05 (% *v*/*v*)	-	[35]
		C. tropicalis	0.01 (% *v*/*v*)	-	[35]
silibinin 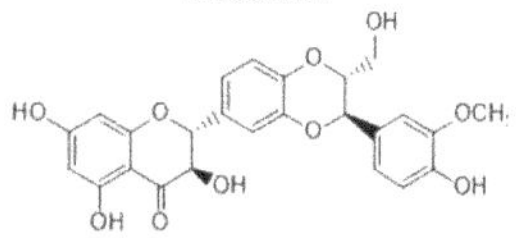	*Silybum marianum*	*C. albicans* (2)	19.3, 1024	-	[36–38]
		C. krusei	1024	-	[36]
		C. tropicalis	1024	-	[36]
		A. flavus	9.6	-	[37]
		C. parapsilosis	9.6	-	[37]
		Malassezia Furfur	19.3	-	[37]
		Trichosporon species	19.3	-	[37]
pinobanksin 	Chinese propolis	*C. albicans*	100	-	[39]
tectochrysin 	Chinese propolis	*C. albicans*	25	-	[39]
chrysin 	Chinese propolis	*C. albicans*	100	-	[39]

Table 1. *Cont.*

Natural Product	Source	Target Fungi	MIC (μg/mL) [a]	FICI [a]	References
3-O-acetylpinobanksin 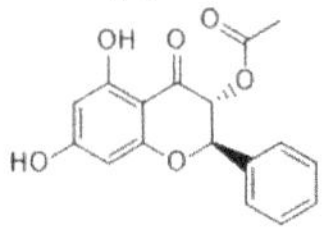	Chinese propolis	*C. albicans*	50	-	[39]
licochalcone A 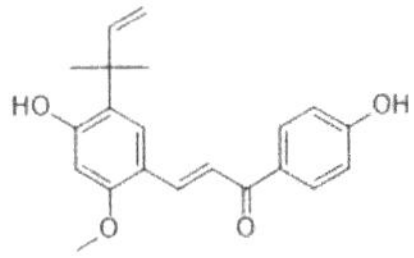	*Glycyrrhiza* family	*C. albicans* (4)	16.92–50.76	-	[40]
baicalein 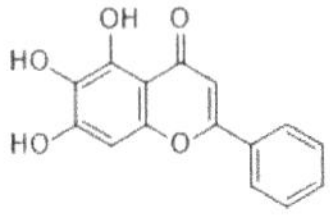	*Scutellaria baicalensis*	*C. albicans*	25	0.039	[41,42]
berberine 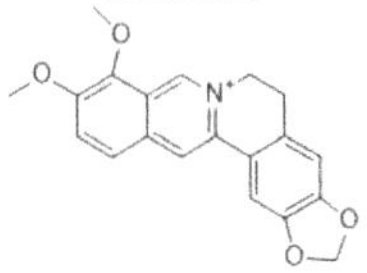	many herbs such as *Coptis chinensis* and *Mahonia aquifolium*	*C. albicans* (4)	80–160	0.017–0.127	[43,44]
		C. krusei (3)	10–20	-	[43]
		C. glabrata (3)	20–160	-	[43]
		C. dubliniensis	40	-	[43]
sanguinarine 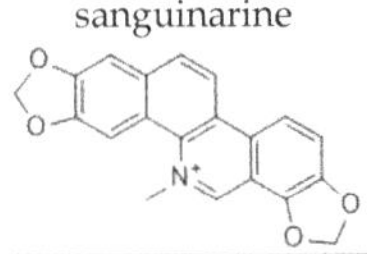	*Papaveraceae* family	*C. albicans* (11)	4, 37.5–50	-	[45,46]
		C. neoformans	64	-	[45]
chelerythrine 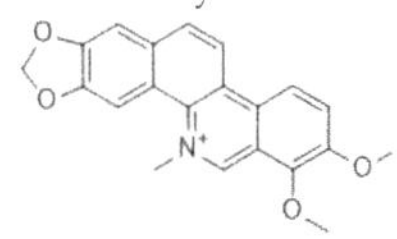	*Papaveraceae* family	*C. albicans* (3)	2–4	-	[45,47]
		C. glabrata (2)	16		[47]
		C. krusei (2)	16		[47]
		C. tropicalis (2)	8		[47]
		C. neoformans	64	-	[45]

Table 1. *Cont.*

Natural Product	Source	Target Fungi	MIC (μg/mL) [a]	FICI [a]	References
plumbagin 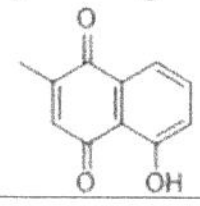	*Plumbago scandens*	*C. albicans*	0.78	-	[48]
		C. neoformans	8	-	[49]
bis-naphthoquinone 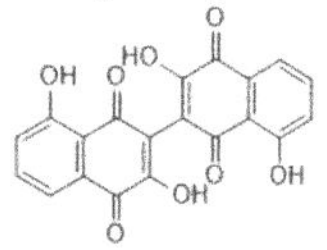	*Ceratostigma plumbaginoides*	*C. albicans*	0.09	-	[50]
shikonin 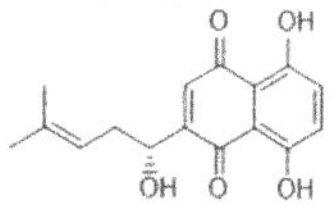	*Lithospermum erythrorhizon*	*C. albicans* (12)	2-8, >64	-	[51,52]
		C. krusei	4	-	[51]
		C. glabrata	8	-	[51]
		C. tropicalis	8	-	[51]
		C. parapsilosis	16	-	[51]
		Saccharomyces cerevisiae	4	-	[51]
		C. neoformans	8	-	[51]
		T. cutaneum	8	-	[51]
		A. fumigatus	>64	-	[51]
osthole 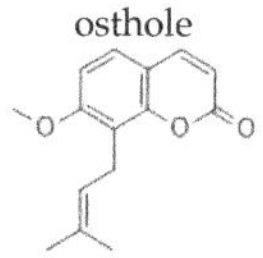	*Cnidii fructus*	*C. albicans* (52)	8–64, >64	0.04–0.37	[6,53]
ethyl caffeate 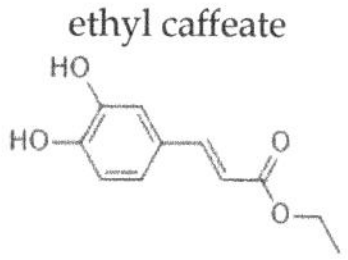	*Elephantopus scaber* L.	*C. albicans* (26)	64–256	0.047–0.375	[54]

[a] MIC, minimum inhibitory concentration. FICI, fractional inhibitory concentration index; the numbers indicate the FICI of natural products combined with fluconazole against fungi. -, not mentioned in the report. [b] Numbers in the brackets indicate the number of strains tested. [c] In this case, MIC is defined as the concentration of the 0.5 mm inhibitory zone produced by the tested compound in a disc diffusion method. [d] In these cases, the unit of MICs is (% v/v) instead of μg/mL, since the agents are in liquid form.

2.1. Terpenoids

Terpenoids are common substances in natural products. Many terpenoids have important biological activities and are important resources for the development of new drugs. Some terpenoids in natural medicines have important biological activities and have been used in clinical practice for many years, such as the antimalarial drug artemisinin, the anticancer drugs paclitaxel and cantharidin, the antibiotic drug lactam, etc. Some terpenoids have antifungal activity and may provide new strategies for antifungal therapy.

For example, dioscin is a steroidal saponin that can be isolated from TCMs of the *Dioscoreaceae* family. It exhibits significant fungicidal effects against the *Candida* species, such as *C. albicans*, *C. glabrata*, and *C. parapsilosis*, with minimum inhibitory concentration (MIC) at 2–4 μg/mL. Moreover, dioscin can also inhibit biofilm formation and even destroy the mature biofilm at high concentrations [14]. The possible antifungal mechanisms of dioscin may be associated with inhibiting the virulence factors of *C. albicans*, including the morphological transformation, the production of phospholipase, and the adherence to the abiotic surface [14]. Another computational study showed that dioscin can bind to the glucoamylase enzyme that is produced in many fungi by molecular docking [55].

Another example of antifungal terpenoids is α-santalol. It can be found in the essential oil of the *Santalum* species, which has been historically used as a noble perfume as well as medicine. It is reported that α-santalol inhibits Trichophyton rubrum at 12.5 μg/disc (using the disc diffusion method), and a preliminary mechanism study revealed that α-santalol influences the fungal cell wall synthesis and mitosis process [16].

Formyl-phloroglucinol meroterpenoids (FPMs) are important secondary metabolites with various biological activities, mainly found in *Eucalyptus* and *Psidium*. FPMs exhibit antifungal effects against many fungal species, including *C. albicans* and *T. mentagrophytes* [56,57]. *Eucalyptus robusta* is widely distributed in Sichuan, Yunnan, and other southern areas of China, and it is commonly used in the traditional medicine of ethnic minorities. Eucalyptus leaves are used to treat respiratory infections, intestinal infections, malaria, trachoma, otitis media, keratitis, dermatitis, etc. The essential oil extracted from the leaves is also used as a preservative. Eucarobustol E, as an FPM from the leaves of *Eucalyptus robusta*, has a broad-spectrum antifungal effect with MIC ranging from 4 to 16 μg/mL. It also inhibits *C. albicans* biofilms in both the formation phase and mature phase, and the mechanism might involve the negative regulation of hyphal growth due to the inhibition of carbon flow towards ergosterol. This mechanism is totally different from the known mechanisms of existing antifungal drugs and may provide a novel strategy for fungal infection treatment [18].

Carvacrol is a phenolic monoterpenoid that can be extracted from *Origanum vulgare*, an herb that is used to treat acute gastroenteritis in TCM. It inhibits the spore germination of *Aspergillus flavus* by 73.3% at the concentration of 100 μg/mL, and 100 μg/mL of carvacrol slows the mycelia growth and reduces mycelia drying weight. The antifungal mechanism involves deficiency in ergosterol production and the alteration of glycerophospholipid metabolism [58].

Thymol is one of the main components of thyme oil, which is used as both pharmacological and cosmetic additives. The MIC of thymol against *C. albicans* is 500 μg/mL [20]. Thymol also exhibits antifungal activity against *C. neoformans*, with MICs ranging from 20–51 μg/mL, and the mode of antifungal action seems to be related to the ergosterol of yeast [23]. Another study demonstrated that thymol regulated several signaling pathways, such as calcineurin, unfolded protein response, and the HOG-MAPK pathway [59].

In addition to the terpenoids with antifungal effects when used alone, there are also some terpenoids that can be used as synergists in combination with existing antifungal drugs to reduce the dosage and increase efficacy. Doke S. K. et al. evaluated the synergistic effects of three terpenoids with fluconazole against *C. albicans*, involving carvacrol, eugenol, and thymol. The results indicated that carvacrol and eugenol showed synergistic effects in combination with fluconazole against planktonic cells and biofilm formation, with FICI ranging from 0.25 to 0.516, whereas thymol showed an indifferent effect with fluconazole.

Using terpenoids in combination with current antifungal drugs, such as fluconazole, may be a feasible strategy to treat fungal infections. The combination can effectively reduce the dosage of antifungal drugs and, therefore, reduce the adverse effects during treatment [20].

Asiatic acid (AA) is a natural ursane-type pentacyclic triterpenoid, mainly found in plants such as *Centella asiatica*. AA shows synergistic effects with fluconazole, and the combination can inhibit fluconazole-resistant *C. albicans* strains in vitro and in vivo, with a fractional inhibitory concentration index (FICI) of 0.25. 32 μg/mL. AA also increases the fluconazole effect against *C. albicans* biofilm at 0.125–0.25 μg/mL. A mechanism study indicated that AA inhibited the drug efflux pump of *C. albicans*. The inhibitory effect of AA on the drug efflux pump may increase the intracellular concentration of fluconazole in *C. albicans* cells, and thereby, AA exhibits synergistic effects with fluconazole. Meanwhile, AA in combination with fluconazole aroused the level of intracellular reactive oxygen species (ROS) and inhibited the hyphal growth of yeast [19].

2.2. *Volatile Oils*

Volatile oils, also known as essential oils, are extracted from plants and animals with various bioactivities. They are an important source for developing antifungal agents. Volatile oils have complex molecular structures and usually have a strong fragrance. According to TCM theories, volatile oils have the functions of inducing diaphoresis, regulating Qi, relieving pain, being antimicrobial, and correcting flavor, whereas modern medical research shows that essential oils have antibacterial, antifungal, antiviral, antioxidant, antitumor, and immunomodulatory activities [60]. It seems that natural medicines with antifungal effects are mostly found in *Lamiaceae, Lauraceae, Myrtaceae*, and *Compositae*, such as *Cinnamomum cassia* Presl, *Rosmarinus officinalis, Ocimum basilicum, Origanum vulgare, Mentha haplocalyx* Briq., *Melaleuca alternifolia, Foeniculum vulgare, Zingiber officinale, Catsia tora Linn, Cinnamomum camphora* (L.) Presl., and *Senecio scandens*.

Abers M. et al. evaluated 19 essential oils for their antifungal activity using a modified disk diffusion test. The results demonstrated that volatiles of lavender, tea tree, cinnamon, peppermint, cassia, and oregano had "moderate" antifungal activity, whereas volatiles of thyme and rosemary had "high" antifungal activity. Clove volatile did not exhibit any antifungal activity in the test [61]. A similar work was conducted by another team. Bona E. et al. tested the sensitivity of 30 vaginal *C. albicans* strains to 12 essential oils, and mint, basil, lavender, tea tree oil, and oregano were more efficient in inhibiting fungal growth and activity than the traditional antifungal drug clotrimazole. The MICs of mint, basil, lavender, and tea tree oil were 0.25–4% v/v for most strains, with values >4% for others. The oregano volatile MICs were lower than 1% v/v in 64% of the strains. They also found that the mechanism of essential oils in antifungal activity was associated with cell wall and membrane damage [25].

Turmeric oil is the extract from *Curcuma longa* Linn. It has long been used as a common household medicine and a yellow spice in Southeast Asia. As for antifungal activities, turmeric oil completely inhibits common dermatophyte growth at the concentration of 0.2% v/v, including *Epidermophyton floccosum, Microsporum gypseum, M. nanum, T. mentagrophyte, T. rubrum*, and *T. violaceum* [29]. Notably, turmeric oil exhibits low toxicity towards humans. It did not show any irritation or adverse effect at a 5% concentration for up to 3 weeks in a clinical trial [29]. Moreover, the main component of turmeric oil, Ar-turmerone, exhibits more potent antifungal activity against dermatophytes than ketoconazole, with MICs ranging from 3.90 to 7.81 μg/mL [30]. Another important substance isolated from *Curcuma longa* L. is curcumin, which also arouses the wide interest of scientists. It is reported to have many bioactivities, such as antitumor, antioxidant, anti-inflammatory, and antimicrobial activities [62–64]. Curcumin shows inhibitory effects against many common fungal pathogens. It inhibits *Candida* species growth with an MIC_{90} ranging from 250 to 650 μg/mL. This includes clinical isolates and fluconazole-resistant strains of *C. albicans, C. glabrata, C. krusei, C. tropicalis*, and *C. guilliermondii* [32]. Curcumin also shows antifungal activity against *C. neoformans* and *C. dubliniensis* with an MIC of 32 μg/mL [34].

In addition, curcumin and its derivatives show synergy effects with fluconazole against resistant *C. albicans*, *C. tropicalis*, and *C. krusei*, with the FICIs ranging from 0.078 to 0.563 [33]. The antifungal mechanisms of curcumin include causing oxidative stress and inhibiting thymidine uptake 1. Curcumin also alters the membrane-associated properties of ATPase activity, ergosterol biosynthesis, and proteinase secretion [33].

Similar to terpenoids, volatile oils can also be used as a synergist with antifungal drugs. Adding tea tree oil at a concentration of 1/4 MIC can effectively reduce MICs of fluconazole against resistant *C. albicans* strains. Among all the 32 tested strains, the average fluconazole MIC is reduced from 244.0 μg/mL to 38.46 μg/mL, and terpinen-4-ol, which is the major bioactive component of tea tree oil, shows an even stronger inhibition effect [27].

Moreover, a combination of different TCMs without antifungal drugs can also have a synergistic effect against fungi. Apart from enhancing antifungal activity and reducing adverse effects, this kind of combination treatment can also reduce the occurrence of drug resistance in fungi. It is reported that an essential oil blend of *Cinnamomum zeylanicum*, *Daucus carota*, *Syzygium aromaticum*, and *Origanum vulgare* shows antifungal activity against *C. albicans*, *C. tropicalis*, and *C. glabrata*, including strains resistant to fluconazole or amphotericin B, with MICs ranging from 0.01 to 0.05% *v*/*v* [35].

2.3. Flavonoids

Some flavonoids from TCMs also have antifungal properties. *Rosmarinus officinalis* L., for example, contains phenols and flavonoids in its extract. The extract shows an antibiofilm effect against *C. albicans*, *C. dubliniensis*, *C. glabrata*, *C. krusei*, and *C. tropicalis*. A 50–200 mg/mL extract can reduce mature biofilms, which are formed 48 h in advance, and the antibiofilm activity of *Rosmarinus officinalis* L. extract is comparable to that of nystatin [26].

Silibinin is a flavonoid that can be found in *Silybum marianum*, a traditional medicine for hepatobiliary diseases in China and Europe [65]. The antifungal activity of silibinin seems not stable or uniform. Dayanne R. O. et al. reported that silibinin inhibits *C. albicans*, *C. krusei*, and *C. tropicalis* with MICs at 1024 μg/mL [36]. In another study, however, the MICs of silibinin against *C. albicans* and *C. parapsilosis* were 19.3 μg/mL and 9.6 μg/mL, respectively [37]. One possible reason for the differences is that the strains used in the tests are different and silibinin does not exhibit stable or uniform antifungal activities against those strains. Silibinin at 100 μg/mL also inhibits *C. albicans* biofilm formation to approximately 50%, and the mode of action involves causing membrane damage to fungal cells [65].

Propolis is a natural product with antimicrobial [66], anti-inflammatory [67], and antioxidant activities [68]. The main bioactive components of propolis are flavonoids. Propolis production is influenced by many environmental factors such as local climate and the plants nearby. Therefore, propolis from different regions contains different kinds and amounts of flavonoids, and it consequently has different bioactivities [69]. In Chinese propolis, the four major flavonoids are pinobanksin, tectochrysin, chrysin, and 3-O-acetylpinobanksin. All the four flavonoids have antifungal activities against *C. albicans*, with MICs of 100 μg/mL, 25 μg/mL, 100 μg/mL, and 50 μg/mL, respectively [39]. The antifungal mechanism of propolis is related to the induction of apoptosis through metacaspase and Ras signaling. It is also reported that propolis disrupts the expression of several genes involving pathogenesis, cell adhesion, biofilm formation, filamentous growth, and phenotypic switching [70].

The *Glycyrrhiza* species is another commonly used ingredient in traditional herbal remedies. Licochalcone A is a bioactive natural product isolated from *Glycyrrhiza* species, and it exhibits antifungal activities against *C. albicans* with MICs ranging from 16.92 μg/mL to 50.76 μg/mL. Licochalcone A also significantly inhibits *C. albicans* biofilm growth at 10 × MIC, and an in vivo experiment conducted in mice with oral candidiasis indicated that licochalcone A decreases numbers of fungal colonies in tongue tissue. The mechanism study revealed that licochalcone A decreased proteinases and phospholipases secreted by *C. albicans* biofilm [40].

The combination of flavonoids and antifungal agents can exhibit synergy. The major active components of TCM *Scutellaria baicalensis* are baicalein and baicalin [71]. Both of the two major components exhibit antifungal activities. The MIC of baicalein is 25 μg/mL [41]. It induces the apoptosis of *C. albicans* cells and has potent synergy with fluconazole against resistant strains, with an FICI of 0.039 [42]. The derivatives of baicalein even show greater synergistic antifungal effects with FICIs < 0.007 [72]. There is also evidence that baicalein reduces the MIC of amphotericin B with an FICI ranging from 0.031 to 0.677, and further mechanism investigation indicated that the combination of the two agents accelerates *C. albicans* apoptosis [73].

2.4. Alkaloids

Alkaloids are a group of cyclic nitrogen-containing small molecules with diverse bioactivities. They are abundant natural products and provide many valuable molecules in the medical field, such as morphine, quinine, vinblastine, and berberine.

Berberine has been used for treating diarrhea for thousands of years in China and is still used widely in contemporary medicine. There are many studies on the antifungal effects of berberine. It is reported that berberine inhibits the *Candida* species, i.e., *C. albicans*, *C. krusei*, *C. glabrata*, and *C. dubliniensis*, with MICs ranging from 10 to 160 μg/mL [43]. The antifungal activity of berberine alone is unstable and weak in some situations. Therefore, researchers have turned their attention to combining berberine with other drugs. Our group members found that berberine showed potent synergy with fluconazole against fluconazole-resistant *C. albicans* strains. We tested 40 clinical isolates, and the median FICI was 0.034 (range, 0.017 to 0.127) [44]. Further mechanism investigation showed that berberine played a major role in killing the fungi in the synergy, and fluconazole increased the intracellular berberine concentration by inhibiting ergosterol synthesis in cell membranes [74,75]. Berberine also exhibits synergy with other antifungal drugs such as amphotericin B and terbinafine. It was reported that the combination of berberine and amphotericin B reduced approximately 75% of the amphotericin B dose in a mouse model [76], and 100 μg/mL of berberine can effectively assist the antifungal potential of terbinafine [77]. Berberine inhibits biofilms of the *Candida* species in a dose-dependent manner, and 40, 5120, 320, 40, and 1280 μg/mL of berberine can inhibit biofilms of *C. albicans* ATCC 10231, *C. albicans* ATCC 90028, *C. krusei* ATCC 6258, *C. glabrata* ATCC 90030, and *C. dubliniensis* MYA 646, with inhibition rates of 43.54 ± 1%, 19.89 ± 0.57%, 96.93 ± 1.37%, 92.36 ± 0.32%, and 21.62 ± 0.51%, respectively [43].

Sanguinarine and chelerythrine are two alkaloids with similar structures. Both of them can be isolated from *Papaveraceae* plants and have diverse bioactivities. They share similar antifungal activities against *C. albicans* and *C. neoformans* with MICs of 4 μg/mL and 64 μg/mL, respectively, and a 1:1 mixture of sanguinarine and chelerythrine exhibits stronger antifungal effects with an MIC of 2 μg/mL for *C. albicans* and 16 μg/mL for *C. neoformans* [45]. Another study reported that sanguinarine inhibited 10 *C. albicans* strains, with the MICs within the range of 37.5–50 μg/mL. Sanguinarine also showed anticandidal effects in a murine model at the dose of 1.5 and 2.5 mg/kg, and the mode of action is related to ergosterol synthesis deficiency in cells [46]. Sanguinarine can also inhibit *C. albicans* biofilms, and 3.2 μg/mL of sanguinarine significantly inhibits *C. albicans* biofilm formation by over 90% and destroys mature biofilms by 68.3%. The possible mechanism of sanguinarine's antibiofilm effect involves its inhibitory effect on adhesion and hypha formation due to cAMP pathway suppression [78]. In addition to *C. albicans* and *C. neoformans*, the antifungal spectrum of chelerythrine consists of *C. glabrata*, *C. krusei*, and *C. tropicalis*. According to Gong et al., the MICs of chelerythrine against *C. albicans*, *C. glabrata*, *C. krusei*, and *C. tropicalis* are 2, 16, 16, and 8 μg/mL, respectively, and the mechanism investigation showed that chelerythrine inhibited hyphal growth, increased intracellular calcium concentration, induced the accumulation of intracellular ROS, and inhibited drug transporter activity [47].

Piper betle L. var. nigra is a plant used as a spice and medicine belonging to the *Piperaceae* family. It contains many bioactive chemicals such as cadinene, caryophyllene, and amide alkaloids [79,80], and the extract of *P. betle* showed inhibitory effects against *C. albicans* according to a disk diffusion test [81].

2.5. Quinones

Quinones are six-membered α,β-dienonic rings that occur in nature. Naphthoquinones is one of the most common classes in quinones. Futuro D.O. et al. reviewed the development of naphthoquinones as antifungal agents against the *Candida* species and identified 30 naphthoquinones with better antifungal activities than those of the existing drugs [82]. Plumbagin, for example, exhibits antifungal effects against *C. albicans* ATCC 10,231, with an MIC of 0.78 μg/mL [48,82]. In a recent study, plumbagin exhibited antifungal effects against *C. neoformans* with an MIC of 8 μg/mL. Further study showed that plumbagin disrupted cell membrane integrity and reduced metabolic activities of the pathogen. It also damaged formation-phase and mature-phase biofilms of *C. neoformans* at concentrations of 64 and 128 μg/mL, respectively. A mechanism study confirmed that plumbagin damaged biofilm by down-regulating FAS1 and FAS2 expression [49], and bis-naphthoquinone, which can be extracted from *Ceratostigma plumbaginoides*, has an MIC of 0.09 μg/mL against *C. albicans* ATCC 25,555 [50].

Another example is shikonin; it has fungicidal activity against *C. albicans* with MICs of 2–8 μg/mL [51,52], and it exhibits antifungal activity against other *Candida* species, *C. neoformans*, *T. cutaneum*, and *Saccharomyces cerevisiae*, with MICs ranging from 4 to 16 μg/mL (see details in Table 1) [51]. Using metabonomics, shikonin was found to boost histone H3 on lysine 56 residue acetylation via *HST3* in *C. albicans* and execute its antifungal activity [83]. Shikonin can also inhibit the formation of biofilms and destroy the maintenance of mature biofilms at concentrations of 1–32 μg/mL and 4–16 μg/mL, respectively. This antibiofilm activity was confirmed in a mouse vulvovaginal candidiasis model. The possible mechanisms involve down-regulating hypha- and adhesion-specific gene expression and inducing farnesol production [52].

2.6. Coumarin

Osthole is a natural coumarin that can be isolated from TCM *Cnidii fructus*. It does not show antifungal activity against *C. albicans* at up to 64 μg/mL. However, 1–16 μg/mL of osthole has significant synergy with fluconazole against fluconazole-resistant *C. albicans*, with FICIs ranging from 0.04 to 0.37 [6,53], and further study indicated that the mechanism of synergy is associated with endogenous ROS accumulation [53].

Designing novel coumarin derivatives is another common way of developing antifungal agents, since coumarin has many modifiable sites. Linking other pharmacophores, such as azole and quinoline, to coumarin molecules can effectively expand the antimicrobial spectrum and enhance antifungal activity [84].

2.7. Others

Ethyl caffeate (EC) can be extracted from *Elephantopus scaber* L., which is widely distributed in the southwest of China and used as TCM, treating fevers and microbial infections [85]. EC exhibits antifungal effects against 26 isolates of *C. albicans*, with MICs ranging from 64 to 256 μg/mL, and the combination of EC and fluconazole shows synergism in 14 out of 26 tested isolates, with FICIs ranging from 0.047 to 0.375. Similar results are observed in *C. albicans* biofilms. EC shows no antibiofilm effects at up to 256 μg/mL. However, EC and fluconazole exhibit synergy against 11 out of 26 isolates, with FICIs ranging from 0.002 to 0.375. The synergy mechanisms may be related to the inhibition of hydrolase secretion and drug efflux function of *C. albicans* [54].

3. Immunomodulatory Ingredients from TCM

The immune modulatory effect of TCM on fungal infection include the inhibition of inflammation and the enhancement of fungal clearance. According to TCM theory, most herbs with anti-inflammatory effects fall under the category of "*heat-clearing*" drugs, including *Fructus Forsythiae*, *Lonicera japonica* Flos, and *Menthae Haplocalycis Herba*. This Chinese medicine is used to treat inflammatory diseases, and active ingredients of these herbs include phenolic acids, flavonoids, volatile oils, lignans, etc. Meanwhile, *Panax ginseng* C. A. Mey., and *Astragalus membranaceus* are herbs that can improve the fungal clearance function of the immune system. The main active ingredients are polysaccharide, glycoside, alkaloid, volatile oil, and so on. In this section, the common TCM components with immunomodulatory effects were classified according to their chemical structures.

3.1. Phenolic Acids

Phenolic acids are widely found in *Lonicerae* and are the main active ingredients in TCM *Lonicera japonica* Flos and *Lonicera japonica Caulis* (both originated from *Lonicera japonica* Thunb.). *Lonicera japonica* Flos is the most important ingredient in many TCM preparations used to treat inflammatory diseases. Its extract shows protective activity against LPS-induced lung inflammatory cytokine release. The mechanism study revealed that the *Lonicera japonica* Flos extract increases nuclear Sp1 binding activity through the incremental phosphorylation of ERK, and it consequently enhances the expression of IL-10. At the same time, the extract suppresses the phosphorylation of IκB, p38, and JNK, thereby inhibits nuclear NF-κB binding activities and down-regulates the expression of TNF-α, IL-1β, and IL-6 in the lung [86].

Chlorogenic acid is one of the main active substances of *Lonicera japonica* Flos (the chemical structures can be found Figure 1). It is reported that chlorogenic acid exhibits anti-proliferation activity against the fibroblast-like synoviocyte cell line (RSC-364), which is stimulated by IL-6. The main mechanism involves JAK/STAT and NF-κB signaling cascades. Chlorogenic acid inhibits these two pathways by suppressing the expression of p-STAT3, JAK1, p50, and IKK, therefore inducing apoptosis in RSC-364 [7]. Interestingly, a recent study revealed that chlorogenic acid also induces apoptosis in fluconazole-resistant *Candida* spp. and exhibits antifungal activity. Molecular docking demonstrates that chlorogenic acid binds to several important drug targets, including thymidylate kinase, CYP51, and ALS3 [87].

Rosmarinic acid (Figure 1) can be found in *Menthae Haplocalycis Herba* and *Prunella vulgaris* [88,89]. Rosmarinic acid has been reported to have anti-inflammatory effects in many diseases, such as arthritis, colitis, atopic dermatitis, asthma, allergic rhinitis, and periodontal disease [90]. It also inhibits LPS-induced inflammation in RAW264.7 cells. The anti-inflammation target of rosmarinic acid involves the NF-κB/MAPK pathway. It suppresses the activation of the NF-κB/MAPK pathway and thereby reduces the production of pro-inflammatory cytokines NO, TNF-α, IL-1β, and IL-6 [91].

3.2. Flavonoids

Flavonoids can be found in many TCMs with anti-inflammatory effects, such as *Taraxacum officinale* L., *Fructus Forsythiae*, and *Menthae Haplocalycis Herba*. Common flavonoids in these herbs are luteolin, quercetin, rutin, etc. The extract of *Taraxacum officinale* L., which contains luteolin (Figure 1), was confirmed to have anti-inflammatory effects in many studies. It can reduce the release of inflammatory cytokines, including NO, PGE2, IL-1β, IL-6, and TNF-α, by suppressing the NF-κB/iNOS and MAPK pathway [8,92–94]. Research on the anti-inflammatory effects of luteolin dates back at least 20 years [95]. Luteolin can reduce LPS-induced inflammation both in vivo and in vitro. A mechanism study showed that luteolin inhibits the pro-inflammatory molecules TNF-α and ICAM-1 expression in mice [95], and in RAW264.7 cells, luteolin also inhibits NO, IL-1β, and IL-6 in addition to TNF-α [96].

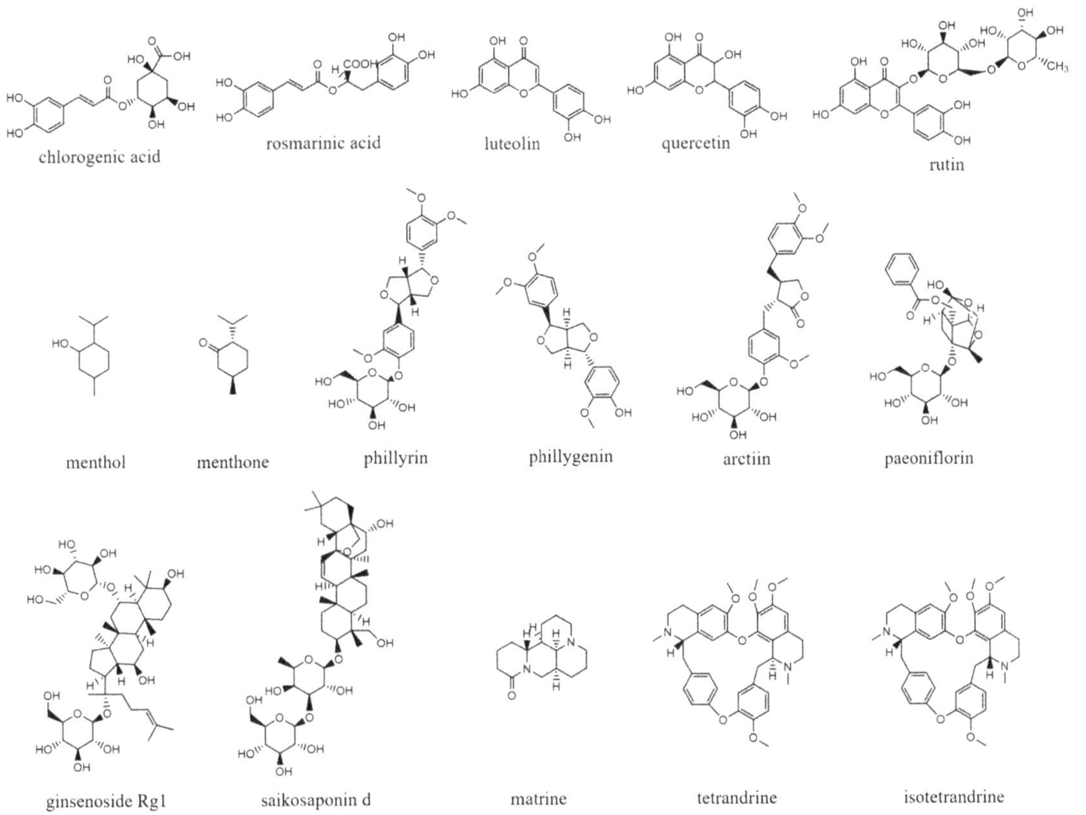

Figure 1. Chemical structures of the natural products isolated from TCMs with immunomodulatory effects.

Quercetin (Figure 1) is another common flavonoid in these "*cold*-natured" herbs. It also can be found in some vegetables and fruits. The anti-inflammatory effect of quercetin has been reported in many diseases such as inflammatory bowel disease, multiple sclerosis, asthma, and atherosclerosis, in vivo or in vitro [97,98]. It was also in a clinical trial for treating rheumatoid arthritis [99]. Quercetin inhibits inflammation via promoting anti-inflammatory cytokine secretion (for example, IL-10), reducing pro-inflammatory cytokine release (TNF-α, IL-1β, and IL-6), inhibiting cyclooxygenase and lipoxygenase expression, and maintaining mast cell stability [97,100].

Rutin (Figure 1) is a flavonoid that exists in *Fructus Forsythiae* [98]. Rutin shows anti-inflammatory effects in activated human neutrophils, through inhibiting TNF-α and NO production, as well as myeloperoxidase (MPO) activity [101]. Rutin also attenuates advanced glycation end product induced inflammation on human chondrocytes. The study revealed that rutin targets BCL-2 and TRAF-6 in the NF-κB/MAPK pathway to inhibit inflammation and treat osteoarthritis [102].

3.3. *Volatile Oils*

Volatile oils are the main active ingredients in *Menthae Haplocalycis Herba*, consisting of menthol, menthone, isomenthone, piperitone, linalool, carvone, limonene, α-pinene, β-pinene, etc. The volatile oils of *Menthae Haplocalycis Herba* are reported to exhibit anti-inflammatory and antimicrobial effects, as well as alleviate mental fatigue [88]. Menthol (Figure 1) can be beneficial in rats with acetic acid-induced acute colitis. The protective effect is related to the inhibition of MPO, as well as the reduction of TNF-α, IL-1β, and

IL-6 [103]. Menthone (Figure 1) is reported to reduce LPS-induced inflammation in mice. The possible mechanism is that menthone inhibits the activation of NLRP3 inflammasome and consequently reduces the release of pro-inflammatory cytokines, including IL-18, IL-1β, IL-5, TNF-α, IFN-γ, G-CSF, GM-CSF, and MIP-1β [104].

3.4. Lignans

Lignans are one of the main active ingredients in *Fructus Forsythiae*. *Fructus Forsythiae* is the dried fruit of *Forsythia suspensa* (Thunb.) Vahl, family *Lignanaceae*. *Fructus Forsythiae* is often used for acute colds belonging to "*heat*", lymphatic tuberculosis, urinary tract infections, etc.

One of its main lignans with an anti-inflammatory effect is phillyrin (Figure 1). Phillyrin inhibits inflammation both in vivo and in vitro. In a zebrafish model, phillyrin reduced inflammation in a dose-dependent manner and improved survival. A mechanism study showed that phillyrin inhibits the MyD88/IκBα/NF-κB signaling pathway, but not ERK1/2 MAPKs or JNK MAPKs. It reduces neutrophil infiltration and down-regulates the release of IκBα, TNF-α, IL-1β, and IL-6 [105]. By regulating NF-κB signaling, phillyrin also alleviates inflammation induced by SARS-CoV-2 in Huh-7 cells. It decreases the release of pro-inflammatory cytokines including TNF-α, IL-6, IL-1β, MCP-1, and IP-10 [106].

Phillygenin (Figure 1) is another lignan in *Fructus Forsythiae* that targets the NF-κB signaling pathway and shows anti-inflammatory effects. It inhibits LPS-induced inflammation in LX2 and RAW 264.7 cell lines. Molecular docking indicated that phillygenin has an affinity for many proteins in the NF-κB pathway, such as IKKβ, p65, IκBα, and TAK1 [107,108].

Arctiin (Figure 1) can be found in *Fructus Forsythiae* and exhibits anti-inflammatory effects as well. It attenuates inflammation in different cells by inhibiting COX-2 expression, which is an essential protein in inflammation [109,110]. Arctiin also activates Nrf2/HO-1 signaling and blocks the RIG-I/JNK MAPK signaling of A549 cells in inflammation induced by H9N2 avian influenza virus [110]. In an LPS-induced acute lung injury mice model, arctiin significantly ameliorated lung histopathological changes and decreased lung MPO activity. The mechanism study suggested that arctiin targets the PI3K/AKT/NF-κB signaling pathway by inhibiting PI3K/Akt phosphorylation and NF-κB activation [111].

3.5. Alkaloids

Some alkaloids have a regulatory effect on the immune function of the body by targeting inflammation-related pathways such as the NF-κB signaling pathway. Matrine, for example, is a tetracyclo-quinolizindine alkaloid (Figure 1) extracted from *Sophora flavescens*. It balanced the Th1/Th2 axis and improved rheumatoid arthritis in a rat model. By regulating the NF-κB pathway, matrine reduced the level of Th1 cytokines (IFN-γ, TNF-α, and IL-1β) and raised Th2 cytokines (IL-4 and IL-10) [112].

Tetrandrine is an isoquinoline alkaloid (Figure 1) that can be isolated from *Radix Stephaniae Tetrandrae*. It is reported to have inhibitory effects on the proliferation of T cells via the NF-κB pathway. Tetrandrine prevents the degradation of IκBα and inhibits nuclear translocation of p65 by blocking IKKα and IKKβ activities, and tetrandrine down-regulates the activation of MAPK including JNK, p38, and ERK, as well as the downstream transcription factor AP-1 [113]. Isotetrandrine has a chemical structure very similar to tetrandrine, differing only in the stereochemistry at the chiral centers (Figure 1). Isotetrandrine has stronger inhibitory effects against the proliferation of T cells than those of tetrandrine [114].

3.6. Polysaccharides

Polysaccharides are sugar chains consisting of at least 10 monosaccharides bound by glycosidic bonds. They are complex mixtures and one of the main active ingredients of tonic herbs in TCM, such as polysaccharides from *Ganoderma*, *Astragalus*, *Ginseng*, *Angelica sinensis*, etc. The pharmacological effects of polysaccharides on immune systems are complicated. Some polysaccharides have an inhibiting effect on the immune system to attenuate excessive immune responses, while others have a stimulating effect on the immune system to help the body fight against infections or tumors. In many cases, some polysaccharides have both of these effects. It seems that these polysaccharides can balance immune cells and restore immune functions from abnormal.

Ganoderma lucidum polysaccharide is extracted from sporoderm-removed spores of the fungus. On one hand, it inhibits inflammation in AOM/DSS-induced colitis. *Ganoderma* polysaccharide suppresses TLR4/MyD88/NF-κB signaling, inhibits macrophage infiltration, and down-regulates IL-1β, iNOS, as well as COX-2 expressions in the colon. It also inhibits LPS-induced inflammation markers and MAPK activation in RAW264.7, HT-29, and NCM460 cells [115]. On the other hand, *Ganoderma* polysaccharide activates the immune responses by binding to dectin-1, TLRs, MR, or CR3 on immune cells including monocytes, macrophages, dendritic cells (DCs), granulocytes, neutrophils, and natural killing (NK) cells. It also can directly activate lymphocytes and neutrophils [116].

3.7. Glycosides

Glycosides can be found in many TCMs such as *Paeonia lactiflora* Pallas, *Radix Ginseng*, *Radix Scutellariae*, *Radix Bupleuri*, etc. As with polysaccharides, the immunomodulatory effects of glycosides are complicated, encompassing both promotive and inhibitory effects.

Paeoniflorin (Figure 1) is a monoterpene glucoside that is the major active component of *Paeonia lactiflora* Pallas. Paeoniflorin has protective effects on many autoimmune diseases in animal models, including arthritis, liver injuries, allergic contact dermatitis, Sjögren syndrome, psoriasis, multiple sclerosis, and asthma [117]. Paeoniflorin regulates the activation of T lymphocytes, B lymphocytes, and macrophages. It also inhibits DC maturation and pro-inflammatory mediator production. As for pathways, paeoniflorin inhibits the MAPK signaling pathway, the JAK2/STAT3 pathway, and the PI3K/Akt/mTOR pathway in immune cells [117].

Ginsenoside Rg1 (Figure 1) is one of the glycosides in *Radix Ginseng*. It has various immune-modulating activities, for example, enhancing the immune activity of Th cells. Ginsenoside Rg1 can help mice fight against disseminated candidiasis. Ginsenoside Rg1 has no antifungal activity against *C. albicans* in vitro. However, it can promote $CD4^{+}$T cell immune response mediated by Th1 cells in infected mice and consequently induces cytokine release including IFN-γ, IL-2, IL-4, and IL-10, exhibiting protective effects in mice [118].

Saikosaponin d (Figure 1) can be found in *Radix Bupleuri* and is one of the major bioactive components of medicine. Saikosaponin d helps generate functional mature neutrophils in cancer chemotherapy-induced neutropenia. The generated neutrophils are capable of resisting infection both in vitro and in vivo. This immune enhancing effect is mediated by the CBL-ERK1/2 pathway, resulting in neutrophil differentiation [119].

4. Discussion

Fungal infections can disrupt homeostasis in a number of ways. On the one hand, the invasive growth of fungi destroys mucous membranes and destroys the structure of tissues and organs. On the other hand, it can induce inflammation and trigger dysfunction in the body, making it harder to recover.

In the theory of TCM, the philosophy of treating fungal diseases does not just focus on antifungal, but on both antifungal and immune regulation. Herbal preparations are commonly used to treat fungal diseases, and these often contain a variety of ingredients that have complex regulatory effects on both the fungus and the host. Some ingredients kill the fungus, and others balance the host's immune response. TCM regulates the host immune response in two directions. Some herbs suppress the immune response by reducing inflammation, while others boost the immune system's ability to clear fungi. This overall regulation of the fungus and the host is very helpful in the treatment of fungal infections.

However, TCM has two drawbacks in treating fungal infections. First, the effective ingredients of TCM are not clear. Second, the mechanism of TCM treatment of fungal infection is still unclear. In recent years, efforts have been made to explore the active ingredients of TCM and elucidate its mechanism of action. The material basis and mechanism of TCM treatment of fungal infection were discussed in this paper.

4.1. Material Basis of TCM for Treating Fungal Infection Diseases

Overall, there are two strategies for TCM to treat fungal infections, antifungal and immunomodulation. There are many chemicals from TCM exhibiting antifungal activities, and most of them can be divided into terpenoids, volatile oils, flavonoids, alkaloids, quinones, coumarin, etc. The terpenoid family of antifungal compounds varies greatly in structure, from thymol, with a molecular weight of less than two hundred, to dioscin, with a molecular weight of more than eight hundred. Interestingly, it seems that the larger the molecular weight of the terpenoid, the better its antifungal activity. Dioscin has the most complex molecular structure in this group, and it has the best antifungal activity against *Candida* spp., with MIC values of 2–4 μg/mL. Volatile oils have their own advantages and disadvantages. Most volatile oils have a pleasant aromatic scent and can be used for environmental sterilization or topical medications. However, their volatile nature makes them difficult to store and transport and are therefore not an ideal antifungal drug preparation. The flavonoids mentioned in this review share a moderate activity against *C. albicans*.

The chemicals with the strongest antifungal activity are from the alkaloid and quinone families, such as chelerythrine and shikonin. They both exhibit broad and potent activities against common *Candida* spp., with MICs ranging from 2 to 16 μg/mL. One potential problem with alkaloids and quinones is their toxicity [120]. Quaternary ammonium alkaloids, such as chelerythrine, bearing a quaternary nitrogen atom, have oxidative effects and are thus usually toxic to cells. An evaluation of the acute hepatotoxicity effect of chelerythrine at a dose of 10 mg/kg/day (i.p.) showed that chelerythrine caused marked parenchymal damage in the liver [121]. Meanwhile, EC_{50} of shikonin against V79 cell lines was 0.4 μg/mL by an MTT assay [122]. Through structural modification, chemicals of these two families are promising antifungal lead compounds.

The single use of Chinese medicine has a significant antifungal effect, and the combination of Chinese medicine and existing antifungal drugs can produce synergistic effects, such as the combination of berberine and fluconazole. Using this synergy, we can enhance the effect of the drug and reduce drug dosage. Therefore, we can reduce the side effects of drugs. In addition, the combination of two or more drugs can also reduce the development of resistance.

Unlike TCM ingredients with antifungal effects, ingredients with immunomodulatory effects have their own unique chemical structure characteristics. Major TCM ingredients with immunomodulatory activity can generally be classified into the categories of phenolic acids, flavonoids, volatile oils, lignans, polysaccharides, glycosides, and alkaloids. There are two types of immunomodulatory effects of these TCM ingredients. Most of the phenolic acids, flavonoids, volatile oils, lignans, and alkaloids show anti-inflammatory effects in the immune response of the host, while polysaccharides and glycosides usually have immune-promoting properties. For example, *Ganoderma lucidum* polysaccharide activates immune response by binding to receptors on immune cells, ginsenoside Rg1 enhance the immune activity of Th cells, and saikosaponin d helps generate functional mature neutrophils in neutropenia hosts.

Panax ginseng is a traditional Chinese valuable herb with a history of application for more than 2000 years. Ginseng is used in Chinese medicine to prolong the life of critically ill patients and is used in folklore as a tonic for strengthening the body. Modern pharmacological studies have shown that it has immune-enhancing and antioxidant properties [123]. Polysaccharides and glycosides are two major active components in ginseng, both of which are very promising immunomodulatory drug candidates [124]. They are relatively safe in vivo [125]. Polysaccharides are a relatively complex mixture of components and are therefore difficult to synthesize in vitro, and biosynthesis may be a good way to develop such drugs.

4.2. Mechanism of Chinese Medicine in Treating Fungal Infectious Diseases

The mechanism of antifungal action of TCM ingredients has been insufficiently studied (Figure 2). Most of the currently reported chemicals do not have a clear target. For the TCM ingredients that act on fungi, the main mechanisms involve inhibiting the virulence factors of fungi (hyphal growth, cell adhesion, production of phospholipase), disturbing fungal cell wall synthesis, damaging the fungal cell membrane, inhibiting the mitosis process, altering the glycerophospholipid metabolism, influencing the drug efflux function, increasing intracellular ROS levels, and inducing apoptosis. Two related pathways in fungal cells are the HOG-MAPK pathway and the Ras-cAMP pathway. Among those chemicals, shikonin is reported to promote histone H3 on lysine 56 residue acetylation via *HST3* in *C. albicans* [83]. Again, shikonin exhibits strong antifungal activities with a broad spectrum and a clear binding target, making it a good candidate to become a lead compound in antifungal drug development.

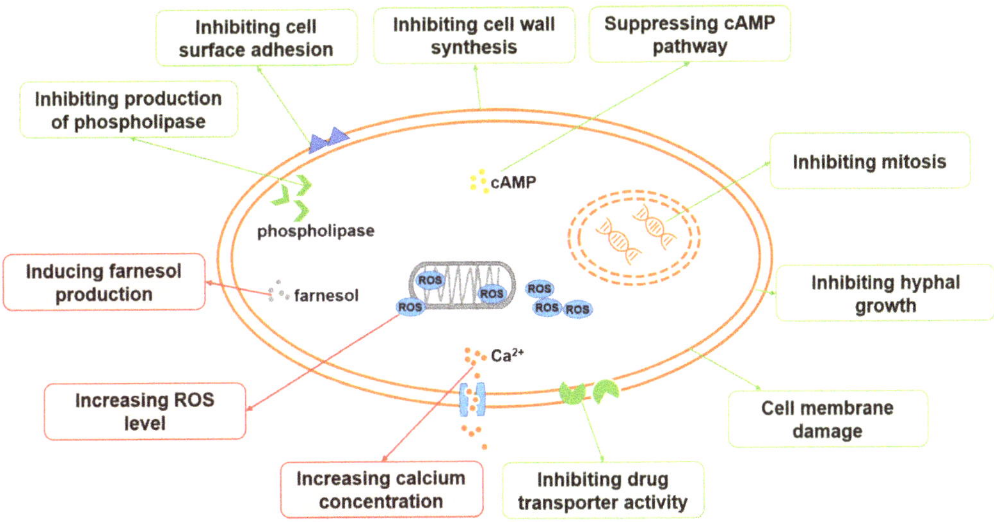

Figure 2. Antifungal mechanisms of some TCM ingredients.

As for immunomodulatory TCM ingredients, the mechanisms involve two aspects: anti-inflammation and immune enhancement. The pathways that anti-inflammatory ingredients related with are the NF-κB pathway, MAPK signaling pathways, and the JAK/STAT pathway. The NF-κB pathway is one of the most important pathways that regulates inflammation responses in hosts. NF-κB remains inactive in the cytoplasm when binding to its inhibitor IκB. When the cell is stimulated, IKK promotes the phosphorylation of IκB and leads to the dissociation of NF-κB. Then, NF-κB is translocated into the nucleus and binds to DNA, resulting in gene transcription and protein synthesis, which are related to inflammation response [126,127]. Nearly all the anti-inflammatory functions of TCMs listed here are related to the NF-κB pathway (Figure 3). MAPK cascades mediate the signal transduction from extracellular signals to intracellular reactions [126]. There are at least three MAPK cascade signal transduction pathways that are involved in the anti-inflammation effects of TCM ingredients: ERK, JNK, and p38. Rosmarinic acid, rutin, and arctiin down-regulate protein expressions in the MAPK pathway. This reduces inflammatory responses. The JAK/STAT pathway is another type of cascade that translates extracellular chemical signals into the permission of JAK phosphorylation and STAT activation. The pathway is related to many cellular processes, including cell proliferation, differentiation, apoptosis, and immune regulation [128]. Chlorogenic acid and paeoniflorin can inhibit the JAK/STAT pathway and alleviate the inflammation response. However, both the MAPK pathway and the JAK/STAT pathway are involved in many essential biological processes, and this prevents the two pathways from being good anti-inflammatory targets. Inhibiting targets in these pathways may cause many adverse effects.

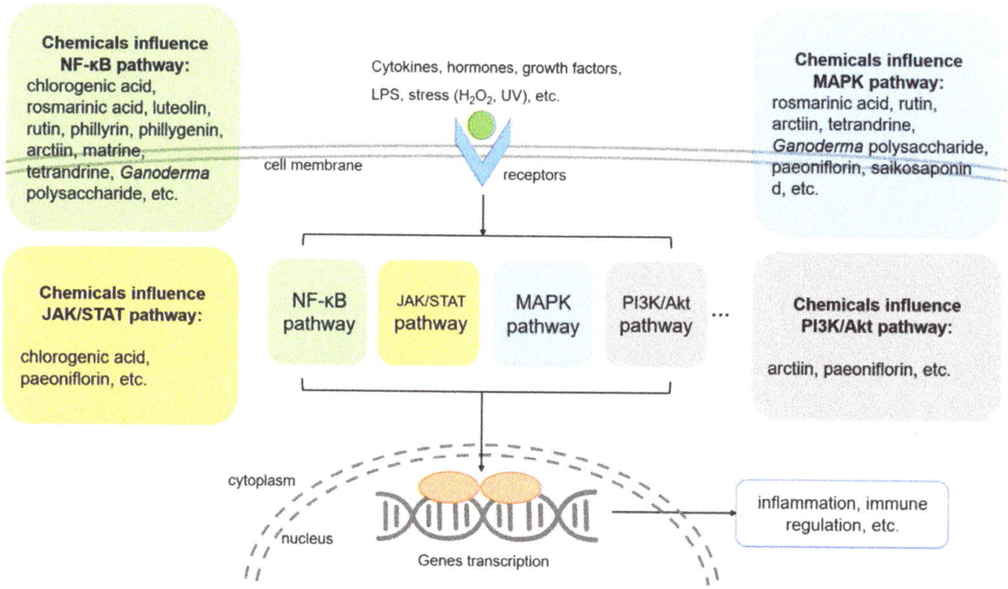

Figure 3. Immunomodulatory mechanisms of some TCM ingredients. When cytokines, hormones, growth factors, or LPS bind to the receptors on the cell membrane, down-stream pathways are stimulated. Transcription factors that are regulated by these pathways enter the nucleus and promote the expression of inflammatory factors or immunomodulatory factors. TCM ingredients affect these pathways and thereby regulate inflammatory processes as well as the immune system.

The immune-enhancing effects are usually induced by stimulating immune cells. For example, *Ganoderma* polysaccharide boosts immune response by binding to dectin-1, TLRs, MR, or CR3 on immune cells. Ginsenoside Rg1 promotes the $CD4^{+}$T cell immune response mediated by Th1 cells, and saikosaponin d helps produce mature neutrophils mediated by the CBL-ERK1/2 pathway.

In addition, there are some interesting molecules, such as chlorogenic acid, that possess both antifungal and anti-inflammatory effects via different mechanisms. On one hand, it possibly binds to CYP51 as well as ALS3 and induces apoptosis in *Candida* spp. On the other hand, it inhibits the JAK/STAT and NF-κB pathways by suppressing p-STAT3, JAK1, p50, and IKK in host cells. This kind of molecule with multiple targets is also a good candidate for drug discovery.

4.3. Exploitation of TCM and Development of New Antifungal Drugs

If antifungal ingredients are combined with immunomodulatory components of TCM, more ideal antifungal therapeutic effects may be obtained. However, there are still some problems in the development of TCM. Many TCM candidates have shown good antifungal activity in vitro, but few have shown consistent efficacy in animal models or clinical trials. This problem makes the application of TCM in antifungal therapy a great challenge.

There may be some reasons for this. First, the content of active ingredients in TCM is relatively low. We can use the technology of synthetic biology to genetically modify medicinal plants or use engineering yeast to increase the yield of effective components. Synthetic biology was applied to improve the production of natural products in many cases. For example, by introducing the *Artemisia annua* Linn. genes that encode the enzymes of the artemisinin biosynthetic pathway into yeast cells, we can improve the artemisinic acid yields from 1% to 25 g/L [129,130], although artemisinin production costs may increase as a result. The application of modern technology to improve the yield and economic benefits of the effective components can effectively solve some difficulties in the application of TCM.

Another possible reason is the relatively low efficacy and poor ADME properties of the active ingredient. We can modify the structure of the active small molecules of TCM by semi-synthesis or total synthesis, explore the structure–activity relationship, and select the molecules with better medicinal properties from the derivatives. TCM is a rich reservoir of biologically active natural products. It can provide us with many implications for the study of novel chemical structures with antifungal or immunomodulatory activities. We can screen for new lead compounds in this different TCM chemical pool. A variety of series of derivatives can be developed based on lead compounds to improve their efficacy, toxicity, and pharmacokinetic properties. By evaluating these derivatives, we can find antifungal candidates with high efficacy, low toxicity, and good ADME properties.

Currently, the development of TCM antifungal drugs is still in the stage of drug screening, and the mechanism of action of these bioactive molecules or mixtures is not clear. Future research should focus on the development of natural drugs with clear ingredients, a clear mechanism of action, and an optimized structure of active ingredients. It is hoped that new antifungal drugs with high efficiency and low toxicity can be developed.

Author Contributions: Conceptualization, Y.W.; resources, L.H. and H.Z.; writing—original draft preparation, H.Z., L.H. and R.-Y.L.; writing—review and editing, Y.W.; supervision, Y.W.; funding acquisition, Y.W. and H.Z. All authors have read and agreed to the published version of the manuscript.

Funding: This research was funded by Hua Zhong and Yan Wang, the National Natural Science Foundation of China (82204470 and 81772124), and the biosafety program (20SWAQX29-1-6).

Institutional Review Board Statement: Not applicable.

Informed Consent Statement: Not applicable.

Data Availability Statement: Data sharing not applicable.

Acknowledgments: Part of the graphical abstract was designed by Freepik.

Conflicts of Interest: The authors declare no conflict of interest.

References

1. Ya-jing, X.; Yong-zhuo, W.; Jia-yin, L.; Rong-jie, Y.; Yun, G.; Dan-qi, D. Retrospective analysis of Chinese antifungal herbal medicines in China. *J. Dermatol. Venereol.* **2019**, *10*, 26–29. [CrossRef]
2. Chen, M.; Xu, Y.; Hong, N.; Yang, Y.; Lei, W.; Du, L.; Zhao, J.; Lei, X.; Xiong, L.; Cai, L.; et al. Epidemiology of fungal infections in China. *Front. Med.* **2018**, *12*, 58–75. [CrossRef] [PubMed]
3. Strickland, A.B.; Shi, M. Mechanisms of fungal dissemination. *Cell. Mol. Life Sci.* **2021**, *78*, 3219–3238. [CrossRef] [PubMed]
4. Kullberg, B.J.; Arendrup, M.C. Invasive Candidiasis. *N. Engl. J. Med.* **2015**, *373*, 1445–1456. [CrossRef]
5. Köhler, J.R.; Hube, B.; Puccia, R.; Casadevall, A.; Perfect, J.R. Fungi that Infect Humans. *Microbiol. Spectr.* **2017**, *5*, 813–843. [CrossRef]
6. Li, L.P.; Wang, X.J.; Zhang, J.Y.; Zhang, L.L.; Cao, Y.B.; Gu, L.Q.; Yu, Y.Q.; Yang, Q.L.; Shen, C.Y.; Han, B.; et al. Antifungal activity of osthol in vitro and enhancement in vivo through Eudragit S100 nanocarriers. *Virulence* **2018**, *9*, 555–562. [CrossRef]
7. Lou, L.; Zhou, J.; Liu, Y.; Wei, Y.I.; Zhao, J.; Deng, J.; Dong, B.; Zhu, L.; Wu, A.; Yang, Y.; et al. Chlorogenic acid induces apoptosis to inhibit inflammatory proliferation of IL-6-induced fibroblast-like synoviocytes through modulating the activation of JAK/STAT and NF-κB signaling pathways. *Exp. Ther. Med.* **2016**, *11*, 2054–2060. [CrossRef]
8. Cen, L.; Xiao, R.; Li, M.; Wang, Z.; Chang, B. Antitumor and anti-inflammatory mechanisms of taraxacum: Research advances. *J. Int. Pharm. Res.* **2020**, *47*, 954–961. [CrossRef]
9. Huang, X.; Yi, Y.; Yong, J.; Sun, J.; Song, Z.; Li, D.; Li, Y. Inhibitory effect of berberine hydrochloride against *Candida albicans* and the role of the HOG-MAPK pathway. *J. Antibiot.* **2021**, *74*, 807–816. [CrossRef]
10. Wang, T.; Shi, G.; Shao, J.; Wu, D.; Yan, Y.; Zhang, M.; Cui, Y.; Wang, C. In vitro antifungal activity of baicalin against *Candida albicans* biofilms via apoptotic induction. *Microb. Pathog.* **2015**, *87*, 21–29. [CrossRef]

11. Dovigo, L.N.; Carmello, J.C.; de Souza Costa, C.A.; Vergani, C.E.; Brunetti, I.L.; Bagnato, V.S.; Pavarina, A.C. Curcumin-mediated photodynamic inactivation of *Candida albicans* in a murine model of oral candidiasis. *Med. Mycol.* **2013**, *51*, 243–251. [CrossRef]
12. El-Saber Batiha, G.; Magdy Beshbishy, A.; Wasef, G.; Elewa, L.; Al-Sagan, A.; Abd El-Hack, M.E.; Taha, A.E.; Abd-Elhakim, Y.M.; Prasad Devkota, H. Chemical Constituents and Pharmacological Activities of Garlic (*Allium sativum* L.): A Review. *Nutrients* **2020**, *12*, 872. [CrossRef] [PubMed]
13. Jia-jia, F.; Yuan-yong, T. Research progress on antifungal effect of traditional Chinese medicine. *J. Pract. Dermatol.* **2019**, *12*, 34–37.
14. Yang, L.; Liu, X.; Zhong, L.; Sui, Y.; Quan, G.; Huang, Y.; Wang, F.; Ma, T. Dioscin Inhibits Virulence Factors of *Candida albicans*. *Biomed. Res. Int.* **2018**, *2018*, 4651726. [CrossRef] [PubMed]
15. Tao, X.; Yin, L.; Xu, L.; Peng, J. Dioscin: A diverse acting natural compound with therapeutic potential in metabolic diseases, cancer, inflammation and infections. *Pharmacol. Res.* **2018**, *137*, 259–269. [CrossRef]
16. Kim, T.H.; Hatano, T.; Okamoto, K.; Yoshida, T.; Kanzaki, H.; Arita, M.; Ito, H. Antifungal and Ichthyotoxic Sesquiterpenoids from *Santalum album* Heartwood. *Molecules* **2017**, *22*, 1139. [CrossRef] [PubMed]
17. Sharifi-Rad, J.; Quispe, C.; Turgumbayeva, A.; Mertdinç, Z.; Tütüncü, S.; Aydar, E.F.; Özçelik, B.; Anna, S.W.; Mariola, S.; Koziróg, A.; et al. Santalum Genus: Phytochemical constituents, biological activities and health promoting-effects. *Z. Fur Naturforschung. C J. Biosci.* **2022**, *12*, 58–75. [CrossRef] [PubMed]
18. Liu, R.H.; Shang, Z.C.; Li, T.X.; Yang, M.H.; Kong, L.Y. In Vitro Antibiofilm Activity of Eucarobustol E against *Candida albicans*. *Antimicrob. Agents Chemother.* **2017**, *61*, e02707. [CrossRef] [PubMed]
19. Wang, Y.; Lu, C.; Zhao, X.; Wang, D.; Liu, Y.; Sun, S. Antifungal activity and potential mechanism of Asiatic acid alone and in combination with fluconazole against *Candida albicans*. *Biomed. Pharmacother.* **2021**, *139*, 111568. [CrossRef]
20. Doke, S.K.; Raut, J.S.; Dhawale, S.; Karuppayil, S.M. Sensitization of *Candida albicans* biofilms to fluconazole by terpenoids of plant origin. *J. Gen. Appl. Microbiol.* **2014**, *60*, 163–168. [CrossRef]
21. Wang, K.; Jiang, S.; Yang, Y.; Fan, L.; Su, F.; Ye, M. Synthesis and antifungal activity of carvacrol and thymol esters with heteroaromatic carboxylic acids. *Nat. Prod. Res.* **2019**, *33*, 1924–1930. [CrossRef] [PubMed]
22. Kowalczyk, A.; Przychodna, M.; Sopata, S.; Bodalska, A.; Fecka, I. Thymol and Thyme Essential Oil-New Insights into Selected Therapeutic Applications. *Molecules* **2020**, *25*, 4125. [CrossRef] [PubMed]
23. Teixeira, A.P.C.; Nóbrega, R.O.; Lima, E.O.; Araújo, W.O.; Lima, I.O. Antifungal activity study of the monoterpene thymol against *Cryptococcus neoformans*. *Nat. Prod. Res.* **2020**, *34*, 2630–2633. [CrossRef] [PubMed]
24. Olea, A.F.; Bravo, A.; Martínez, R.; Thomas, M.; Sedan, C.; Espinoza, L.; Zambrano, E.; Carvajal, D.; Silva-Moreno, E.; Carrasco, H. Antifungal Activity of Eugenol Derivatives against *Botrytis cinerea*. *Molecules* **2019**, *24*, 1239. [CrossRef]
25. Bona, E.; Cantamessa, S.; Pavan, M.; Novello, G.; Massa, N.; Rocchetti, A.; Berta, G.; Gamalero, E. Sensitivity of *Candida albicans* to essential oils: Are they an alternative to antifungal agents? *J. Appl. Microbiol.* **2016**, *121*, 1530–1545. [CrossRef]
26. Meccatti, V.M.; Oliveira, J.R.; Figueira, L.W.; Lagareiro Netto, A.A.; Zamarioli, L.S.; Marcucci, M.C.; Camargo, S.E.A.; Carvalho, C.A.T.; Oliveira, L.D. *Rosmarinus officinalis* L. (rosemary) extract has antibiofilm effect similar to the antifungal nystatin on Candida samples. *An. Acad. Bras. Ciênc.* **2021**, *93*, e20190366. [CrossRef]
27. Mertas, A.; Garbusińska, A.; Szliszka, E.; Jureczko, A.; Kowalska, M.; Król, W. The influence of tea tree oil (*Melaleuca alternifolia*) on fluconazole activity against fluconazole-resistant *Candida albicans* strains. *Biomed. Res. Int.* **2015**, *2015*, 590470. [CrossRef]
28. Roana, J.; Mandras, N.; Scalas, D.; Campagna, P.; Tullio, V. Antifungal Activity of Melaleuca alternifolia Essential Oil (TTO) and Its Synergy with Itraconazole or Ketoconazole against *Trichophyton rubrum*. *Molecules* **2021**, *26*, 461. [CrossRef]
29. Pandey, K.P.; Mishra, R.K.; Kamran, A.; Mishra, P.; Bajaj, A.K.; Dikshit, A. Studies on antidermatophytic activity of waste leaves of *Curcuma longa* L. *Physiol. Mol. Biol. Plants* **2010**, *16*, 177–185. [CrossRef]
30. Jankasem, M.; Wuthi-Udomlert, M.; Gritsanapan, W. Antidermatophytic Properties of Ar-Turmerone, Turmeric Oil, and Curcuma longa Preparations. *ISRN Dermatol.* **2013**, *2013*, 250597. [CrossRef]
31. Achimón, F.; Brito, V.D.; Pizzolitto, R.P.; Ramirez Sanchez, A.; Gómez, E.A.; Zygadlo, J.A. Chemical composition and antifungal properties of commercial essential oils against the maize phytopathogenic fungus *Fusarium verticillioides*. *Rev. Argent. Microbiol.* **2021**, *53*, 292–303. [CrossRef] [PubMed]
32. Khan, N.; Shreaz, S.; Bhatia, R.; Ahmad, S.I.; Muralidhar, S.; Manzoor, N.; Khan, L.A. Anticandidal activity of curcumin and methyl cinnamaldehyde. *Fitoterapia* **2012**, *83*, 434–440. [CrossRef]
33. Dong, H.H.; Wang, Y.H.; Peng, X.M.; Zhou, H.Y.; Zhao, F.; Jiang, Y.Y.; Zhang, D.Z.; Jin, Y.S. Synergistic antifungal effects of curcumin derivatives as fungal biofilm inhibitors with fluconazole. *Chem. Biol. Drug Des.* **2021**, *97*, 1079–1088. [CrossRef] [PubMed]
34. Martins, C.V.; da Silva, D.L.; Neres, A.T.; Magalhães, T.F.; Watanabe, G.A.; Modolo, L.V.; Sabino, A.A.; de Fátima, A.; de Resende, M.A. Curcumin as a promising antifungal of clinical interest. *J. Antimicrob. Chemother.* **2009**, *63*, 337–339. [CrossRef]
35. Brochot, A.; Guilbot, A.; Haddioui, L.; Roques, C. Antibacterial, antifungal, and antiviral effects of three essential oil blends. *Microbiologyopen* **2017**, *6*, e00459. [CrossRef] [PubMed]
36. de Oliveira, D.R.; Tintino, S.R.; Braga, M.F.; Boligon, A.A.; Athayde, M.L.; Coutinho, H.D.; de Menezes, I.R.; Fachinetto, R. In vitro antimicrobial and modulatory activity of the natural products silymarin and silibinin. *Biomed. Res. Int.* **2015**, *2015*, 292797. [CrossRef]
37. Yun, D.G.; Lee, D.G. Silibinin triggers yeast apoptosis related to mitochondrial Ca(2+) influx in *Candida albicans*. *Int. J. Biochem. Cell Biol.* **2016**, *80*, 1–9. [CrossRef] [PubMed]

38. Tuli, H.S.; Mittal, S.; Aggarwal, D.; Parashar, G.; Parashar, N.C.; Upadhyay, S.K.; Barwal, T.S.; Jain, A.; Kaur, G.; Savla, R.; et al. Path of Silibinin from diet to medicine: A dietary polyphenolic flavonoid having potential anti-cancer therapeutic significance. *Semin. Cancer Biol.* **2021**, *73*, 196–218. [CrossRef]
39. Jingwen, T.; Siyu, L.; Yiyang, W.; Zhiqin, G.; Hong, Y.; Lianjuan, Y. Flavonoids constitutes and in vitro anti-*Candida albicans* activity in propolis. *World Clin. Drugs* **2021**, *42*, 957–961.
40. Seleem, D.; Benso, B.; Noguti, J.; Pardi, V.; Murata, R.M. In Vitro and In Vivo Antifungal Activity of Lichochalcone-A against *Candida albicans* Biofilms. *PLoS ONE* **2016**, *11*, e0157188. [CrossRef]
41. Wong, K.S.; Tsang, W.K. In vitro antifungal activity of the aqueous extract of *Scutellaria baicalensis* Georgi root against *Candida albicans*. *Int. J. Antimicrob. Agents* **2009**, *34*, 284–285. [CrossRef] [PubMed]
42. Huang, S.; Cao, Y.Y.; Dai, B.D.; Sun, X.R.; Zhu, Z.Y.; Cao, Y.B.; Wang, Y.; Gao, P.H.; Jiang, Y.Y. In vitro synergism of fluconazole and baicalein against clinical isolates of *Candida albicans* resistant to fluconazole. *Biol. Pharm. Bull.* **2008**, *31*, 2234–2236. [CrossRef] [PubMed]
43. Xie, Y.; Liu, X.; Zhou, P. In vitro Antifungal Effects of Berberine against *Candida* spp. In Planktonic and Biofilm Conditions. *Drug Des. Dev. Ther.* **2020**, *14*, 87–101. [CrossRef] [PubMed]
44. Quan, H.; Cao, Y.Y.; Xu, Z.; Zhao, J.X.; Gao, P.H.; Qin, X.F.; Jiang, Y.Y. Potent in vitro synergism of fluconazole and berberine chloride against clinical isolates of *Candida albicans* resistant to fluconazole. *Antimicrob. Agents Chemother.* **2006**, *50*, 1096–1099. [CrossRef]
45. Qian, W.; Yang, M.; Li, X.; Sun, Z.; Li, Y.; Wang, X.; Wang, T. Anti-microbial and anti-biofilm activities of combined chelerythrine-sanguinarine and mode of action against *Candida albicans* and *Cryptococcus neoformans* in vitro. *Colloids Surf. B Biointerfaces* **2020**, *191*, 111003. [CrossRef]
46. Hu, Z.; Hu, H.; Hu, Z.; Zhong, X.; Guan, Y.; Zhao, Y.; Wang, L.; Ye, L.; Ming, L.; Riaz Rajoka, M.S.; et al. Sanguinarine, Isolated From Macleaya cordata, Exhibits Potent Antifungal Efficacy Against Candida albicans Through Inhibiting Ergosterol Synthesis. *Front. Microbiol.* **2022**, *13*, 908461. [CrossRef]
47. Gong, Y.; Li, S.; Wang, W.; Li, Y.; Ma, W.; Sun, S. In vitro and in vivo activity of chelerythrine against *Candida albicans* and underlying mechanisms. *Future Microbiol.* **2019**, *14*, 1545–1557. [CrossRef]
48. de Paiva, S.R.; Figueiredo, M.R.; Aragão, T.V.; Kaplan, M.A. Antimicrobial activity in vitro of plumbagin isolated from Plumbago species. *Mem. Inst. Oswaldo Cruz* **2003**, *98*, 959–961. [CrossRef]
49. Qian, W.; Wang, W.; Zhang, J.; Fu, Y.; Liu, Q.; Li, X.; Wang, T.; Zhang, Q. Exploitation of the antifungal and antibiofilm activities of plumbagin against *Cryptococcus neoformans*. *Biofouling* **2022**, *38*, 558–574. [CrossRef]
50. Elansary, H.O.; Yessoufou, K.; Mahmoud, E.A.; Skalicka-Woźniak, K. In vitro Antioxidant and Antimicrobial Effects of Ceratostigma plumbaginoides. *Nat. Prod. Commun.* **2016**, *11*, 1455–1458. [CrossRef]
51. Sasaki, K.; Abe, H.; Yoshizaki, F. In vitro antifungal activity of naphthoquinone derivatives. *Biol. Pharm. Bull.* **2002**, *25*, 669–670. [CrossRef] [PubMed]
52. Yan, Y.; Tan, F.; Miao, H.; Wang, H.; Cao, Y. Effect of Shikonin Against *Candida albicans* Biofilms. *Front. Microbiol.* **2019**, *10*, 1085. [CrossRef] [PubMed]
53. Li, D.D.; Chai, D.; Huang, X.W.; Guan, S.X.; Du, J.; Zhang, H.Y.; Sun, Y.; Jiang, Y.Y. Potent In Vitro Synergism of Fluconazole and Osthole against Fluconazole-Resistant *Candida albicans*. *Antimicrob. Agents Chemother.* **2017**, *61*, e00436-17. [CrossRef] [PubMed]
54. Wang, T.; Pan, M.; Xiao, N.; Wu, J.; Wang, Q.; Cheng, T.; Yan, G.; Wu, D.; Li, N.; Shao, J. In vitro and in vivo analysis of monotherapy and dual therapy with ethyl caffeate and fluconazole on virulence factors of *Candida albicans* and systemic candidiasis. *J. Glob. Antimicrob. Resist.* **2021**, *27*, 253–266. [CrossRef] [PubMed]
55. Hamaamin Hussen, N.; Hameed Hasan, A.; Jamalis, J.; Shakya, S.; Chander, S.; Kharkwal, H.; Murugesan, S.; Ajit Bastikar, V.; Pyarelal Gupta, P. Potential inhibitory activity of phytoconstituents against black fungus: In silico ADMET, molecular docking and MD simulation studies. *Comput. Toxicol.* **2022**, *24*, 100247. [CrossRef] [PubMed]
56. Wong, J.H.; Lau, K.M.; Wu, Y.O.; Cheng, L.; Wong, C.W.; Yew, D.T.; Leung, P.C.; Fung, K.P.; Hui, M.; Ng, T.B.; et al. Antifungal mode of action of macrocarpal C extracted from *Eucalyptus globulus* Labill (Lan An) towards the dermatophyte *Trichophyton mentagrophytes*. *Chin. Med.* **2015**, *10*, 34. [CrossRef]
57. Bharate, S.B.; Khan, S.I.; Yunus, N.A.; Chauthe, S.K.; Jacob, M.R.; Tekwani, B.L.; Khan, I.A.; Singh, I.P. Antiprotozoal and antimicrobial activities of O-alkylated and formylated acylphloroglucinols. *Bioorganic Med. Chem.* **2007**, *15*, 87–96. [CrossRef]
58. Qu, C.; Li, Z.; Wang, X. UHPLC-HRMS-Based Untargeted Lipidomics Reveal Mechanism of Antifungal Activity of Carvacrol against *Aspergillus flavus*. *Foods* **2021**, *11*, 93. [CrossRef]
59. Jung, K.W.; Chung, M.S.; Bai, H.W.; Chung, B.Y.; Lee, S. Investigation of Antifungal Mechanisms of Thymol in the Human Fungal Pathogen, *Cryptococcus neoformans*. *Molecules* **2021**, *26*, 3476. [CrossRef]
60. Guo-hui, Y.; Jin-ping, C.; Li-li, W.; Xiao-wen, H. Analysis of Chemical Compositions and Antifungal Activity of Plant Essential Oils Against *Candida* spp. *Chin. J. Exp. Tradit. Med. Formulae* **2018**, *24*, 215–227.
61. Abers, M.; Schroeder, S.; Goelz, L.; Sulser, A.; St Rose, T.; Puchalski, K.; Langland, J. Antimicrobial activity of the volatile substances from essential oils. *BMC Complement. Med. Ther.* **2021**, *21*, 124. [CrossRef] [PubMed]
62. Chen, M.; Li, L.; Xia, L.; Jiang, S.; Kong, Y.; Chen, X.; Wang, H. The kinetics and release behaviour of curcumin loaded pH-responsive PLGA/chitosan fibers with antitumor activity against HT-29 cells. *Carbohydr. Polym.* **2021**, *265*, 118077. [CrossRef]

63. Pontes-Quero, G.M.; Benito-Garzón, L.; Pérez Cano, J.; Aguilar, M.R.; Vázquez-Lasa, B. Amphiphilic polymeric nanoparticles encapsulating curcumin: Antioxidant, anti-inflammatory and biocompatibility studies. *Mater. Sci. Eng. C Mater. Biol. Appl.* **2021**, *121*, 111793. [CrossRef]
64. Fakhrullina, G.; Khakimova, E.; Akhatova, F.; Lazzara, G.; Parisi, F.; Fakhrullin, R. Selective Antimicrobial Effects of Curcumin@Halloysite Nanoformulation: A *Caenorhabditis elegans* Study. *ACS Appl. Mater. Interfaces* **2019**, *11*, 23050–23064. [CrossRef]
65. Yun, D.G.; Lee, D.G. Assessment of silibinin as a potential antifungal agent and investigation of its mechanism of action. *IUBMB Life* **2017**, *69*, 631–637. [CrossRef] [PubMed]
66. Navarro-Pérez, M.L.; Vadillo-Rodríguez, V.; Fernández-Babiano, I.; Pérez-Giraldo, C.; Fernández-Calderón, M.C. Antimicrobial activity of a novel Spanish propolis against planktonic and sessile oral *Streptococcus* spp. *Sci. Rep.* **2021**, *11*, 23860. [CrossRef] [PubMed]
67. Nani, B.D.; Sardi, J.C.O.; Lazarini, J.G.; Silva, D.R.; Massariolli, A.P.; Cunha, T.M.; de Alencar, S.M.; Franchin, M.; Rosalen, P.L. Anti-inflammatory and anti-Candida Effects of Brazilian Organic Propolis, a Promising Source of Bioactive Molecules and Functional Food. *J. Agric. Food Chem.* **2020**, *68*, 2861–2871. [CrossRef] [PubMed]
68. Laaroussi, H.; Bakour, M.; Ousaaid, D.; Aboulghazi, A.; Ferreira-Santos, P.; Genisheva, Z.; Teixeira, J.A.; Lyoussi, B. Effect of antioxidant-rich propolis and bee pollen extracts against D-glucose induced type 2 diabetes in rats. *Food Res. Int.* **2020**, *138*, 109802. [CrossRef]
69. Yanshan, L.; Wenjun, P.; Yongguang, B. Ultrasonic extraction of flavonoids from propolis. *Tech. Acoust.* **2020**, *39*, 190–194. [CrossRef]
70. Zulhendri, F.; Chandrasekaran, K.; Kowacz, M.; Ravalia, M.; Kripal, K.; Fearnley, J.; Perera, C.O. Antiviral, Antibacterial, Antifungal, and Antiparasitic Properties of Propolis: A Review. *Foods* **2021**, *10*, 1360. [CrossRef]
71. Su, H.; Song, S.; Yan, X.; Fang, L.; Zeng, B.; Zhu, Y. Endogenous salicylic acid shows different correlation with baicalin and baicalein in the medicinal plant *Scutellaria baicalensis* Georgi subjected to stress and exogenous salicylic acid. *PLoS ONE* **2018**, *13*, e0192114. [CrossRef] [PubMed]
72. Wang, Y.-H.; Zhou, H.-Y.; Yang, N.; Zhao, F.; Jin, Y.-S. Synthesis and synergistic antifungal effects of flavonoids against drug resistant *Candida albicans*. *Bioorganic Med. Chem. Lett.* **2016**, *26*, 3098–3102. [CrossRef]
73. Fu, Z.; Lu, H.; Zhu, Z.; Yan, L.; Jiang, Y.; Cao, Y. Combination of baicalein and Amphotericin B accelerates *Candida albicans* apoptosis. *Biol. Pharm. Bull.* **2011**, *34*, 214–218. [CrossRef]
74. Li, D.D.; Xu, Y.; Zhang, D.Z.; Quan, H.; Mylonakis, E.; Hu, D.D.; Li, M.B.; Zhao, L.X.; Zhu, L.H.; Wang, Y.; et al. Fluconazole assists berberine to kill fluconazole-resistant *Candida albicans*. *Antimicrob. Agents Chemother.* **2013**, *57*, 6016–6027. [CrossRef]
75. Xu, Y.; Quan, H.; Wang, Y.; Zhong, H.; Sun, J.; Xu, J.; Jia, N.; Jiang, Y. Requirement for Ergosterol in Berberine Tolerance Underlies Synergism of Fluconazole and Berberine against Fluconazole-Resistant *Candida albicans* Isolates. *Front. Cell. Infect. Microbiol.* **2017**, *7*, 491. [CrossRef] [PubMed]
76. Han, Y.; Lee, J.H. Berberine synergy with amphotericin B against disseminated candidiasis in mice. *Biol. Pharm. Bull.* **2005**, *28*, 541–544. [CrossRef] [PubMed]
77. Lam, P.; Kok, S.H.; Lee, K.K.; Lam, K.H.; Hau, D.K.; Wong, W.Y.; Bian, Z.; Gambari, R.; Chui, C.H. Sensitization of *Candida albicans* to terbinafine by berberine and berberrubine. *Biomed. Rep.* **2016**, *4*, 449–452. [CrossRef]
78. Zhong, H.; Hu, D.D.; Hu, G.H.; Su, J.; Bi, S.; Zhang, Z.E.; Wang, Z.; Zhang, R.L.; Xu, Z.; Jiang, Y.Y.; et al. Activity of Sanguinarine against *Candida albicans* Biofilms. *Antimicrob. Agents Chemother.* **2017**, *61*, e02259-16. [CrossRef]
79. Ghosh, K.; Bhattacharya, T.K. Chemical constituents of *Piper betle* Linn. (Piperaceae) roots. *Molecules* **2005**, *10*, 798–802. [CrossRef]
80. Tang, G.H.; Chen, D.M.; Qiu, B.Y.; Sheng, L.; Wang, Y.H.; Hu, G.W.; Zhao, F.W.; Ma, L.J.; Wang, H.; Huang, Q.Q.; et al. Cytotoxic amide alkaloids from *Piper boehmeriaefolium*. *J. Nat. Prod.* **2011**, *74*, 45–49. [CrossRef]
81. Prasetya, F.; Salam, S.; Rahmadani, A.; Haikal, K.; Febrina, L.; Anshory, H.; Arifuddin, M.; Siregar, V.O.; Narsa, A.C.; Herman, H.; et al. Novel Amides Derivative with Antimicrobial Activity of *Piper betle* var. nigra Leaves from Indonesia. *Molecules* **2021**, *26*, 335. [CrossRef]
82. Futuro, D.O.; Ferreira, P.G.; Nicoletti, C.D.; Borba-Santos, L.P.; Silva, F.C.D.; Rozental, S.; Ferreira, V.F. The Antifungal Activity of Naphthoquinones: An Integrative Review. *An. Acad. Bras. Ciências* **2018**, *90*, 1187–1214. [CrossRef] [PubMed]
83. Liao, Z.; Zhu, Z.; Li, L.; Wang, L.; Wang, H.; Jiang, Y.; Cao, Y. Metabonomics on *Candida albicans* indicate the excessive H3K56ac is involved in the antifungal activity of Shikonin. *Emerg. Microbes Infect.* **2019**, *8*, 1243–1253. [CrossRef] [PubMed]
84. Lin, S.; Zhi, X. Xiangdousu Zaheti Jiqi Kangzhenjun Huoxing [Coumarin derivatives and their antifungal activities]. *Guowai Yiyao Kangshengsu Fence* **2018**, *39*, 327–335.
85. Guo, Y.; Li, M.; Chen, P.; Wu, Q.; Gao, C.; Lu, Y.; Zhang, L.; Yuan, D.; Fu, H. A pair of new elemanolide sesquiterpene lactones from *Elephantopus scaber* L. *Magn. Reson. Chem.* **2017**, *55*, 677–681. [CrossRef] [PubMed]
86. Kao, S.T.; Liu, C.J.; Yeh, C.C. Protective and immunomodulatory effect of flos *Lonicerae japonicae* by augmenting IL-10 expression in a murine model of acute lung inflammation. *J. Ethnopharmacol.* **2015**, *168*, 108–115. [CrossRef] [PubMed]
87. Rocha da Silva, C.; Sá, L.; Dos Santos, E.V.; Ferreira, T.L.; Coutinho, T.; Moreira, L.E.A.; de Sousa Campos, R.; de Andrade, C.R.; Barbosa da Silva, W.M.; de Sá Carneiro, I.; et al. Evaluation of the antifungal effect of chlorogenic acid against strains of *Candida* spp. resistant to fluconazole: Apoptosis induction and in silico analysis of the possible mechanisms of action. *J. Med. Microbiol.* **2022**, *71*, 001526. [CrossRef] [PubMed]

88. Rui, Y.; Xuanhao, C.; Jin, L.; Jun, H.; Yanxu, C. Research development of Menthae Haplocalycis Herba on chemical composition and pharmacological activity. *J. Tianjin Univ. Tradit. Chin. Med.* **2022**, *41*, 4–13.
89. Wang, S.J.; Wang, X.H.; Dai, Y.Y.; Ma, M.H.; Rahman, K.; Nian, H.; Zhang, H. Prunella vulgaris: A Comprehensive Review of Chemical Constituents, Pharmacological Effects and Clinical Applications. *Curr. Pharm. Des.* **2019**, *25*, 359–369. [CrossRef] [PubMed]
90. Luo, C.; Zou, L.; Sun, H.; Peng, J.; Gao, C.; Bao, L.; Ji, R.; Jin, Y.; Sun, S. A Review of the Anti-Inflammatory Effects of Rosmarinic Acid on Inflammatory Diseases. *Front. Pharmacol.* **2020**, *11*, 153. [CrossRef]
91. Xiangyang, C. Study on Chemical Constituents and Anti-Inflammatory Activity of Phenolic Fraction of Menthae Haplocalycis Herba. Ph.D. Thesis, Beijing University of Chinese Medicine, Beijing, China, 2016.
92. Kikuchi, T.; Tanaka, A.; Uriuda, M.; Yamada, T.; Tanaka, R. Three Novel Triterpenoids from Taraxacum officinale Roots. *Molecules* **2016**, *21*, 1121. [CrossRef] [PubMed]
93. Koh, Y.J.; Cha, D.S.; Ko, J.S.; Park, H.J.; Choi, H.D. Anti-inflammatory effect of Taraxacum officinale leaves on lipopolysaccharide-induced inflammatory responses in RAW 264.7 cells. *J. Med. Food* **2010**, *13*, 870–878. [CrossRef] [PubMed]
94. Park, C.M.; Park, J.Y.; Noh, K.H.; Shin, J.H.; Song, Y.S. Taraxacum officinale Weber extracts inhibit LPS-induced oxidative stress and nitric oxide production via the NF-κB modulation in RAW 264.7 cells. *J. Ethnopharmacol.* **2011**, *133*, 834–842. [CrossRef] [PubMed]
95. Kotanidou, A.; Xagorari, A.; Bagli, E.; Kitsanta, P.; Fotsis, T.; Papapetropoulos, A.; Roussos, C. Luteolin reduces lipopolysaccharide-induced lethal toxicity and expression of proinflammatory molecules in mice. *Am. J. Respir. Crit. Care Med.* **2002**, *165*, 818–823. [CrossRef] [PubMed]
96. Zhou, X.N.; Han, C.; Song, P.Y.; Zhao, X.H.; Zhong, X.H. Anti-inflammatory Effects of Luteolin and Quercetin in Vitro. *Prog. Vet. Med.* **2017**, *38*, 56–61. [CrossRef]
97. Shen, P.; Lin, W.; Deng, X.; Ba, X.; Han, L.; Chen, Z.; Qin, K.; Huang, Y.; Tu, S. Potential Implications of Quercetin in Autoimmune Diseases. *Front. Immunol.* **2021**, *12*, 689044. [CrossRef]
98. Wang, Y.; Wang, M.; Li, N.; Wu, Y.; Yan, N. Research Progress on Anti-inflammatory Active Components of Lianqiao (*Fructus forsythiae*) and its Action Mechanism. *Chin. Arch. Tradit. Chin. Med.* **2022**, *40*, 115–120. [CrossRef]
99. Javadi, F.; Ahmadzadeh, A.; Eghtesadi, S.; Aryaeian, N.; Zabihiyeganeh, M.; Rahimi Foroushani, A.; Jazayeri, S. The Effect of Quercetin on Inflammatory Factors and Clinical Symptoms in Women with Rheumatoid Arthritis: A Double-Blind, Randomized Controlled Trial. *J. Am. Coll. Nutr.* **2017**, *36*, 9–15. [CrossRef]
100. Carullo, G.; Cappello, A.R.; Frattaruolo, L.; Badolato, M.; Armentano, B.; Aiello, F. Quercetin and derivatives: Useful tools in inflammation and pain management. *Future Med. Chem.* **2017**, *9*, 79–93. [CrossRef]
101. Nikfarjam, B.A.; Adineh, M.; Hajiali, F.; Nassiri-Asl, M. Treatment with Rutin—A Therapeutic Strategy for Neutrophil-Mediated Inflammatory and Autoimmune Diseases: Anti-inflammatory Effects of Rutin on Neutrophils. *J. Pharmacopunct.* **2017**, *20*, 52–56. [CrossRef]
102. Chen, X.; Yu, M.; Xu, W.; Zou, L.; Ye, J.; Liu, Y.; Xiao, Y.; Luo, J. Rutin inhibited the advanced glycation end products-stimulated inflammatory response and extra-cellular matrix degeneration via targeting TRAF-6 and BCL-2 proteins in mouse model of osteoarthritis. *Aging* **2021**, *13*, 22134–22147. [CrossRef] [PubMed]
103. Ghasemi-Pirbaluti, M.; Motaghi, E.; Bozorgi, H. The effect of menthol on acute experimental colitis in rats. *Eur. J. Pharmacol.* **2017**, *805*, 101–107. [CrossRef] [PubMed]
104. Wang, F.; Wen, T.; Xu, F.; Sang, W.; Chen, R.; Nan, Z. The protective effects of menthone on endotoxin-induced infalmmation in mice. *Chin. Pharmacol. Bull.* **2017**, *33*, 227–234.
105. Yang, L.; Zhou, X.; Huang, W.; Fang, Q.; Hu, J.; Yu, L.; Ma, N.; Zhang, W. Protective Effect of Phillyrin on Lethal LPS-Induced Neutrophil Inflammation in Zebrafish. *Cell. Physiol. Biochem.* **2017**, *43*, 2074–2087. [CrossRef] [PubMed]
106. Ma, Q.; Li, R.; Pan, W.; Huang, W.; Liu, B.; Xie, Y.; Wang, Z.; Li, C.; Jiang, H.; Huang, J.; et al. Phillyrin (KD-1) exerts anti-viral and anti-inflammatory activities against novel coronavirus (SARS-CoV-2) and human coronavirus 229E (HCoV-229E) by suppressing the nuclear factor kappa B (NF-κB) signaling pathway. *Phytomedicine* **2020**, *78*, 153296. [CrossRef]
107. Hu, N.; Wang, C.; Dai, X.; Zhou, M.; Gong, L.; Yu, L.; Peng, C.; Li, Y. Phillygenin inhibits LPS-induced activation and inflammation of LX2 cells by TLR4/MyD88/NF-κB signaling pathway. *J. Ethnopharmacol.* **2020**, *248*, 112361. [CrossRef]
108. Zhou, M.; Tang, Y.; Liao, L.; Liu, M.; Deng, Y.; Zhao, X.; Li, Y. Phillygenin inhibited LPS-induced RAW 264.7 cell inflammation by NF-κB pathway. *Eur. J. Pharmacol.* **2021**, *899*, 174043. [CrossRef]
109. Lee, S.; Shin, S.; Kim, H.; Han, S.; Kim, K.; Kwon, J.; Kwak, J.H.; Lee, C.K.; Ha, N.J.; Yim, D.; et al. Anti-inflammatory function of arctiin by inhibiting COX-2 expression via NF-κB pathways. *J. Inflamm.* **2011**, *8*, 16. [CrossRef]
110. Zhou, B.; Wang, L.; Liang, Y.; Li, J.; Pan, X. Arctiin suppresses H9N2 avian influenza virus-mediated inflammation via activation of Nrf2/HO-1 signaling. *BMC Complement. Med. Ther.* **2021**, *21*, 289. [CrossRef]
111. Zhou, B.; Weng, G.; Huang, Z.; Liu, T.; Dai, F. Arctiin Prevents LPS-Induced Acute Lung Injury via Inhibition of PI3K/AKT Signaling Pathway in Mice. *Inflammation* **2018**, *41*, 2129–2135. [CrossRef]
112. Niu, Y.; Dong, Q.; Li, R. Matrine regulates Th1/Th2 cytokine responses in rheumatoid arthritis by attenuating the NF-κB signaling. *Cell Biol. Int.* **2017**, *41*, 611–621. [CrossRef] [PubMed]

113. Ho, L.J.; Juan, T.Y.; Chao, P.; Wu, W.L.; Chang, D.M.; Chang, S.Y.; Lai, J.H. Plant alkaloid tetrandrine downregulates IkappaBalpha kinases-IkappaBalpha-NF-kappaB signaling pathway in human peripheral blood T cell. *Br. J. Pharmacol.* **2004**, *143*, 919–927. [CrossRef] [PubMed]
114. Xu, W.; Kusano, J.; Chen, S.; Yamamoto, R.; Matsuda, H.; Hara, Y.; Fujii, Y.; Hayashi, S.; Tanaka, S.; Sugiyama, K.; et al. Absolute configuration of tetrandrine and isotetrandrine influences their anti-proliferation effects in human T cells via different regulation of NF-κB. *Z. Fur Naturforschung. C J. Biosci.* **2021**, *76*, 21–25. [CrossRef] [PubMed]
115. Guo, C.; Guo, D.; Fang, L.; Sang, T.; Wu, J.; Guo, C.; Wang, Y.; Wang, Y.; Chen, C.; Chen, J.; et al. Ganoderma lucidum polysaccharide modulates gut microbiota and immune cell function to inhibit inflammation and tumorigenesis in colon. *Carbohydr. Polym.* **2021**, *267*, 118231. [CrossRef]
116. Ren, L.; Zhang, J.; Zhang, T. Immunomodulatory activities of polysaccharides from Ganoderma on immune effector cells. *Food Chem.* **2021**, *340*, 127933. [CrossRef]
117. Zhang, L.; Wei, W. Anti-inflammatory and immunoregulatory effects of paeoniflorin and total glucosides of paeony. *Pharmacol. Ther.* **2020**, *207*, 107452. [CrossRef]
118. Lee, J.H.; Han, Y. Ginsenoside Rg1 helps mice resist to disseminated candidiasis by Th1 type differentiation of CD4+ T cell. *Int. Immunopharmacol.* **2006**, *6*, 1424–1430. [CrossRef]
119. Qi, X.; Fan, M.; Huang, N.; Zhang, X.; Liu, J.; Li, X.; Sun, R. Saikosaponin d contributed to cancer chemotherapy induced neutropenia therapy by promoting neutrophil differentiation via activation CBL-dependent ERK pathway. *Pharmacol. Res.* **2020**, *160*, 105149. [CrossRef]
120. Yadav, S.; Sharma, A.; Nayik, G.A.; Cooper, R.; Bhardwaj, G.; Sohal, H.S.; Mutreja, V.; Kaur, R.; Areche, F.O.; AlOudat, M.; et al. Review of Shikonin and Derivatives: Isolation, Chemistry, Biosynthesis, Pharmacology and Toxicology. *Front. Pharmacol.* **2022**, *13*, 905755. [CrossRef]
121. Valipour, M.; Zarghi, A.; Ebrahimzadeh, M.A.; Irannejad, H. Therapeutic potential of chelerythrine as a multi-purpose adjuvant for the treatment of COVID-19. *Cell Cycle* **2021**, *20*, 2321–2336. [CrossRef]
122. Figat, R.; Zgadzaj, A.; Geschke, S.; Sieczka, P.; Pietrosiuk, A.; Sommer, S.; Skrzypczak, A. Cytotoxicity and antigenotoxicity evaluation of acetylshikonin and shikonin. *Drug Chem. Toxicol.* **2021**, *44*, 140–147. [CrossRef] [PubMed]
123. Liu, S. The Effectiveness Study of Polysaccharides and Its Componets from Ginseng on Alzheimer's Disease. Ph.D. Thesis, Jilin University, Changchun, China, 2014.
124. Nan, M.; Zhao, Y.; Lv, N.; He, Y.; Si, X.; Zhao, Q. Advances in Studies on the Chemical Structure and Hypoglycemic Activity of Ginseng Polysaccharide. *China Pharm.* **2014**, *25*, 4506–4508. [CrossRef]
125. Phu, H.T.; Thuan, D.T.B.; Nguyen, T.H.D.; Posadino, A.M.; Eid, A.H.; Pintus, G. Herbal Medicine for Slowing Aging and Aging-associated Conditions: Efficacy, Mechanisms and Safety. *Curr. Vasc. Pharmacol.* **2020**, *18*, 369–393. [CrossRef] [PubMed]
126. Lin, Y.; He, F.; Wu, L.; Xu, Y.; Du, Q. Matrine Exerts Pharmacological Effects Through Multiple Signaling Pathways: A Comprehensive Review. *Drug Des. Dev. Ther.* **2022**, *16*, 533–569. [CrossRef]
127. Zhang, Q.; Lenardo, M.J.; Baltimore, D. 30 Years of NF-κB: A Blossoming of Relevance to Human Pathobiology. *Cell* **2017**, *168*, 37–57. [CrossRef]
128. Müller, P.; Kuttenkeuler, D.; Gesellchen, V.; Zeidler, M.P.; Boutros, M. Identification of JAK/STAT signalling components by genome-wide RNA interference. *Nature* **2005**, *436*, 871–875. [CrossRef]
129. Paddon, C.J.; Westfall, P.J.; Pitera, D.J.; Benjamin, K.; Fisher, K.; McPhee, D.; Leavell, M.D.; Tai, A.; Main, A.; Eng, D.; et al. High-level semi-synthetic production of the potent antimalarial artemisinin. *Nature* **2013**, *496*, 528–532. [CrossRef]
130. Fu, R.; Li, J.; Yu, H.; Zhang, Y.; Xu, Z.; Martin, C. The Yin and Yang of traditional Chinese and Western medicine. *Med. Res. Rev.* **2021**, *41*, 3182–3200. [CrossRef]

antibiotics

Article

Identification and Characterization of Genes Related to Ampicillin Antibiotic Resistance in *Zymomonas mobilis*

Binan Geng †, **Xingyu Huang** †, **Yalun Wu**, **Qiaoning He** * and **Shihui Yang** *

State Key Laboratory of Biocatalysis and Enzyme Engineering, Environmental Microbial Technology Center of Hubei Province, School of Life Sciences, Hubei University, Wuhan 430062, China
* Correspondence: qiaoninghe@hubu.edu.cn (Q.H.); shihui.yang@hubu.edu.cn (S.Y.)
† These authors contributed equally to this work.

Abstract: Antibiotics can inhibit or kill microorganisms, while microorganisms have evolved antibiotic resistance strategies to survive antibiotics. *Zymomonas mobilis* is an ideal industrial microbial chassis and can tolerate multiple antibiotics. However, the mechanisms of antibiotic resistance and genes associated with antibiotic resistance have not been fully analyzed and characterized. In this study, we investigated genes associated with antibiotic resistance using bioinformatic approaches and examined genes associated with ampicillin resistance using CRISPR/Cas12a−based genome−editing technology. Six ampicillin−resistant genes (*ZMO0103*, *ZMO0893*, *ZMO1094*, *ZMO1650*, *ZMO1866*, and *ZMO1967*) were identified, and five mutant strains ZM4Δ0103, ZM4Δ0893, ZM4Δ1094, ZM4Δ1650, and ZM4Δ1866 were constructed. Additionally, a four−gene mutant ZM4ΔARs was constructed by knocking out *ZMO0103*, *ZMO0893*, *ZMO1094*, and *ZMO1650* continuously. Cell growth, morphology, and transformation efficiency of mutant strains were examined. Our results show that the cell growth of ZM4Δ0103 and ZM4ΔARs was significantly inhibited with 150 μg/mL ampicillin, and cells changed to a long filament shape from a short rod shape. Moreover, the transformation efficiencies of ZM4Δ0103 and ZM4ΔARs were decreased. Our results indicate that ZMO0103 is the key to ampicillin resistance in *Z. mobilis*, and other ampicillin−resistant genes may have a synergetic effect with it. In summary, this study identified and characterized genes related to ampicillin resistance in *Z. mobilis* and laid a foundation for further study of other antibiotic resistance mechanisms.

Keywords: *Zymomonas mobilis*; antibiotic resistance; ampicillin; genome editing; CRISPR−Cas12a; resistance selection markers

Citation: Geng, B.; Huang, X.; Wu, Y.; He, Q.; Yang, S. Identification and Characterization of Genes Related to Ampicillin Antibiotic Resistance in *Zymomonas mobilis*. *Antibiotics* **2022**, *11*, 1476. https://doi.org/10.3390/antibiotics11111476

Academic Editor: John E. Gustafson

Received: 10 September 2022
Accepted: 24 October 2022
Published: 25 October 2022

1. Introduction

Antibiotics are natural secondary metabolites or artificially synthesized analogs produced by organisms, such as bacteria, animals, and plants during the metabolic process to kill pathogens [1]. They are commonly used during plasmid and strain construction in genetic engineering and are usually used as the feed additive for the growth and disease resistance of plants and animals [2]. In the environment, competition exists among microorganisms for living space and nutrients. Some microorganisms compete with others by producing antibiotics that are sensitive to antibiotics. For example, *streptomyces* can produce 80% of the antibiotics currently known [3]. At the same time, various microorganisms naturally have certain resistance to different antibiotics, which brings challenges to the genetic engineering of strains [4].

At present, four classes of antibiotics are mainly classified according to their inhibition pathway: (i) inhibition of cell wall synthesis, such as β−lactam antibiotics and glycopeptides; (ii) inhibition of protein synthesis including macrolides, oxazolidinones, amphenicols, lincosamides, tetracyclines, and aminoglycosides [5]; (iii) inhibition of DNA synthesis by targeting gyrase or DNA, such as fluoroquinolones and nitroimidazoles [6,7]; and (iv) inhibition of membrane integrity, such as lipopeptides (e.g., daptomycin) [8] and polymyxins (e.g., colistin) [9]. β−Lactams are the most widely used antibiotics with the potential to interrupt bacterial cell wall formation as a result of covalent binding to essential penicillin−binding protein (PBP) enzymes that are involved in the terminal steps of peptidoglycan cross−linking in both Gram−negative and Gram−positive bacteria [10]. Ampicillin is the representative β−lactam antibiotic and commonly used in genetic engineering to screen colonies with ampicillin resistance.

To survive in an environment containing antibiotics, microorganisms have developed a variety of antibiotic resistance (AR) mechanisms during evolution. Based on the biochemical route involved in resistance, the mechanisms of antibiotic resistance currently are classified into four types. The first one is to modify the antimicrobial molecule. Some enzymes are capable of chemical alterations to inactivate the antibiotics with acetylation, phosphorylation, and adenylation of the antibiotics, such as aminoglycosides, chloramphenicol, and streptogramins [11]. Some enzymes can destroy the antibiotic molecule, such as β−lactamases, rendering the antibiotic unable to interact with the target sites [11]. The second is to prevent antibiotics from reaching their target by decreasing penetration or actively extruding the antimicrobial compound. A reduced number or differential expression of porins, such as the OprD porin protein in some microorganisms, prevented the entry of carbapenems [12]. Extruding the toxic compound out of the cell through an efflux pump [13,14] and the formation of biofilm [15] are also effective to prevent the entrance of antibiotics. The third one is to change or bypass target sites by avoiding the antibiotic to reach its binding site or to modify the target sites that results in decreased affinity for the antibiotic molecule. The last type is a global cell adaptive response to the antibacterial attack.

Z. mobilis is a facultative anaerobic Gram−negative bacterium with a unique Entner–Doudoroff (ED) pathway and many excellent physiological characteristics for industrial bioethanol production, such as the highly efficient utilization of sugar and high ethanol yield and ethanol tolerance [16,17]. Currently, the available antibiotics used for genetic engineering in *Z. mobilis* include ampicillin, kanamycin, spectinomycin, chloramphenicol, tetracycline, streptomycin, and gentamicin [18,19]. Among them, some antibiotics are naturally resisted by *Z. mobilis* at low concentrations, such as ampicillin < 300 μg/mL, kanamycin < 350 μg/mL, streptomycin < 300 μg/mL, gentamicin < 100 μg/mL, and tetracycline and chloramphenicol < 25 μg/mL [18,19]. In addition, different subspecies of *Z. mobilis* have variable susceptibility to different antibiotics. For example, the working concentrations of ampicillin, chloramphenicol, tetracycline, and kanamycin that are used for genetics studies were 300 vs. 500, 100 vs. 100, 25 vs. 25, and 350 vs. 250 μg/mL in *Z. mobilis* ZM4 and CP4, respectively [19].

Although *Z. mobilis* is tolerant to ampicillin, only one work reported that *ZMO0103* probably is an ampicillin−resistant gene, which reported the results of a heterologous protein expression and an enzymatic kinetic analysis [20]. The genome sequence of *Z. mobilis* ZM4 was published, and the genome annotation was further improved [21–23]. Moreover, the genome−editing tools including the native type I−F CRISPR−Cas system and the CRISPR−Cas12 system as well as the platform to identify and characterize biological parts have been established in *Z. mobilis* [24–26]. Therefore, we attempted to explore the ampicillin tolerance mechanism of *Z. mobilis* by identifying potential resistance genes using bioinformatics approaches and constructing ampicillin−sensitive mutant strains by genome engineering to verify its function.

2. Results

2.1. *In Silico Analysis of the AR Genes of Z. mobilis ZM4*

A total of 100 candidate AR genes in *Z. mobilis* ZM4 were predicted using the databases of CARD and MEGARes. The results demonstrate that *Z. mobilis* ZM4 contains 9 putative lactamase−related genes (*ZMO0103, ZMO0108, ZMO0598, ZMO0675, ZMO0781, ZMO1336, ZMO1574, ZMO1914*, and *ZMO1967*), 7 putative transferase−related genes (*ZMO0111, ZMO0183, ZMO1143, ZMO1306, ZMO1355, ZMO1452*, and *ZMO1577*), 8 putative porin−related genes (*ZMO0079, ZMO0257, ZMO0478, ZMO1124, ZMO1164, ZMO1177, ZMO1322,* and *ZMO1387*), and 76 putative efflux pump−related genes (See Supplementary Materials Table S1). In addition, 68 β−lactamase genes were predicted in ZM4 by BLASTP with the gene sequences of the β−lactam class in the UniProt database (See Supplementary Materials Table S2). Among them, six genes were annotated as β−lactamase−encoding genes (*ZMO0103, ZMO0893, ZMO1094, ZMO1650, ZMO1866, ZMO1967*), two candidate genes in list 1 (*ZMO0103, ZMO1967*), and five genes in list 2 (*ZMO0103, ZMO0893, ZMO1650, ZMO1866,* and *ZMO1967*).

Five putative ampicillin−resistant (AR) candidate genes of *ZMO0103, ZMO0893, ZMO1650, ZMO1866,* and *ZMO1967* were predicted as the β−lactamase genes, and *ZMO1094* was annotated as metallo−beta−lactamase−like protein−encoding gene in ZM4. Multiple sequence alignment revealed that ZMO0103, ZMO0893, and ZMO1650 belong to the AmpC superfamily (Supplementary Materials Figure S1); ZMO1967 and ZMO1094 belong to the PenP superfamily (β−lactamase class A), while ZMO1866 belongs to the RnjA superfamily (Supplementary Materials Figure S1) according to the conserved domain search [27]. Moreover, ZMO0103, ZMO0893, and ZMO1650 have similar conserved structures based on the multisequence alignment results (Supplementary Materials Figure S2).

2.2. Ampicillin Resistance−Related Gene Knockout in Z. mobilis ZM4

Subsequently, six ampicillin−resistant (AR) candidate genes based on the above bioinformatics study were selected for knockout using the CRISPR−Cas12a genome−editing system. Except for *ZMO1967*, which may be an essential gene and cannot be knocked out, five other genes of *ZMO0103, ZMO0893, ZMO1094, ZMO1650,* and *ZMO1866* were successfully knocked out with the CRISPR−Cas12a system. As demonstrated in Figure 1, the expected sizes of the amplicon of *ZMO0103, ZMO0893, ZMO1094, ZMO1650,* and *ZMO1866* in the wild type (WT) strain were obtained, which were ~4021, 3314, 3122, 3746, and 3768 bp, respectively. As a contrast, the corresponding amplified fragments in the knockout strains were ~2247, 2232, 2361, 2221, and 2080 bp, respectively (Figure 1). The inactivation of the AR genes was further confirmed by Sanger sequencing of the PCR products. Then, the knockout strains harboring the editing plasmids were cultured in RMG5 for several passages to obtain the final strains with the editing plasmid cured. Except for *ZMO1866, ZMO0103, ZMO0893, ZMO1094,* and *ZMO1650,* the knockout strains lost the editing plasmid successfully, which can only grow in RMG5 after 24 h cultivation, but not in the medium with chloramphenicol (Supplementary Materials Figure S2).

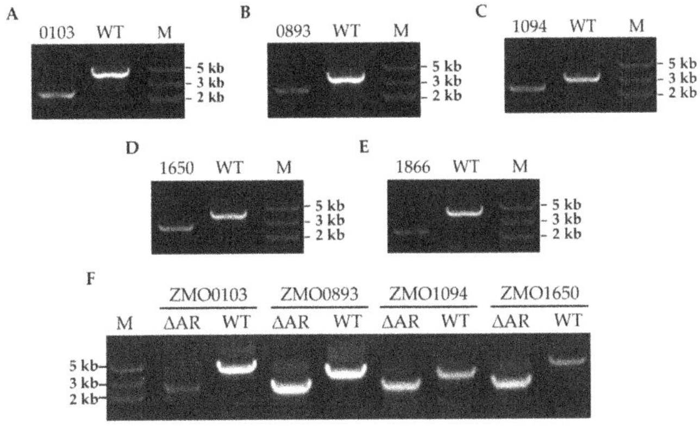

Figure 1. Confirmation of ampicillin−resistant (AR) knockout strains in *Z. mobilis* ZM4 by PCR. The mutants of ZM4Δ0103 (**A**), ZM4Δ0893 (**B**), ZM4Δ1650 (**C**), ZM4Δ1094 (**D**), ZM4Δ1866 (**E**), and ZM4ΔARs (**F**) were confirmed by colony PCR using their corresponding primers. The sizes of PCR products (bp) of WT and knockout strains were 4021, 2247 (ZM4Δ0103); 3314, 2232 (ZM4Δ0893); 3122, 2361 (ZM4Δ1650); 3746, 2221 (ZM4Δ1094); 3768, 2080 (ZM4Δ1866).

To obtain the strain with all the ampicillin−resistant (AR) candidate genes knocked out, we further conducted six rounds of genome editing continuously with the CRISPR−Cas12a system. Consistent with the above single−gene deletion experiments, we only obtained a mutant strain, ZM4ΔArs, with four ampicillin−resistant (AR) genes of *ZMO0103*, *ZMO0893*, *ZMO1094*, and *ZMO1650* knocked out continuously, while ZMO1967 and ZMO1866 were not able to be deleted. The ZM4ΔARs was further identified by colony PCR using the primers for each gene. The results of the correct PCR products of four genes indicate that these four ampicillin−resistant (AR) genes were knocked out successfully in ZM4ΔARs (Figure 1F).

2.3. Antibiotic Tolerance of Ampicillin−Resistant (AR) Gene Knockout Strains

Previous studies reported that the concentration of ampicillin required for plate screening and liquid culture of the transformants in genetic engineering manipulation of *Z. mobilis* ZM4 was 300 μg/mL [19]. In this study, we set three ampicillin concentration gradients of 0, 150, and 300 μg/mL to cultivate the ampicillin−resistant (AR) gene knockout strains. As demonstrated in the Supplementary Materials Figure S4, the deletion of *ZMO1094* and *ZMO1650* individually did not decrease the ampicillin resistance. Specifically, the growth of ZM4Δ1094 and ZM4Δ1650 under different ampicillin concentrations was not inhibited compared with ZM4, whereas the deletion of *ZMO0893* and *ZMO1866* individually were effective. The growth of ZM4Δ0893 and ZM4Δ1866 was inhibited under 150 μg/mL ampicillin with a slower growth rate of $0.15 \pm 0.03\ h^{-1}$ and $0.05 \pm 0.01\ h^{-1}$ compared with the value of $0.24 \pm 0.02\ h^{-1}$ for wild−type ZM4 (Figure 2). The growth of the *ZMO0103* mutant ZM4Δ0103 was the slowest among all strains with the lowest growth rate of $0.21 \pm 0.01\ h^{-1}$ under 0 μg/mL ampicillin, and ZM4Δ0103 was inhibited by all concentrations of ampicillin, such as 150 μg/mL (Figure 2, Supplementary Materials Figure S4).

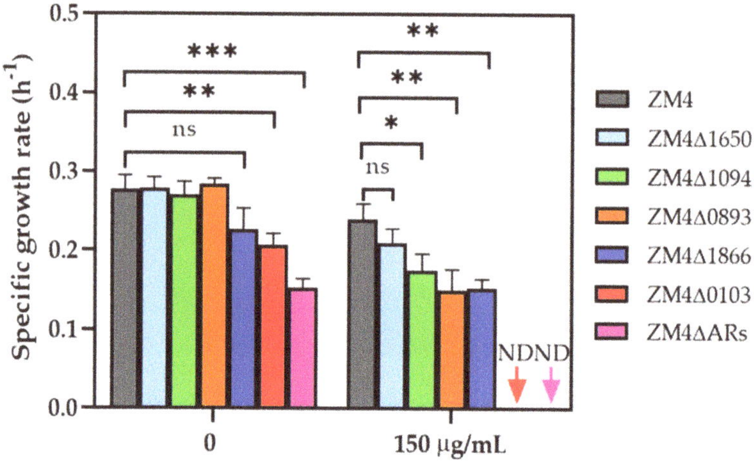

Figure 2. The specific growth rate of ampicillin−resistant (AR) gene knockout strains cultured under 0 and 150 μg/mL of ampicillin. Three replicates were performed for the experiment. When the mutant could not grow under the condition, the sample is marked "ND" (not detected). * represents a significant difference with p−value < 0.05. ** represents a significant difference with p−value < 0.01. *** represents a significant difference with p−value < 0.001. ns represents no significant difference.

The same experiment was also carried out for ZM4ΔARs, a mutant strain with four ampicillin−resistant (AR) genes deleted. Under 0 μg/mL ampicillin, a long lag phase was observed with a lower growth rate of $0.15 \pm 0.01\ h^{-1}$ in ZM4ΔARs compared to ZM4Δ0103 (Figure 2, Supplementary Materials Figure S4), and ZM4ΔARs also cannot grow at 150 μg/mL ampicillin. Combined with the growth of single−gene knockout strains under RMG5, we speculated that the deletion of *ZMO0103* is one of the major reasons for the poor growth of ZM4ΔARs, and other ampicillin−resistant (AR) genes may have a synergetic effect on ampicillin resistance. However, as previously reported, when the ampicillin concentration increased to 300 μg/mL, nearly all the strains could not grow. These results also demonstrate that the mutants ZM4Δ0103 and ZM4ΔARs had a lower ampicillin tolerance concentration—150 μg/mL.

2.4. Cell Morphology of Mutants Treated with Ampicillin

Previous studies demonstrated that cell cultures in the presence of antibiotics resulted in abnormal cellular morphology in various degrees, such as elongated or distorted cell shapes [28,29]. To evaluate the morphological changes of ampicillin−resistant (AR) gene knockout strains during the ampicillin treatment, cell morphologies of three *Z. mobilis* strains (ZM4, ZM4Δ0103, and ZM4ΔArs) were observed under light microscopy. The results show that ZM4Δ0103 and ZM4ΔARs had longer rod shapes with various cell lengths and widths. The lengths of ZM4Δ0103 and ZM4ΔARs were around 10.6 ± 7.2 μm and 11.3 ± 5.9 μm compared to that of wild−type ZM4 (3.2 ± 0.9 μm) in RMG5 without ampicillin (Figure 3). However, when strains were cultured in RMG5 with 100 μg/mL ampicillin (RMA100), the cellular morphology of ZM4Δ0103 (16.1 ± 9.3 μm) and ZM4ΔARs (16.0 ± 7.8 μm) changed and was longer with a filament shape, while the length of ZM4 did not change much (4.1 ± 1.0 μm). This demonstrated that ampicillin was more stressful to ZM4Δ0103 and ZM4ΔARs during cell growth, and ZM4Δ0103 and ZM4ΔARs were more sensitive to ampicillin than ZM4. Similar morphological changes of *Z. mobilis* have been previously described when the cells were exposed to different stresses, such as high temperature [30], lignocellulosic hydrolysate inhibitory [31], high concentration of xylose [32], and salt conditions [33].

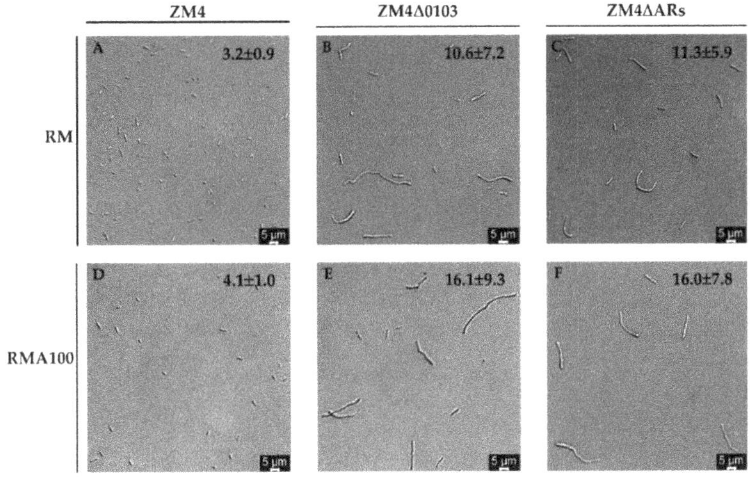

Figure 3. Cell morphology of strains ZM4, ZM4Δ0103, and ZM4ΔARs cultured in RM and RMA100 was observed by light microscopy. The numbers with error value in each image represent the average cell size (μm) analyzed with ImageJ software. Numbers in the lower right corner of each represent the scale. RM and RMA100 represent the different RMG5 media with 0 and 100 μg/mL of ampicillin.

2.5. Effects of Ampicillin−Resistant (AR) Gene Mutagenesis on Genetic Transformation Efficiency

The transformation efficiencies of both ZM4Δ0103 and ZM4ΔARs were significantly reduced using the plasmids of pEZ15A (~3 kb) and pE39−MVA (~10 kb). The electroporation efficiency of plasmid pEZ15A in ZM4 was (1.50 ± 0.08) × 10^5 CFU/μg DNA, which was decreased to (1.66 ± 1.98) × 10^4 and (2.66 ± 1.11) × 10^3 CFU/μg DNA in ZM4Δ0103 and ZM4ΔARs, respectively (Figure 4, Supplementary Materials Figure S5), with the transformation efficiencies reduced ca 10~100 fold. Few colonies grew on the plate after the electroporation of pE39−MVA to ZM4Δ0103 and ZM4ΔARs, while the efficiency of $(1.35 \pm 0.01) \times 10^4$ CFU/μg DNA in ZM4 was observed (Figure 4, Supplementary Materials Figure S6). The transformation efficiency of ZM4Δ0103 and ZM4ΔARs cannot be calculated when the pE39−MVA plasmid with a large size was used. These results suggest that the predicted β−lactamase genes influenced the genetic transformation efficiency and significantly reduced the electroporation efficiency especially with the larger plasmid.

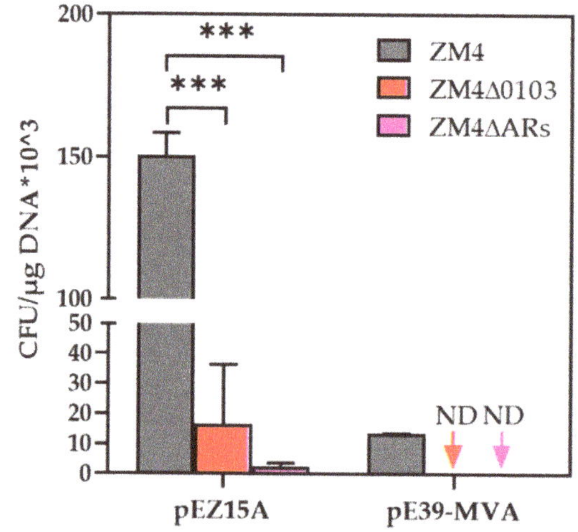

Figure 4. Electroporation efficiency of ZM4, ZM4Δ0103, and ZM4ΔARs using plasmids of pEZ15A (~3 kb) and pE39−MVA (~10 kb). Three replicates were performed for the experiment. The error bar represents standard deviation (SD). When transformation of a plasmid was below the limit of detection (0.00001), the sample is marked "ND" (not detected). *** represents a very significant difference (p−value < 0.001).

3. Discussion

In this study, we first predicted six functional β−lactamase genes that may cause ampicillin antibiotic resistance in *Z. mobilis* ZM4 by bioinformatics analysis. ZMO0103, ZMO0893, and ZMO1650 belong to the AmpC superfamily containing similar conserved structures, indicating that they may have similar functions in cellular processes. ZMO1967 belongs to the PenP superfamily, which is associated with β−lactamase class A. ZMO1094 and ZMO1866 are not closely related to the function of β−lactams. We attempted to knock out all six of them in ZM4; unfortunately, only 5 of them were deleted individually using the CRISPR−Cas12a genome−editing system. The ZMO1967 protein has a transmembrane domain at the N−terminus with a probability of 0.98 by TMHMM−2.0 (https://dtu.biolib.com/DeepTMHMM, accessed on 24 August 2022). In addition, *ZMO1967* is probably an essential gene according to the results of the genome−wide CRISPRi (unpublished data), which may be the reason that it failed to be knocked out in this study. We successfully deleted five β−lactamase genes individually in ZM4, but the editing plasmid for *ZMO1866* deletion was not able to be eliminated.

Our study demonstrated that in addition to ampicillin resistance, these genes annotated as ampicillin−resistant (AR) genes had other effects on cell morphology, cell growth, and transformation efficiency. The knockouts *ZMO0893*, *ZMO1094*, and *ZMO1866* slightly inhibited the growth under the addition of 150 μg/mL. Especially, when *ZMO0103* was deleted, the growth of the strain was inhibited mostly with different concentrations of ampicillin. The biomass of ZM4Δ0103 hardly increased under ampicillin ≥ 150 μg/mL. In addition, similar growth inhibition was also observed in four ampicillin−resistant (AR) genes of *ZMO0103*, *ZMO0893*, *ZMO1094* and the *ZMO1650* knockout strain ZM4ΔARs. The previous study reported that *ZMO0103* was a β−lactamase gene, which contains a 55%

amino acid sequence identity with class C β−lactamase genes [20]. A higher expression level [30,34–36] of *ZMO0103* in ZM4 may be ascribed to the absence of the AmpR that can inhibit the expression of *ZMO0103*. The result of multiple sequence alignment shows that the homologous protein of β−lactamase ZMO0103 was only found in *Sphingomonas* (Supplementary Materials Figures S7 and S8). Combining the result of the improved sensitivity of ZM4Δ0103 to ampicillin in this study, we speculated that ZMO0103 is a unique protein and the most important β−lactamase in *Z. mobilis* ZM4, resulting in the high resistance of ZM4 to β−lactam antibiotics, such as ampicillin [17].

In addition, we also found that the growth of ZM4Δ0103 and ZM4ΔARs was inhibited even without the addition of ampicillin. Considering the existence of transmembrane helices and signal peptides at the N−terminus (https://services.healthtech.dtu.dk/service.php?SignalP, accessed on 28 August 2022) of ZMO0103, ZMO0103 located on the cell membrane may directly hydrolyze ampicillin in the periplasmic space, and the integrity of the membrane could be disrupted by knocking out *ZMO0103* leading to the defective growth of ZM4Δ0103 and ZM4ΔARs. Furthermore, cell sizes of ZM4Δ0103 and ZM4ΔARs became longer and further lengthened with the addition of ampicillin indicating that the AmpC family lactamase protein ZMO0103 is related to cell wall biosynthesis and deconstruction and crucial for cell morphology and growth. Microscopic observations under RMG5 showed that the longest cells in ZM4Δ0103 and ZM4ΔARs increased in length by 13.7 μm and 13.1 μm, respectively, compared to the longest cells in the wild−type ZM4. In addition, when 100 μg/mL of ampicillin (RMA100) was added, the cell lengths of the longest cells in ZM4Δ0103 and ZM4ΔARs could increase by 7.6 μm and 6.6 μm, respectively, compared to the lengths of ZM4Δ0103 and ZM4ΔARs under RM. These results suggest that the deletion of ZMO0103 affected the cell wall structure and the cell membrane and therefore led to changes in the intracellular osmotic pressure, thus enlarging the cells similar to the result of the previous study [37].

Based on our speculation that the deletion of *ZMO0103* may affect cell membrane and cell wall structures, we expected that the exogenous DNA could be more easily transferred into the cells. So, we further tested the transform efficiency of ZM4Δ0103 and ZM4ΔARs using two plasmids of pEZ15A and pE39−MVA with different sizes. However, lower transformation efficiencies were observed when pEZ15A and pE39−MVA were electroporated into ZM4Δ0103 and ZM4ΔARs compared with ZM4. Particularly, in the case of pE39−MVA, almost no transformants were obtained, which suggested that the transformation efficiency of the knockout strains was affected by plasmids sizes. However, the mechanisms of how *ZMO0103* influences the transformation efficiency remain to be investigated including the constructing a truncated mutant of *ZMO0103* by deleting the catalytic domain (amino acids from position 52 to 405) or the C−terminal domain (amino acids from position 406 to 520) that may influence the transformation of exogenous DNA. In addition, the restriction–modification (R–M) system genes (*ZMO0028*, *ZMO1933*, and *ZMOp32x025_028*) could be knocked out in ampicillin−resistant gene knockout strains to further improve the transformation efficiency [38–40].

4. Materials and Methods

4.1. Strains and Cultural Conditions

E. coli DH5α was stored in our laboratory and used for plasmid maintenance and construction. During culturing, 50 μg/mL of chloramphenicol was added to Luria−Bertani (LB) medium (10 g/L tryptone, 5 g/L yeast extract, and 10 g/L NaCl, pH 7.0) for *E. coli* and cultured at 37 °C. *Z. mobilis* ZM4 was used as the parental strain for the construction of derived mutants and cultured with Rich Medium (RMG5) (50 g/L glucose, 10 g/L yeast extract, and 2 g/L KH_2PO_4) at 30 °C. When required, 50 μg/mL of chloramphenicol and 100 μg/mL of spectinomycin were added to the LB and RMG5. All *E. coli*, *Z. mobilis*, and their derivative strains used in this study are listed in the Supplementary Materials Table S3.

4.2. *In Silico Analysis of the AR Genes of Z. mobilis ZM4*

A BLASTP analysis of the *Z. mobilis* ZM4 proteome was performed using two databases for resistance genes: CARD (Comprehensive antibiotic resistance database) and MEGARes database. In addition to using BLASTP to investigate the relatedness of the sequences to those contained within the CARD and MEGARes databases, we further aligned with the gene sequences of the β−lactamase class in the UniProt database with the *Z. mobilis* ZM4 protein sequences. We chose the common genes of candidate β−lactamase genes after BLASTP with CARD and MEGARes databases and genes annotated as β−lactamase as candidate gene list 1. Then, we chose the common genes of candidate β−lactamase genes after BLASTP with the UniProt database and genes annotated as β−lactamase as candidate gene list 2. The common genes within both candidate gene list 1 and list 2 were the final candidate β−lactamase genes used for investigation in this work. The β−lactamase genes with a significant E−value of 2×10^{-10} were selected.

4.3. *Construction of Editing Plasmids of Ampicillin−Resistant (AR) Genes in Z. mobilis ZM4*

The spacers were designed to bear the entire 23 bp sequences containing a 5′−NTTN−3′ PAM for six genes of *ZMO0103*, *ZMO0893*, *ZMO1094*, *ZMO1358*, *ZMO1866*, and *ZMO1967*. The oligonucleotides of spacers were ordered from TsingKe Biotechnology Co., Ltd. (Beijing, China). CRISPR−Cas12a−editing plasmids were constructed following the previous description [25]. The Cas12a−targeting gRNA sequence was annealed using two single−stranded oligonucleotides by first heating the reaction mixture to 95 °C for 5 min and subsequently cooling down gradually to room temperature. Then, the annealed spacer was ligated into *Bsa* I−linearized pEZ−sgr by T4 ligase at 22 °C for 3 h. The resulting plasmids were named as pEZ−sgr−0103, pEZ−sgr−0893, pEZ−sgr−1094, pEZ−sgr−1385, pEZ−sgr−1866, and pEZ−sgr−1967.

Gibson assembly method was utilized for donor construction. Donor sequences including extra ~800 bp upstream and downstream flank sequences of the candidate gene were amplified using Primer STAR polymerase (Takara, Japan) from the genomic DNA of *Z. mobilis* ZM4 and then cloned into pEZ−sgr vector by T5 exonuclease (NEB, WA, USA). The resulting plasmids were named as pEZ−sgr−0103−D, pEZ−sgr−0893−D, pEZ−sgr−1094−D, pEZ−sgr−1385−D, pEZ−sgr−1866−D, and pEZ−sgr−1967−D. All plasmids used in this work are provided in the Supplementary Materials Table S4. The sequences of primers used in this study are shown in the Supplementary Materials Table S5.

4.4. *Curing of Editing Plasmids*

Transformants harboring editing plasmids were cultured in RMG5 broth without the supplement of antibiotics. After 6 consecutive passages in the nonresistant RMG5 liquid medium, 100 μL cultures were diluted and plated on nonresistant RMG5 plates. Then, single colonies were picked to conduct colony PCR using primers pEZ15A−F/R for amplifying the editing plasmid. Editing plasmids were lost if no PCR product was amplified compared with the control. Single colonies with correct PCR results were then inoculated on RMG5 with or without chloramphenicol for further verification. The knockout strains losing the editing plasmids can only grow in RMG5, but not in the medium with the supplementation of chloramphenicol.

4.5. *Electroporation of Editing Plasmids to Z. mobilis ZM4*

Transformation of *Z. mobilis* ZM4 with the editing plasmid was achieved by preparing electrocompetent cells using a modified protocol as previously described. *Z. mobilis* ZM4 was inoculated with 40 mL RMG5 in a 100 mL flask and was grown to an OD600 nm = 0.4~0.6. Cells were harvested by centrifugation at 4000× *g* for 10 min at room temperature. The supernatant was discarded, and the cell pellet was washed with Milli−Q® ultrapure water and 10% (*v*/*v*) glycerol before being resuspended in a final volume of 400 μL of 10% (*v*/*v*) glycerol. Cells were stored as 50 μL aliquots on ice for immediate use in electroporation experiments. Then, 500 ng plasmid DNAs was used for electroporation

(Bio−Rad, CA, USA). After pulsing at 16 kV/cm, 25 μF and 200 Ω, 1 mL of RMG5 was added to the electroporated solutions and then incubated at 30 °C for 4~6 h. Finally, the cells were plated and selected on RM plates supplemented with chloramphenicol until colonies were visible (≤2 d).

4.6. Construction of ZM4ΔARs by Continuous Gene Editing

We constructed ZM4ΔARs with four genes of *ZMO0103*, *ZMO0893*, *ZMO1094*, and *ZMO1650* knocked out following these steps:

Step 1. A CRISPR−Cas12a−editing plasmid pEZ−sgr−0103−D was transformed into ZM4−Cas12a. Colony PCR was then performed with the primers 0103−out−F and 0103−out−R.

Step 2. The positive single clone with the correct PCR size was cultured in RMG5, and the editing plasmid was then cured following Section 4.4.

Step 3. After curing the editing plasmid, the electrocompetent cells of knockout strain ZM4Δ0103 were then prepared for the next−round knockout.

Knockout of *ZMO0893*, *ZMO1094*, *ZMO1650*, *ZMO1866*, and *ZMO1967* followed the above steps 1 to 3. After each gene was edited and the corresponding editing plasmid was lost, the competent cells of the corresponding mutant were prepared for transferring to the next editing plasmid.

4.7. Genetic Transformation Efficiency of β−lactamase Mutants

To calculate the electroporation efficiencies of the β−lactamase mutants, a 3−kb shuttle plasmid pEZ15A and a 10−kb shuttle plasmid pE39−MVA were prepared from *E. coli* DH5α and introduced into ZM4, ZM4Δ0103, and ZM4ΔArs. Electroporation efficiency was presented by the colony forming units (CFUs) on selective plates when 50 μg plasmid DNA was introduced and 100 μL recovery culture was plated. The calculating formula is described below:

$$\text{CFU}/\mu\text{g}^{-1}\ \text{DNA} = (\text{Cp}/\text{Tp}) \times (\text{Vt}/\text{Vp});$$

where Cp is the colony number counted on selective plates; Tp is the total amount of plasmid DNA (μg) used here; Vt is the total transformation volume (μL); Vp is the volume (μL) plated.

4.8. Confirmation of Ampicillin−Resistant Gene Deletion in Z. mobilis ZM4

A pair of primers was used to confirm the disruption of each AR gene in *Z. mobilis* transformants: the primers were used to amplify the region including upstream of the corresponding gene, targeting gene, and downstream of the corresponding gene (see primer pairs in the Supplementary Materials Table S5, e.g., 0103−out−F and 0103−out−R).

4.9. Growth Studies and Analysis of Ampicillin−Resistant Gene Mutants

Cell growth was monitored by measuring the cell optical density (OD) values using a Bioscreen C high−throughput growth measurement instrument (Bioscreen C MBR, Helsinki, Finland). The single colony was inoculated in 1 mL of RMG5 at 30 °C overnight as the seed culture. Then, the seed culture of *Z. mobilis* was transferred to RMG5 until reaching the exponential phase. Subsequently, the cultures were inoculated to the well of the Bioscreen C plate containing 300 μL bacterium suspension with an initial $OD_{600\ nm}$ value of 0.05. Three technical replicates were used for each condition.

4.10. Cell Morphology Observation

Z. mobilis strains were cultured in RMG5 or RMG5 with 150 μg/mL ampicillin overnight at 30 °C. Bacterial pellets were collected at the exponential phase, washed twice with 1× phosphate buffer saline (1× PBS), resuspended in the same buffer, and observed under a light microscope (Leica Dmi8, Buffalo Grove, IL, USA) at 400 magnifications. Each image was taken using the image−based autofocus system LAS X software of the Leica Dmi8 system, and the cell size was measured using ImageJ software [41].

5. Conclusions

In summary, genes associated with antibiotic resistance in *Z. mobilis* were systematically investigated, and six ampicillin−resistant genes in *Z. mobilis* were identified. Five of them were knocked out individually, and four ampicillin−resistant gene deletion mutant ZM4ΔArs was also constructed to verify their functions. The strains of ZM4Δ0103 and ZM4ΔArs were sensitive to ampicillin at a lower concentration of 150 μg/mL with a long and filament shape. The putative membrane protein ZMO0103 is probably the essential β−lactamase, and other ampicillin−resistant genes may have a synergetic effect with it. This study not only identified the ampicillin−resistant genes in *Z. mobilis* ZM4, but also verified their functions on cell growth, morphology, and transformation efficiency. In addition, ampicillin−sensitive mutants can serve as the parental strains for metabolic engineering practices in *Z. mobilis* enabling the usage of antibiotics, such as ampicillin, that are routinely used for genetic engineering.

Supplementary Materials: The following supporting information can be downloaded at: https://www.mdpi.com/article/10.3390/antibiotics11111476/s1, Figure S1: Secondary structure prediction of ZMO0103, ZMO0893, ZMO1094, ZMO1650, ZMO1866, and ZMO1967 in NCBI database; Figure S2: Multiple sequence alignment of the AmpC superfamily and β−lactamase class A protein in Z. mobilis; Figure S3: Loss of editing plasmids in ampicillin−resistant (AR) gene knockout strains; Figure S4: Cell growth of single and multiple ampicillin−resistant gene knockout strains cultured under 0 (A), 150 (B), and 300 (C) μg/mL of ampicillin; Figure S5: Electroporation efficiency of ZM4Δ0103, ZM4ΔArs, and ZM4 by pEZ15A; Figure S6: Electroporation efficiency of ZM4Δ0103, ZM4ΔArs, and ZM4 by pEZ15A and pE39−MVA; Figure S7: Distance tree of ZMO0103 after multiple sequence alignment with other proteins collected in NCBI using BALSTP; Figure S8: The detailed result of multiple sequence alignment with other proteins collected in NCBI using BALSTP; Table S1: List of the putative efflux pump−type antibiotic−resistant genes by BLASTP with CARD and MEGARes databases; Table S2: List of the putative β−lactamase genes by BLASTP in the UniProt database; Table S3: Strains used in this study; Table S4: Plasmids used in this study; Table S5: Primers used in this study.

Author Contributions: Conceptualization, S.Y.; methodology, B.G., X.H. and Y.W.; validation, X.H.; formal analysis, B.G. and X.H.; investigation, B.G., X.H., Y.W., Q.H. and S.Y.; data curation, B.G. and X.H.; writing—original draft preparation, B.G. and X.H.; writing— review and editing, Q.H. and S.Y.; visualization, Y.W.; supervision and project administration, S.Y. All authors have read and agreed to the published version of the manuscript.

Funding: Supported by National Key Research and Development Program of China (2018YFC1801200), the National Natural Science Foundation of China (21978071 and U1932141), Leading Innovative and Entrepreneur Team Introduction Program of Zhejiang Province (2018R01014), and the Innovation Base for Introducing Talents of Discipline of Hubei Province (2019BJH021).

Institutional Review Board Statement: Ethical review and approval were waived for this study as the study did not involve animal interventions.

Informed Consent Statement: Not applicable.

Data Availability Statement: No new data were created or analyzed in this study. Data sharing does not apply to this article.

Acknowledgments: We acknowledge the support from the State Key Laboratory of Biocatalysis and Enzyme Engineering.

Conflicts of Interest: The authors declare no conflict of interest.

References

1. Awad, H.; El-Shahed, K.Y.I.; Aziz, R.; Sarmidi, M.R.; El-Enshasy, H.A. Antibiotics as microbial secondary metabolites: Production and application. *J. Teknol.* **2012**, *59*, 1–11. [CrossRef]
2. Parker, D.S.; Armstrong, D.G. Antibiotic feed additives and livestock production. *Proc. Nutr. Soc.* **1987**, *46*, 415–421. [CrossRef]
3. Procopio, R.E.; Silva, I.R.; Martins, M.K.; Azevedo, J.L.; Araujo, J.M. Antibiotics produced by *Streptomyces*. *Braz. J. Infect. Dis.* **2012**, *16*, 466–471. [CrossRef]

4. Urban−Chmiel, R.; Marek, A.; Stepien−Pysniak, D.; Wieczorek, K.; Dec, M.; Nowaczek, A.; Osek, J. Antibiotic resistance in bacteria−a review. *Antibiotics* **2022**, *11*, 1079. [CrossRef]
5. Singh, S.B.; Young, K.; Silver, L.L. What is an "ideal" antibiotic? Discovery challenges and path forward. *Biochem. Pharmacol.* **2017**, *133*, 63–73. [CrossRef]
6. Ezelarab, H.A.A.; Abbas, S.H.; Hassan, H.A.; Abuo−Rahma, G.E.A. Recent updates of fluoroquinolones as antibacterial agents. *Arch Pharm* **2018**, *351*, e1800141. [CrossRef]
7. Ang, C.W.; Jarrad, A.M.; Cooper, M.A.; Blaskovich, M.A.T. Nitroimidazoles: Molecular fireworks that combat a broad spectrum of infectious diseases. *J. Med. Chem.* **2017**, *60*, 7636–7657. [CrossRef]
8. Vilhena, C.; Bettencourt, A. Daptomycin: A review of properties, clinical use, drug delivery and resistance. *Mini Rev. Med. Chem.* **2012**, *12*, 202–209. [CrossRef]
9. El−Sayed Ahmed, M.A.E.; Zhong, L.L.; Shen, C.; Yang, Y.; Doi, Y.; Tian, G.B. Colistin and its role in the Era of antibiotic resistance: An extended review (2000–2019). *Emerg. Microbes Infect.* **2020**, *9*, 868–885. [CrossRef]
10. Bush, K.; Bradford, P.A. β−Lactams and β−Lactamase Inhibitors: An Overview. *Cold Spring Harb. Perspect. Med.* **2016**, *6*, a025247. [CrossRef]
11. Wilson, D.N. Ribosome−targeting antibiotics and mechanisms of bacterial resistance. *Nat. Rev. Microbiol.* **2014**, *12*, 35–48. [CrossRef]
12. Fernandez, L.; Hancock, R.E. Adaptive and mutational resistance: Role of porins and efflux pumps in drug resistance. *Clin. Microbiol. Rev.* **2012**, *25*, 661–681. [CrossRef] [PubMed]
13. McMurry, L.; Petrucci, R.E., Jr.; Levy, S.B. Active efflux of tetracycline encoded by four genetically different tetracycline resistance determinants in *Escherichia coli*. *Proc. Natl. Acad. Sci. USA* **1980**, *77*, 3974–3977. [CrossRef]
14. Poole, K. Efflux−mediated antimicrobial resistance. *J. Antimicrob. Chemother.* **2005**, *56*, 20–51. [CrossRef] [PubMed]
15. Hoiby, N.; Bjarnsholt, T.; Givskov, M.; Molin, S.; Ciofu, O. Antibiotic resistance of bacterial biofilms. *Int. J. Antimicrob. Agents* **2010**, *35*, 322–332. [CrossRef]
16. Wang, X.; He, Q.N.; Yang, Y.F.; Wang, J.W.; Haning, K.; Hu, Y.; Wu, B.; He, M.X.; Zhang, Y.P.; Bao, J.; et al. Advances and prospects in metabolic engineering of *Zymomonas mobilis*. *Metab. Eng.* **2018**, *50*, 57–73. [CrossRef]
17. Yang, Y.F.; Geng, B.N.; Song, H.Y.; He, Q.N.; He, M.X.; Bao, J.; Bai, F.W.; Yang, S.H. Progress and perspectives on developing *Zymomonas mobilis* as a chassis cell. *Synth. Biol. J.* **2021**, *2*, 59–90. [CrossRef]
18. Yang, S.H.; Pelletier, D.A.; Lu, T.Y.; Brown, S.D. The *Zymomonas mobilis* regulator hfq contributes to tolerance against multiple lignocellulosic pretreatment inhibitors. *BMC Microbiol.* **2010**, *10*, 135. [CrossRef]
19. Zou, S.L.; Zhang, K.; Jing, X.; Wang, Z.R.; Zhang, M.H. Studies on selectable marker for genetic engineering of *Zymomonas mobilis* ZM4 and CP4 strain. *Ind. Microbiol.* **2012**, *42*, 72–77.
20. Rajnish, K.N.; Asraf, S.A.; Manju, N.; Gunasekaran, P. Functional characterization of a putative beta−lactamase gene in the genome of *Zymomonas mobilis*. *Biotechnol. Lett.* **2011**, *33*, 2425–2430. [CrossRef]
21. Yang, S.; Pappas, K.M.; Hauser, L.J.; Land, M.L.; Chen, G.L.; Hurst, G.B.; Pan, C.; Kouvelis, V.N.; Typas, M.A.; Pelletier, D.A.; et al. Improved genome annotation for *Zymomonas mobilis*. *Nat. Biotechnol.* **2009**, *27*, 893–894. [CrossRef]
22. Yang, S.; Vera, J.M.; Grass, J.; Savvakis, G.; Moskvin, O.V.; Yang, Y.; McIlwain, S.J.; Lyu, Y.; Zinonos, I.; Hebert, A.S.; et al. Complete genome sequence and the expression pattern of plasmids of the model ethanologen *Zymomonas mobilis* ZM4 and its xylose−utilizing derivatives 8b and 2032. *Biotechnol. Biofuels* **2018**, *11*, 125. [CrossRef]
23. Seo, J.S.; Chong, H.; Park, H.S.; Yoon, K.O.; Jung, C.; Kim, J.J.; Hong, J.H.; Kim, H.; Kim, J.H.; Kil, J.I.; et al. The genome sequence of the ethanologenic bacterium *Zymomonas mobilis* ZM4. *Nat. Biotechnol.* **2005**, *23*, 63–68. [CrossRef]
24. Zheng, Y.L.; Han, J.M.; Wang, B.Y.; Hu, X.Y.; Li, R.X.; Shen, W.; Ma, X.D.; Ma, L.X.; Yi, L.; Yang, S.H.; et al. Characterization and repurposing of the endogenous Type I−F CRISPR−Cas system of *Zymomonas mobilis* for genome engineering. *Nucleic Acids Res.* **2019**, *47*, 11461–11475. [CrossRef]
25. Shen, W.; Zhang, J.; Geng, B.N.; Qiu, M.Y.; Hu, M.M.; Yang, Q.; Bao, W.W.; Xiao, Y.B.; Zheng, Y.L.; Peng, W.F.; et al. Establishment and application of a CRISPR−Cas12a assisted genome−editing system in *Zymomonas mobilis*. *Microb. Cell Fact.* **2019**, *18*, 162. [CrossRef]
26. Yang, Y.F.; Shen, W.; Huang, J.; Li, R.X.; Xiao, Y.B.; Wei, H.; Chou, Y.C.; Zhang, M.; Himmel, M.E.; Chen, S.W.; et al. Prediction and characterization of promoters and ribosomal binding sites of *Zymomonas mobilis* in system biology era. *Biotechnol. Biofuels* **2019**, *12*, 52. [CrossRef]
27. Lu, S.; Wang, J.; Chitsaz, F.; Derbyshire, M.K.; Geer, R.C.; Gonzales, N.R.; Gwadz, M.; Hurwitz, D.I.; Marchler, G.H.; Song, J.S.; et al. CDD/SPARCLE: The conserved domain database in 2020. *Nucleic Acids Res.* **2020**, *48*, D265–D268. [CrossRef]
28. Chan, E.L.; Harris, R.C.; Dalton, H.P. The effect of antibiotics on the cell morphology of *Legionella pneumophila*. *J. Med. Microbiol.* **1987**, *23*, 149–154. [CrossRef]
29. Rodgers, F.G.; Tzianabos, A.O.; Elliott, T.S. The effect of antibiotics that inhibit cell−wall, protein, and DNA synthesis on the growth and morphology of *Legionella pneumophila*. *J. Med. Microbiol.* **1990**, *31*, 37–44. [CrossRef]
30. Li, R.X.; Shen, W.; Yang, Y.F.; Du, J.; Li, M.; Yang, S.H. Investigation of the impact of a broad range of temperatures on the physiological and transcriptional profiles of *Zymomonas mobilis* ZM4 for high−temperature−tolerant recombinant strain development. *Biotechnol. Biofuels* **2021**, *14*, 146. [CrossRef]

31. Mohagheghi, A.; Linger, J.G.; Yang, S.H.; Smith, H.; Dowe, N.; Zhang, M.; Pienkos, P.T. Improving a recombinant *Zymomonas mobilis* strain 8b through continuous adaptation on dilute acid pretreated corn stover hydrolysate. *Biotechnol. Biofuels* **2015**, *8*, 55. [CrossRef]
32. Lou, J.Y.; Wang, J.W.; Yang, Y.F.; Yang, Q.; Li, R.X.; Hu, M.M.; He, Q.N.; Du, J.; Wang, X.; Li, M.; et al. Development and characterization of efficient xylose utilization strains of *Zymomonas mobilis*. *Biotechnol. Biofuels* **2021**, *14*, 231. [CrossRef]
33. Fuchino, K.; Bruheim, P. Increased salt tolerance in *Zymomonas mobilis* strain generated by adaptative evolution. *Microb. Cell Fact.* **2020**, *19*, 147. [CrossRef]
34. Yang, S.; Pan, C.; Hurst, G.B.; Dice, L.; Davison, B.H.; Brown, S.D. Elucidation of *Zymomonas mobilis* physiology and stress responses by quantitative proteomics and transcriptomics. *Front. Microbiol.* **2014**, *5*, 246. [CrossRef]
35. Yang, Q.; Yang, Y.; Tang, Y.; Wang, X.; Chen, Y.; Shen, W.; Zhan, Y.; Gao, J.; Wu, B.; He, M.; et al. Development and characterization of acidic−pH−tolerant mutants of *Zymomonas mobilis* through adaptation and next−generation sequencing−based genome resequencing and RNA−Seq. *Biotechnol. Biofuels* **2020**, *13*, 144. [CrossRef]
36. He, M.X.; Wu, B.; Shui, Z.X.; Hu, Q.C.; Wang, W.G.; Tan, F.R.; Tang, X.Y.; Zhu, Q.L.; Pan, K.; Li, Q.; et al. Transcriptome profiling of *Zymomonas mobilis* under furfural stress. *Appl. Microbiol. Biotechnol.* **2012**, *95*, 189–199. [CrossRef]
37. Mack, K.; Titball, R.W. Transformation of *Burkholderia pseudomallei* by electroporation. *Anal. Biochem.* **1996**, *242*, 73–76. [CrossRef]
38. Wu, B.; He, M.; Feng, H.; Zhang, Y.; Hu, Q.; Zhang, Y. Construction and characterization of restriction−modification deficient mutants in *Zymomonas mobilis* ZM4. *Chin. J. Appplied Environ. Biol.* **2013**, *19*, 189–197. [CrossRef]
39. Kerr, A.L.; Jeon, Y.J.; Svenson, C.J.; Rogers, P.L.; Neilan, B.A. DNA restriction−modification systems in the ethanologen *Zymomonas mobilis* ZM4. *Appl. Microbiol. Biotechnol.* **2011**, *89*, 761–769. [CrossRef]
40. Lal, P.B.; Wells, F.; Myers, K.S.; Banerjee, R.; Guss, A.M.; Kiley, P.J. Improving mobilization of foreign DNA into *Zymomonas mobilis* strain ZM4 by removal of multiple restriction systems. *Appl. Environ. Microbiol.* **2021**, *87*, e0080821. [CrossRef]
41. Rueden, C.T.; Schindelin, J.; Hiner, M.C.; DeZonia, B.E.; Walter, A.E.; Arena, E.T.; Eliceiri, K.W. ImageJ2: ImageJ for the next generation of scientific image data. *BMC Bioinform.* **2017**, *18*, 529. [CrossRef]

antibiotics

Article

SspH, a Novel HATPase Family Regulator, Controls Antibiotic Biosynthesis in *Streptomyces*

Xue Yang [1,†], Yanyan Zhang [1,*,†], Shanshan Li [1], Lan Ye [1,2], Xiangjing Wang [2] and Wensheng Xiang [1,2,*]

1 State Key Laboratory for Biology of Plant Diseases and Insect Pests, Institute of Plant Protection, Chinese Academy of Agricultural Sciences, Haidian District, Beijing 100193, China; yangxue504499816@163.com (X.Y.); ssli@ippcaas.cn (S.L.); yelan_617@163.com (L.Y.)

2 School of Life Science, Northeast Agricultural University, Harbin 150030, China; wangneau2013@163.com

* Correspondence: yyzhang@ippcaas.cn (Y.Z.); xiangwensheng@neau.edu.cn (W.X.)

† These authors contributed equally to this work.

Abstract: *Streptomyces* can produce a wealth of pharmaceutically valuable antibiotics and other bioactive compounds. Production of most antibiotics is generally low due to the rigorously controlled regulatory networks, in which global/pleiotropic and cluster-situated regulatory proteins coordinate with various intra- and extracellular signals. Thus, mining new antibiotic regulatory proteins, particularly the ones that are widespread, is essential for understanding the regulation of antibiotic biosynthesis. Here, in the biopesticide milbemycin producing strain *Streptomyces bingchenggensis*, a novel global/pleiotropic regulatory protein, SspH, a single domain protein containing only the HATPase domain, was identified as being involved in controlling antibiotic biosynthesis. The *sspH* overexpression inhibited milbemycin production by repressing the expression of milbemycin biosynthetic genes. The *sspH* overexpression also differentially influenced the expression of various antibiotic biosynthetic core genes. Site-directed mutagenesis revealed that the HATPase domain was essential for SspH's function, and mutation of the conserved amino acid residues N54A and D84A led to the loss of SspH function. Moreover, cross-overexpression experiments showed that SspH and its orthologs, SCO1241 from *Streptomyces coelicolor* and SAVERM_07097 from *Streptomyces avermitilis*, shared identical functionality, and all exerted a positive effect on actinorhodin production but a negative effect on avermectin production, indicating that SspH-mediated differential control of antibiotic biosynthesis may be widespread in *Streptomyces*. This study extended our understanding of the regulatory network of antibiotic biosynthesis and provided effective targets for future antibiotic discovery and overproduction.

Keywords: global regulator; milbemycins; *Streptomyces bingchenggensis*; SspH; actinorhodin; avermectin

Citation: Yang, X.; Zhang, Y.; Li, S.; Ye, L.; Wang, X.; Xiang, W. SspH, a Novel HATPase Family Regulator, Controls Antibiotic Biosynthesis in *Streptomyces*. *Antibiotics* **2022**, *11*, 538. https://doi.org/10.3390/antibiotics11050538

Academic Editor: Jesús F. Aparicio

Received: 30 March 2022
Accepted: 15 April 2022
Published: 19 April 2022

1. Introduction

Streptomyces can produce numerous pharmaceutically valuable antibiotics, many of which have been widely used in veterinary, agricultural and medical areas. On the other hand, sequencing of *Streptomyces* genomes also revealed the existence of a large number of cryptic and uncharacterized antibiotic biosynthetic gene clusters (BGCs) other than the known antibiotic BGCs [1]. The expression of antibiotic BGCs is rigorously controlled by pyramidal regulatory cascades, in which global/pleiotropic and cluster-situated transcriptional regulatory proteins from different families coordinate with environmental and physiological cues [2–4]. Meanwhile, this rigorous control plays a key role in determining the production level of antibiotics under specific fermentation conditions. To date, many different families of regulatory proteins have been identified and their role in the biosynthesis of antibiotics have been characterized in depth, thus generating regulator-based strategies—such as activator overexpression, repressor deletion, promoter replacement, BGC refactoring and duplication—to achieve antibiotic overproduction and activation [5,6]. However, there still exist a large number of transcriptional regulatory factors with unknown functions in the genome of *Streptomyces*, and our understanding of the antibiotic production regulatory network is very poor; discovering new regulators, particularly the highly conserved ones, is necessary, and will enrich our knowledge of the antibiotic regulatory network and provide new strategies for antibiotic overproduction and activation.

In the natural world, bacteria live in diverse environments and are frequently exposed to nutrient scarcity and challenging environmental conditions. To survive in such unfavorable conditions, they have evolved complex adaptive response systems including the two-component system (TCS). *Streptomyces* possesses a large number of TCSs to cope with complicated, changing and competitive environments, and many TCSs have been shown to impact morphological development and (or) antibiotic production under laboratory culture conditions [7,8]. A typical TCS consists of a membrane-bound histidine kinase (HK) and a cytoplasmic response regulator (RR). In response to a specific signal, the HK autophosphorylates at a conserved His residue, and the phosphoryl group is subsequently transferred to an Asp residue of the RR; the RR is then activated and interacts with target promoter regions to switch on the downstream cell responses [9]. Generally, a typical HK is mainly composed of the periplasmic sensing domain, the dimerization domain (also known as the HK acceptor [HisKA] domain) and the HK-like ATPase (HATPase) domain. The HisKA domain possesses the conserved histidine residue that can be phosphorylated by the HK itself, and the HATPase domain is characterized by five conserved amino acid motifs (N, G1, F, G2 and G3 boxes) [10,11]. Due to the presence of the HATPase domain, the HK also belongs to the GHKL (Gyrase, Hsp90, HK and MutL) ATPase/Kinase superfamily, and the conserved F Motif is unique to HK, distinguishing the HK from the other members of the GHKL family. Interestingly, bioinformatics analysis of the *Streptomyces* genomes revealed the existence of a unique class of one-component signal transduction systems, which shared high similarity with the HK, but contained only the HATPase domain and did not contain the HisKA autophosphorylation site (here, these single domain proteins with only the HATPase are named HATPase-ol proteins). For example, there are at least 26 HATPase-ol proteins in the model strain *Streptomyces coelicolor*, 24 in the avermectin producing strain *Streptomyces avermitilis* and 21 in the milbemycin producing strain *Streptomyces bingchenggensis* (Supplementary Table S1), indicating that HATPase-ol proteins may be widespread in *Streptomyces*; however, little is known about their biological functions.

Polyketide milbemycins are a group of 16-membered macrolide chemicals. The main active components, milbemycin A3/A4 and the derivatives of milbemycin A3/A4 (lepimectin, latidectin and milbemycin oxime), have been commercialized as biopesticides and veterinary drugs [12]. *S. bingchenggensis* is an important industrial producer of milbemycins. Its genome contains hundreds of regulatory genes, but the functions of only a few of them are known, including the milbemycin cluster-situated regulators MilR and MilR2 [12,13], the quorum-sensing system SbbR/SbbA [14], the pleiotropic SARP family regulator KelR [15], and the four-component system SbrH1-R [16]. Thus, our knowledge of the regulatory network of milbemycin biosynthesis is still limited; mining novel transcriptional regulators affecting milbemycin production is necessary to improve the overall understanding of the milbemycin biosynthesis regulatory network and will contribute to the full exploitation of potential targets influencing milbemycin production.

In this report, we discovered a novel regulator of milbemycin biosynthesis, SBI_08867 (also named *S. bingchenggensis* small protein containing only the HATPase domain [SspH]). SspH consists of 142 amino acid residues and contains only the HATPase domain, thus it belongs to the HATPase-ol subfamily. We demonstrated that SspH may control milbemycin production by repressing the expression of milbemycin BGC. SspH also played differential roles in the control of multiple other antibiotic BGCs present in the *S. bingchenggensis* genome. SspH and its orthologs (SCO1241 and SAVERM_07097) shared the same function; both were positive regulators of actinorhodin biosynthesis in *S. coelicolor* and negative regulators of avermectin production in *S. avermitilis*. These results enriched our understanding of the antibiotic regulatory network and provided a useful method for antibiotic activation and overproduction.

2. Results

2.1. Identification of the Novel HATPase-ol Family Regulator SspH in S. bingchenggensis

In our recent work, the two hypothetical proteins (SbrH1H2) and the typical TCS (SbrKR) that together comprise the four-component system SbrH1-R (also known as SbrH1/H2/K/R) have been identified to be negative regulators of milbemycin biosynthesis. SbrH1H2 and SbrKR act synergistically to repress milbemycin biosynthesis [16]. Remarkably, of all the detected genes, the FPKM values of *sbrH1* and *sbrH2* were the highest during the entire milbemycin production period, suggesting that regulatory factors with relatively high expression may play a role in milbemycin biosynthesis. Based on this speculation, we extracted regulatory genes with FPKM values higher than 2000 from the transcriptomic data of *S. bingchenggensis* BC04 collected on day 3 (when milbemycin production begins) to mine new regulators influencing milbemycin production. A total of eleven candidate transcriptional regulatory genes were obtained (Figure 1a), of which two (*kelR* and *sbrK*) have been shown to be involved in milbemycin biosynthesis [15,16], eight (e.g., *sbi_05734*, *sbi_03959*, *sbi_05953*, *sbi_03799*, *sbi_06268*, *sbi_08867*, *sbi_05811* and *sbi_04164*) may be widespread in *Streptomyces* and even actinobacteria (KEGG homolog analysis), and one (*sbi_05271*) encodes a MarR family regulator whose homologs are only found in several *Streptomyces* species (Table 1 and Table S2). Among the eight widespread regulatory genes, orthologs of seven genes have been reported in other *Streptomyces* species to be involved in developmental differentiation and/or antibiotic biosynthesis [17–23], but nothing is known about the function of *sbi_08867* and its orthologs. The *sbi_08867* gene encodes a 15.2-kDa polypeptide containing only the HATPase domain (amino acids 39–139) (SMART No. SM000387) (Figure 1b), which is present in several functionally diverse proteins including the gyrase class of DNA topoisomerases, the heat shock protein HSP90, HK and MutL-like DNA mismatch repair proteins (collectively known as GHKL superfamily proteins) [10]. The HATPase domain contains four conserved amino acid motifs, namely, N, G1, G2 and G3 boxes, which form the nucleotide binding site [11]. As for SBI_08867 and its orthologs, the corresponding assumed conserved motifs are shown in Figure 1c. Furthermore, PSI-BLAST was used to search the nonredundant protein sequence (nr) database for SBI_08867 homologs (query coverage 90% to 100% and percent identity 60% to 100%). A total of 1395 hits were obtained, and almost 99% of the hits were from *Streptomyces*, indicating that SBI_08867 and its homologs were mainly present in *Streptomyces* species (Table S3). Thus, determining the roles of SBI_08867 (hereafter named SspH) and its homologs is necessary, strengthening our knowledge about *Streptomyces* genetics.

2.2. SspH Negatively Influences Milbemycin Production

To determine whether SspH affected the biosynthesis of milbemycins, an *sspH* overexpression strain (BC04/OsspH) and repression strain (BC04/RsspH) were constructed using pSET152::P_{hrdB} (with this plasmid, target gene overexpression is controlled by the *hrdB* promoter) and the ddCpf1-based CRISPRi system pSETddCpf1, respectively (Figure 2a). Moreover, pSET152::P_{hrdB} and pSETddCpf1 were also introduced into BC04 to obtain control strains BC04-C1 and BC04-C2, respectively. Then the four strains and the parental strain BC04 were cultured and tested for milbemycin A3/A4 production. Compared with BC04 and the control strains (BC04-C1 and BC04-C2), BC04/OsspH showed an obvious decrease in milbemycin production while BC04/RsspH showed a slight increase in milbemycin titer (Figure 2b), indicating that SspH played a negative role in milbemycin production.

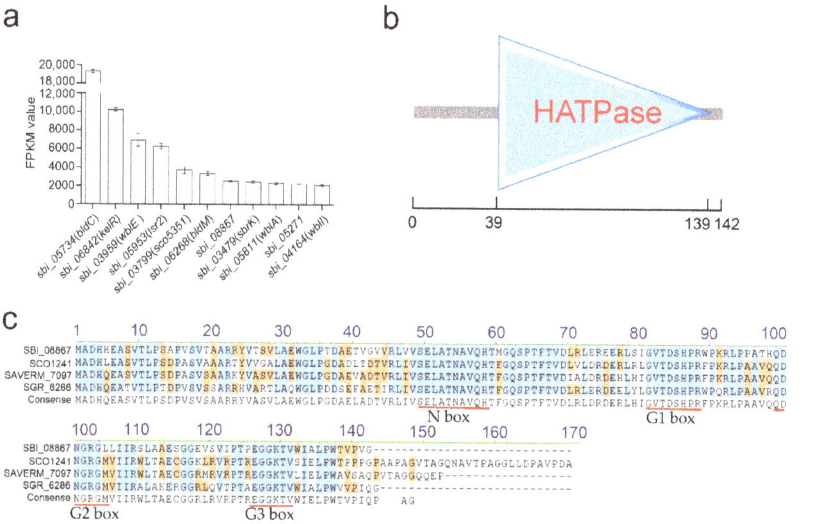

Figure 1. Identification of SBI_08867 in *S. bingchenggensis*. (**a**) Transcript abundances of genes with FPKM values higher than 2000 measured by RNA-seq on day 3. (**b**) Predicted domain structure of SBI_08867. HATPase, histidine kinase-like ATPase. (**c**) Amino acid sequence alignment of SBI_08867 and its orthologs. Identical or similar residues in all sequences are highlighted in blue and orange, respectively. The corresponding assumed conserved motifs are marked in red. SCO1241, SBI_08867 ortholog from *S. coelicolor*; SAVERM_7097, SBI_08867 ortholog from *S. avermitilis*; SGR_6286, SBI_08867 ortholog from *S. griseus*.

Table 1. 11 candidate transcriptional regulatory genes with high FPKM values (FPKM > 2000) on day 3.

Gene ID	Type of Product	Reported Homolog	Predicted Function	Reference(s)
sbi_05734	MerR	BldC (SCO4091)	Widespread; a repressor to maintain vegetative growth and to delay entry into development.	[17]
sbi_06842	SARP	-	A cluster-situated activator of type II PKS cluster, also essential for milbemycin production.	[15]
sbi_03959	Wbl	WblE (SAVERM_3016)	Widespread; an important regulator of morphological differentiation.	[18]
sbi_05953	Lsr2	Lsr2 (SVEN_3225)	Widespread; a nucleoid-associated protein that influences chromosome segregation, DNA replication, transcription and secondary metabolism.	[19]
sbi_03799	TCS (RR)	SCO5351	Widespread; a pleiotropic regulator involved in secondary metabolism and development.	[20]
sbi_06268	TCS (RR)	BldM (SCO4768)	Widespread; an essential regulator for aerial hyphae formation.	[21]
sbi_08867	HATPase-ol	-	Widespread; function unknown.	This work
sbi_03479	TCS (HK)	SbrK	A repressor of milbemycin biosynthesis.	[16]
sbi_05811	Wbl	WblA (SCO3579)	Widespread; a global regulator involved in differentiation and secondary metabolism.	[22]
sbi_05271	MarR	-	Function unknown.	-
sbi_04164	Wbl	WblI (SCO5046)	Widespread; a positive impact on antibiotic production.	[23]

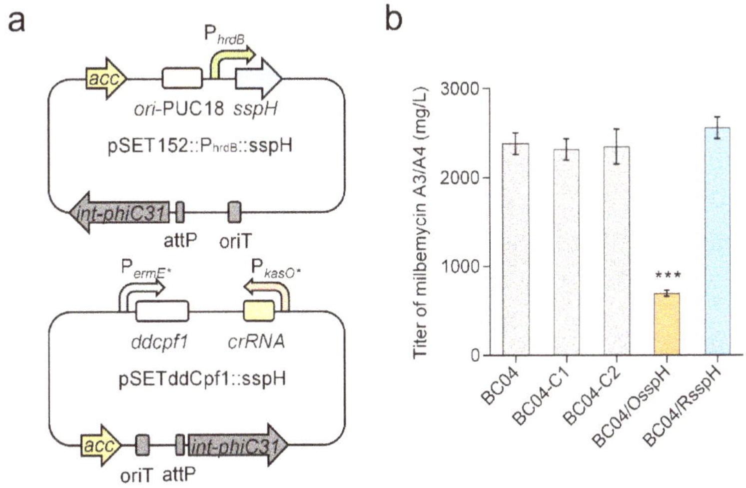

Figure 2. Effects of *sspH* overexpression and repression on milbemycin production. (**a**) Schematic diagrams of *sspH* overexpression and repression plasmids. The *hrdB* promoter P_{hrdB} was used to drive the expression of *sspH*. Two strong promoters, P_{ermE^*} and P_{kasO^*}, were used to drive the expression of *ddcpf1* and *crRNA*, respectively. (**b**) Titer of milbemycin A3/A4 in strains BC04, BC04-C1, BC04-C2, BC04/OsspH and BC04/RsspH. The milbemycin titer was determined after fermentation for 9 days. BC04, the parental strain; BC04-C1, BC04 with pSET152::P_{hrdB}; BC04-C2, BC04 with pSETddCpf1; BC04/OsspH, *sspH* overexpression strain; BC04/RsspH, *sspH* repression strain. Data are presented as the averages of three independent experiments conducted in triplicate. Error bars show standard deviations from three replicates. *p*-values were determined by Student's *t*-test. *** $p < 0.001$.

2.3. SspH Represses the Transcription of milR, milF and milA1

To determine the effects of SspH on the expression of milbemycin BGC genes, qRT-PCR analysis was carried out to assess the transcriptional levels of *milR* (the milbemycin cluster-situated activator gene) [12], *milF* (the tailoring enzyme gene controlled by MilR) [24] and *milA1* (type I polyketide synthase gene) [25] (Table S4). Total RNAs were extracted from the mycelia of BC04, BC04/OsspH and BC04/RsspH cultivated for various days (2, 3, 4, 6 and 8 days). Transcription of *sspH* in BC04 peaked on day 3 (when milbemycin production began) and decreased (but remained relatively high) thereafter. This transcriptional profile was similar to that of *milR*. As expected, the transcript level of *sspH* was significantly higher in BC04/OsspH than in BC04 at most time points, suggesting the successful overexpression of *sspH* (Figure 3). In contrast, transcription of *milR* was obviously lower in BC04/OsspH compared with BC04 at most time points. Similar results were also observed for *milF* and *milA1* (Figure 3), suggesting that SspH may control milbemycin production by repressing milbemycin biosynthetic genes. In BC04/RsspH, the expression level of *sspH* decreased obviously and was only 30–40% of that in BC04, indicating a partial inhibition by the CRISPRi system. However, no significant change was observed for *milR* and *milA1*, and only *milF* showed a little increase in BC04/RsspH at the middle and late fermentation stages (6 to 8 days) (Figure 3). Perhaps it is not surprising that the titer of milbemycins and the expression of milbemycin BGC showed no significant change, because although the transcript level of *sspH* was significantly downregulated, the resultant transcript level was still very high and could effectively inhibit the production of milbemycins.

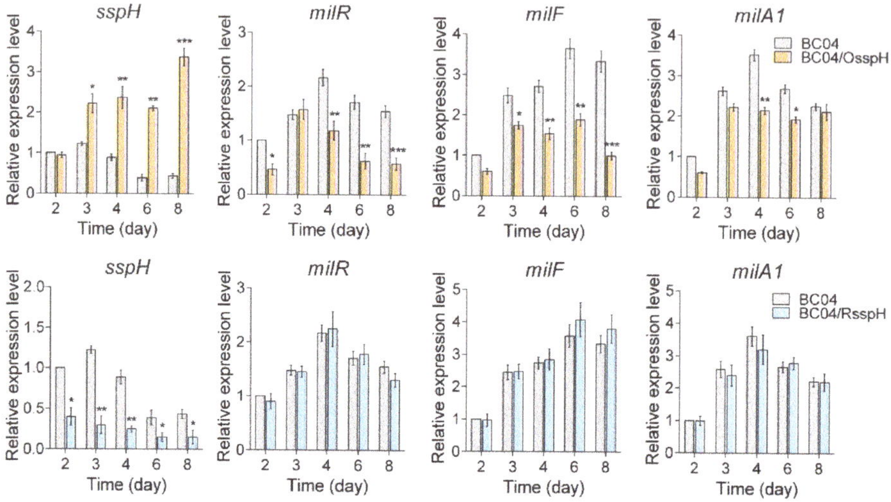

Figure 3. The qRT-PCR analysis of *sspH*, *milR*, *milF* and *milA1* in BC04, BC04/OsspH and BC04/RsspH. The transcriptional levels of related genes are presented relative to that of BC04 on day 2. 16S rRNA was used as the internal control. Data are presented as the averages of three triplicates. Error bars show standard deviations. * $p < 0.05$, ** $p < 0.01$, *** $p < 0.001$.

2.4. The sspH Overexpression in BC04 Differentially Affects Expression of Multiple PKS/NRPS/PKS-NRPS Genes

PKS, NRPS and PKS-NRPS genes are the three main types of antibiotic BGCs within the genome of *Streptomyces*. In *S. bingchenggensis*, there are about 31 such gene clusters. To determine whether *sspH* overexpression affected the expression of PKS/NRPS/PKS-NRPS genes, the transcriptional levels of 31 biosynthetic core genes from these PKS/NRPS/PKS-NRPS clusters were detected by qRT-PCR in strains BC04 and BC04/OsspH (Table S5). Transcription of most biosynthetic core genes was downregulated, and only one gene (*sbi_09249*) showed about 2-fold upregulation in BC04/OsspH on day 3 (Figure 4a). In BC04/OsspH on day 6, eight genes (*sbi_00522*, *sbi_00625*, *sbi_01029*, *sbi_02988*, *sbi_06052*, *sbi_09195*, *sbi_09057* and *sbi_09652*) showed 2- to 6-fold upregulation (Figure 4b), while the other genes exhibited similar or lower expression levels than in BC04. These results suggested that SspH may be a pleiotropic regulator differentially influencing the expression of multiple antibiotic BGCs, and *sspH* overexpression was effective in upregulating some cryptic and uncharacterized antibiotic BGCs.

2.5. The Conserved HATPase Domain Is Essential for SspH Function

SspH is a single domain protein containing only the HATPase domain. To verify the importance of the HATPase domain, we intended to perform site-directed mutation of certain conserved amino acid residues. Among the four conserved binding motifs, the Asn residue (essential for chelating Mg^{2+}) from the N box and the Asp residue (which interacts directly with ATP and forms a hydrogen bond with an adenine base) from the G1 box are absolutely conserved in HATPase domain containing proteins [10,26], so we decided to perform point mutations of these two key residues. First we identified the two conserved amino acids from the SspH amino acid sequence. As for SspH, the conserved Asn residue of the N box should be Asn54. This is easy to determine because SspH's N box shares the core sequence "NA" with the N box of the HK HATPase domain. However, the G1 box of the HATPase domain of SspH seemed to have a unique characteristic that distinguished it from the HK and most other members of the GHKL family. Secondary and three-dimensional (3D) structure analysis of SspH showed that the Asp84 residue at the C-terminal end of sheet β3 might be the conserved Asp residue of the G1 motif that could directly interact with ATP, but its downstream flanking sequence (the loop region between sheet β3 and helix α3) did not contain Gly (Figure S1), which was different from the G1 motif core sequence (core sequence: DxGxG) of most GHKL family members; instead, a Gly81 upstream of Asp84 was found within the β3 sheet (Figure S1). We speculate that this arrangement of Gly81 and Asp84 might be a unique feature of the G1 motif of SspH-like proteins. To determine whether this arrangement of Asp and Gly in the G1 motif was conserved among SspH-like proteins, the G1 motifs of nine orthologs of SspH were analyzed. As expected, the relative positions of the conserved Asp and Gly in the G1 motifs of these proteins were the same as in the G1 motif of SspH (Figure S2).

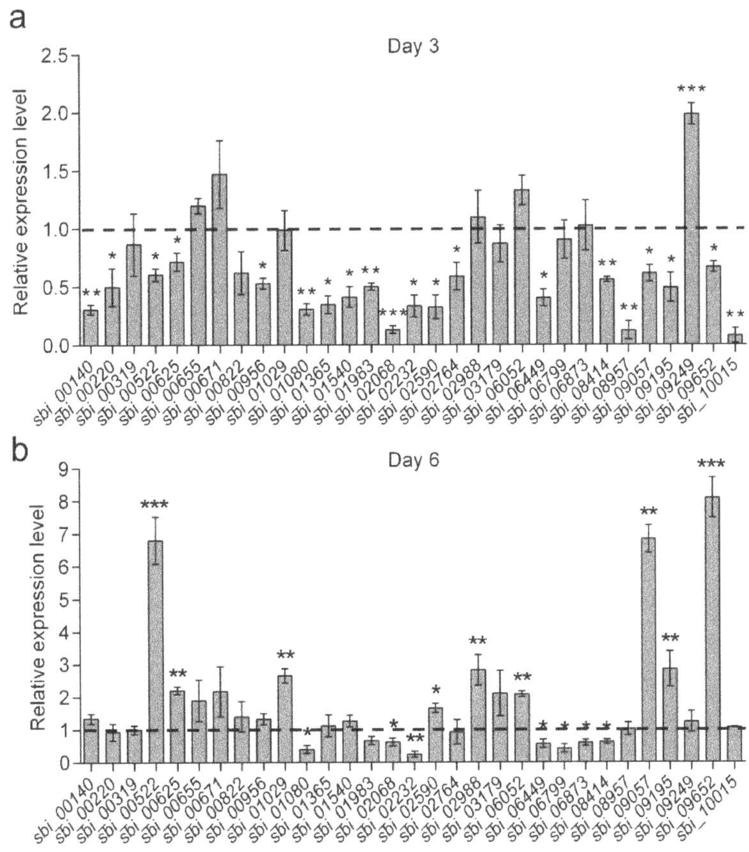

Figure 4. The qRT-PCR analysis of 31 PKS, NRPS and PKS-NRPS biosynthetic core genes in BC04 and BC04/OsspH (**a**) on day 3 and (**b**) on day 6. RNA samples were isolated from 3- and 6-day cultures. The expression level of each biosynthetic core gene in BC04 was assigned a value of 1 (represented by the dotted line). 16S rRNA was used as the internal control. Data are presented as the averages of three triplicates. Error bars show standard deviations. * $p < 0.05$, ** $p < 0.01$, *** $p < 0.001$.

Next, after the conserved Asn54 and Asp84 in SspH were identified, the two sites were mutated to alanine to verify their importance (Figure 5a). Two mutagenized *sspH* plasmids pSET152::N54A and pSET152::D84A were constructed and further introduced into BC04 to generate strains BC04/N54A and BC04/D84A. The two strains, together with BC04 and BC04/OsspH, were compared for milbemycin production. The results showed that the titers of milbemycin A3/A4 in the two *sspH* mutant overexpression strains were similar to that in BC04 (Figure 5b). To determine the influence of the mutant *sspHs* on the expression of the milbemycin BGC, qRT-PCR was performed again to detect expression changes of *milR*, *milF* and *milA1*. As expected, the expression levels of the three genes in BC04/N54A and BC04/D84A were comparable to those in BC04 (Figure 5c). These results strongly suggested that Asn54 and Asp84 were the key active residues and that the HATPase domain was essential for the function of SspH.

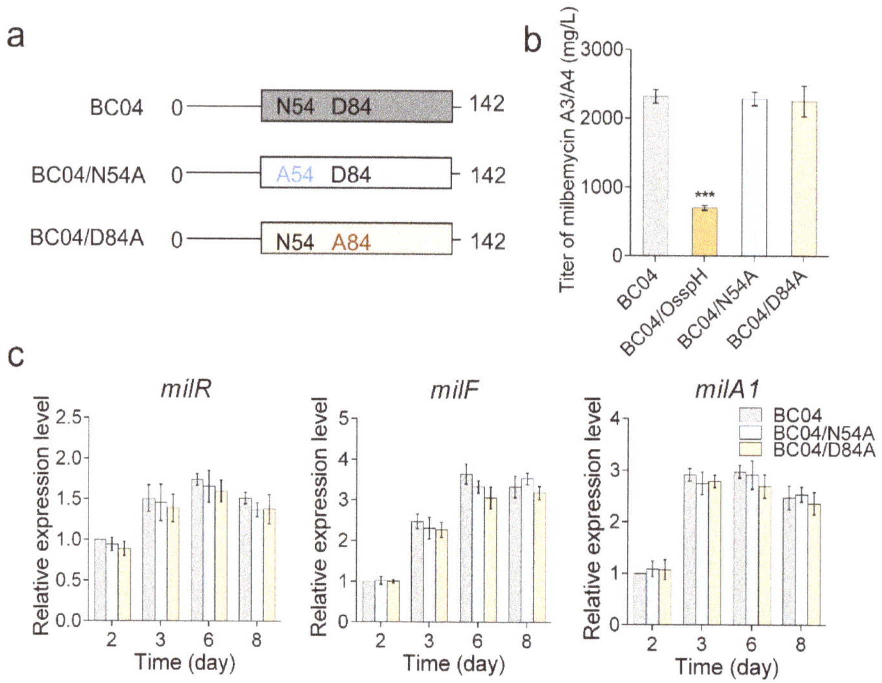

Figure 5. Effects of site-directed mutagenesis in the HATPase domain of SspH on its function. (**a**) Diagrams of site-directed mutation in the HATPase domain. A54, Asn54 was mutated to Ala; A84, Asp84 was changed to Ala. (**b**) Titers of milbemycin A3/A4 in strains BC04, BC04/OsspH, BC04/N54A and BC04/D84A. (**c**) qRT-PCR analysis of the transcriptional levels of *milR*, *milF* and *milA1* in BC04, BC04/N54A and BC04/D84A. The expression of each gene in BC04 on day 2 was assigned a value of 1. The *p*-values were determined by Student's *t*-test. *** $p < 0.001$.

2.6. SspH Is Commonly Involved in the Control of Antibiotic Production in *Streptomyces*

As mentioned above, SspH homologs are widely distributed in *Streptomyces*. To assess whether SspH and its homologs were involved in modulating antibiotic production in other *Streptomyces* species, gene overexpression tests were also performed in the model strain *S. coelicolor* M145 and in the avermectin industrial producer *S. avermitilis* NEAU12 (Figure 6a). First, we introduced the *sspH* overexpression plasmid pSET152::P_{hrdB}::sspH into M145 and NEAU12 to obtain strains M145/OsspH and NEAU12/OsspH, respectively. After flask fermentation, M145/OsspH exhibited a marked increase in actinorhodin production compared with the parental strain M145, while NEAU12/OsspH greatly reduced avermectin B1a production compared with NEAU12, indicating that SspH is a positive regulator of actinorhodin production in *S. coelicolor* but a repressor of avermectin biosynthesis in *S. avermitilis*. Second, genes encoding the orthologs of SspH, *sco1241* (protein product shares 72.5% sequence identity with SspH) from *S. coelicolor* and *saverm_07097* (protein product shares 74.1% sequence identity with SspH) from *S. avermitilis*, were both cloned into the integrative plasmid pSET152::P_{hrdB}, generating the overexpression plasmids pSET152::P_{hrdB}::1241 and pSET152::P_{hrdB}::7097. The two plasmids were then introduced into *S. bingchenggensis* BC04, *S. coelicolor* M145 and *S. avermitilis* NEAU12 to obtain another six overexpression strains BC04/O1241, BC04/O7097, M145/O1241, M145/O7097, NEAU12/O1241 and NEAU12/O7097 (Figure 6a). As expected, overexpression of *sco1241* or *saverm_07097* in BC04 resulted in a significant decrease of milbemycin production; overexpression of *sco1241* or *saverm_07097* in M145 increased the titer of actinorhodin; and overexpression of *sco1241* or *saverm_07097* in NEAU12 led to marked reduction of avermectin production (Figure 6b). These results suggested that SspH and its orthologs were important regulators of antibiotic biosynthesis, sharing the same function; their effects on antibiotic production may be different from species to species.

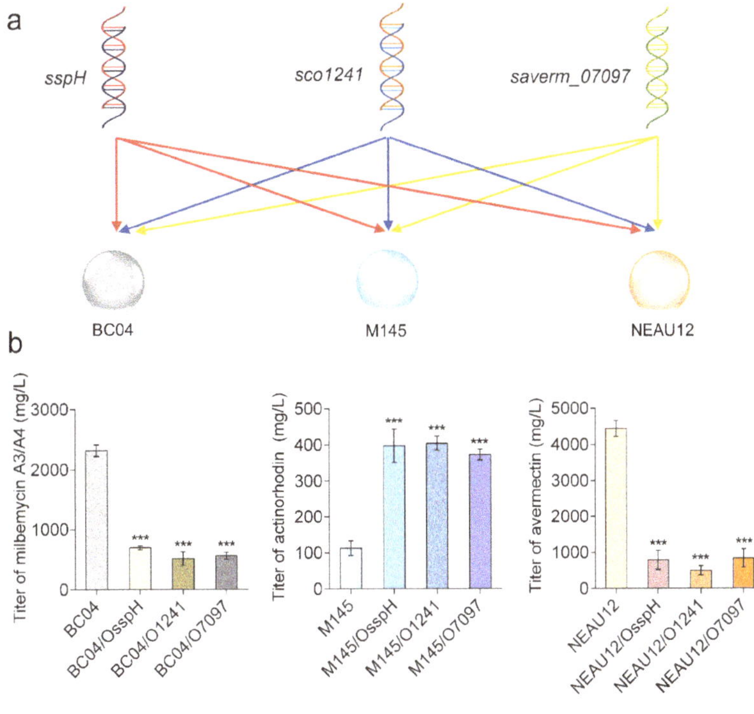

Figure 6. SspH and its orthologs are involved in modulating antibiotic production in three *Streptomyces* species. (**a**) Schematic diagrams of cross-overexpression of *sspH*, *sco1241* and *saverm_07097* in BC04, M145 and NEAU12. (**b**) Effects of overexpression of *sspH*, *sco1241* and *saverm_07097* on the titers of milbemycin A3/A4, actinorhodin and avermectin B1a in BC04, M145 and NEAU12. *p*-values were determined by Student's *t*-test. *** $p < 0.001$.

3. Discussion

In recent years, with the growing emergence of drug-resistant pathogens, the occurrence of new diseases, and the demands for energy conservation, pollution reduction and high yield, the creation of new drugs and green biomanufacturing have become extremely urgent. *Streptomyces* is one of the prominent sources of natural antibiotics for drug discovery and development [27]. However, the production of most antibiotics under laboratory culture conditions is usually very low or non-existent due to tightly controlled regulatory networks, in which both global/pleiotropic and cluster-situated regulatory proteins are involved. Thus, mining new regulators and understanding their functions and molecular mechanisms in antibiotic biosynthesis is of great importance for developing regulation-based methods to increase or activate antibiotic production. In this study, a novel and highly conserved HATPase domain containing regulator, SspH, was identified to be involved in antibiotic biosynthesis in *Streptomyces*. The *sspH* overexpression could not only significantly repress milbemycin production, but also differentially influence the expression of many other potential antibiotic biosynthetic core genes. Moreover, SspH and its orthologs could also promote the high yield of actinorhodin in *S. coelicolor* and inhibit avermectin production in *S. avermitilis*, indicating that SspH and its orthologs play a universal role in controlling antibiotic production in *Streptomyces*.

Many reports have shown that global/pleiotropic regulatory protein-mediated differential control of antibiotic biosynthesis was widespread in *Streptomyces*. We took the well-studied global/pleiotropic regulators AdpA and AtrA and the quorum-sensing system as the examples. In most *Streptomyces* species, the absence of *adpA* leads to the loss of antibiotic biosynthesis, but in *Streptomyces ansochromogenes*, although *adpA* disruption led to the failure of nikkomycin production, a cryptic oviedomycin BGC was activated, leading to the detectable production of oviedomycin [28]. The regulator AtrA and its ortholog AveI have the identical functionality; they play a positive role in actinorhodin production in *S. coelicolor* and a negative role in avermectin production in *S. avermitilis*, reflecting species-differential regulation by the same regulator [29]. The quorum-sensing signal receptor ArpA mostly acts as a repressor of antibiotic production; however, some ArpA homologs also exhibit positive effects on antibiotic biosynthesis, such as *S. bingchenggensis* SbbR (positive effect on milbemycin production) [14] and *S. venezuelae* JadR3 (positive effect on jadomycin production) [30]. Therefore, the manipulation of global/pleiotropic regulatory genes has become one of the major strategies for antibiotic overproduction and activation. Here, we identified a novel and highly conserved HATPase-ol family regulator, SspH. In *S. bingchenggensis*, SspH could not only inhibit the production of milbemycins, but also promote the expression of several cryptic biosynthetic core genes. SspH and its orthologs (SCO1241 and SAVERM_07097) shared the same function; both could promote the production of actinorhodin in *S. coelicolor* but inhibited the production of avermectin in *S. avermitilis*. Thus, genetic manipulation of *sspH* and its orthologs could be applied as an effective method for future antibiotic discovery and overproduction in *Streptomyces*.

SspH has been verified to be an important regulator of antibiotic production. However, the molecular mechanism for SspH-mediated control of antibiotic biosynthesis is still not known. Luckily, detailed studies of RsbW in *Bacillus cereus* and SpoIIAB of *Bacillus subtilis* may provide some clues. Similar to SspH, RsbW and SpoIIAB are small proteins with only an HATPase domain. RsbW, an anti-σ factor possessing kinase activity, is involved in the σ^B-mediated stress response in *Bacillus cereus*. When there was no stress, RsbW formed complexes with σ^B to make σ^B unable to interact with RNA polymerase; meanwhile, RsbW could phosphorylate a ser residue of RsbV (an anti-σ factor antagonist of RsbW), making RsbV unable to bind RsbW. Under stress, RsbV is dephosphorylated by the phosphatase so that it can form complexes with RsbW, leading to the release of σ^B [31]. Like RsbW, SpoIIAB from *Bacillus subtilis* is also an anti-σ factor that possesses kinase activity. SpoIIAB negatively modulates the activity of the sporulation factor σ^F and can phosphorylate the anti-anti-σ factor SpoIIAA. The molecular mechanism of SpoIIAB–SpoIIAA modulating the activity of σ^F is basically similar to that of RsbW–RsbV in the control of σ^B activity [32]. Here, although SspH displayed very low overall sequence similarity with RsbW and SpoIIAB, the 3D structure of SspH is highly similar to those of RsbW and SpoIIAB, thus SspH may also function by interacting with unknown proteins and by phosphorylation of certain proteins. The search for putative proteins interacting with SspH is currently under way in our laboratory, which may provide clues to improve our understanding of the HATPase-ol protein regulatory cascade in *Streptomyces*.

4. Materials and Methods

4.1. Strains, Plasmids and Culture Conditions

All strains used in this work were listed in Table S6. *S. bingchenggensis* BC04 is an industrial milbemycin producer generated from *S. bingchenggensis* CGMCC 1734 after random mutagenesis (GenBank Accession No. CP002047.1) [12]. *S. coelicolor* M145 is a prototrophic derivative of strain *S. coelicolor* A3(2) lacking the endogenous plasmid SCP1 and SCP2 (GenBank Accession No. AL645882.2) [33]. *S. avermitilis* NEAU12 is an avermectin B1a producing strain, and its genome shares high similarity with the genome sequence of *S. avermitilis* MA-4680 (GenBank Accession No. NC_003155.5) [34]. All *E. coli* strains were grown on Luria–Bertani (LB) medium supplemented with antibiotics as required at 37 °C [35]. For conjugation, *E. coli* ET12567 (pUZ8002) was used for transferring plasmids from *E. coli* to *Streptomyces*, and *Streptomyces* strains were grown on MS agar at 28 °C [36]. For spore collection, *S. bingchenggensis*, *S. coelicolor*, *S. avermitilis* and their derivatives were grown at 28 °C on SKYM agar plates [12] and MS agar plates [36,37], respectively. Flask fermentation for the production of milbemycins, actinorhodin and avermectins were the same as previously reported [12,36,37].

All plasmids and primers used in this work were listed in Tables S7 and S8, respectively. A quantity of pSET152, which can integrate into the *Streptomyces* chromosome by site-specific recombination at the phage ΦC31, was used to create recombinant plasmids for overexpressed mutant strains [38]. Meanwhile, pSETddCpf1 was used to construct plasmids for CRISPRi gene suppression mutant strains [39]. Additionally, pBluescript KS (+) was used for site-directed mutagenesis as previously reported [12].

4.2. Gene Overexpression and Repression

For gene overexpression, three fragments containing the coding region of *sspH*, *sco1241* and *saverm_7097* were individually amplified from the genomic DNAs of *S. bingchenggensis* BC04, *S. coelicolor* M145 and *S. avermitilis* NEAU12 using primers SspH-F/R, sco1241-F/R and saverm_7097-F/R. The three fragments were then cloned into the EcoRI/XbaI sites of pSET152::P_{hrdB} to generate gene overexpression plasmids pSET152::P_{hrdB}::sspH, pSET152::P_{hrdB}::1241 and pSET152::P_{hrdB}::7097, respectively. The three plasmids were separately introduced into three strains including *S. bingchenggensis* BC04, *S. coelicolor* M145 and *S. avermitilis* NEAU12, and nine overexpression strains BC04/OsspH, BC04/O1241, BC04/O7097, M145/OsspH, M145/O1241, M145/O7097, NEAU12/OsspH, NEAU12/O1241 and NEAU12/O7097 were obtained (Table S7).

The *sspH* repression strain was constructed based on pSETddCpf1 [39]. Meanwhile, *sspH* specific crRNA expression cassette was amplified using pSETddCpf1 as the template, and ddCpf1-sspH-F and the ddCpf1-R (crRNA-rev) as the primer pair. Additionally, *sspH* specific crRNA cassette was digested with NdeI/SpeI and ligated into the NdeI/SpeI sites of pSETddCpf1, generating *sspH* repression plasmid pSETddCpf1::sspH. Next, pSETdd-Cpf1::sspH was introduced into *S. bingchenggensis* BC04 to obtain the repression strain BC04/RsspH.

4.3. Site-Directed Mutagenesis of sspHpH

In order to construct the *sspH* mutant plasmids, the fragment containing the constitutive promoter P_{hrdB} and *sspH* coding region was amplified using pSET152::P_{hrdB}::sspH as the template. This fragment was then cloned into the EcoRI/BamHI sites of pBluescriptKS (+) to obtain pBluescriptKS (+)::P_{hrdB}::sspH, which was used as the template to conduct subsequent site-directed mutation experiments. The primers for site-directed mutagenesis were phosphorylated at the 5′ end with T4 polynucleotide kinase, respectively, and were further used for amplification of mutant *sspH*. The PCR products were purified by gel extraction and ligated by self-ligation to generate mutagenized *sspH* plasmids. The SspH mutants were N54A (AAC to GCC) and D84A (GAC to GCC). Finally, the DNA fragments containing mutant *sspH* were cut from pBluescriptKS (+) and ligated into the EcoRI/BamHI sites of pSET152, generating pSET152::P_{hrdB}::N54A and pSET152::P_{hrdB}::D84A, respectively. Subsequently, the two plasmids were integrated into BC04 to obtain the *sspH* mutant overexpression strains BC04/N54A and BC04/D84A.

4.4. Detection Methods of Milbemycins, Actinorhodin and Avermectins

For milbemycin A3/A4 and avermectin B1a analysis, 0.5 mL fermented cell cultures of *S. bingchenggensis* was mixed with three volumes of ethanol to extract milbemycin A3/A4, and 0.5 mL fermentation cultures of *S. avermitilis* was mixed with three volumes of methanol. HPLC analysis was carried out by Agilent 1260 II system with an SB-C18 column (Zorbax, 4.6 mm × 260 mm, 5 mm) at a flow rate of 1.0 mL/min. For milbemycin A3/A4, a linear gradient of solvent B was applied from 0 to 100% in 15 min (Solvent A: CH_3CN:H_2O:CH_3OH = 350:50:100, $v/v/v$; Solvent B: CH_3OH), and samples were detected at 242 nm. For avermectin B1a, samples were detected at 246 nm with 90% methyl alcohol. For the detection of actinorhodin, 0.5 mL of fermented cell cultures was treated with 0.5 mL NaOH (1 M NaOH), centrifuged, and the OD_{608nm} of the supernatant was determined [40].

4.5. RNA Isolation and Quantitative Real-Time PCR (qRT-PCR) Assay

RNAs were isolated from the fermentation cultures of *S. bingchenggensis* grown at 28 °C at different time points (2, 3, 4, 6 and 8 days). The detailed steps for RNA extraction were described as previously [12]. Each RNA sample was treated with RQ1 RNase-free DNase I (Promega) to exclude the possibility of genomic DNA contamination. UV spectroscopy and agarose gel electrophoresis were used to detect the RNA quality and quantity. For qRT-PCR, 1 μg of total RNAs was used for first-strand cDNA synthesis, and primers were designed to amplify fragments of 100–200 bp (Table S8). The PCR procedures were the same as previously reported [12]. After PCR amplification, the data were analyzed using LightCycler®-96 Series Software (v1.4.1, Roche Diagnostics, Rotkreuz, Switzerland) and the $2^{-\Delta\Delta CT}$ Method [41].

4.6. Sequence Analysis

Protein sequence alignment and domain architectures were analyzed by publicly available databases and their application tools, including BLAST (http://www.blast.ncbi.nlm.nih.gov/Blast.cgi, accessed on 18 June 2021) and SMART (http://www.smart.embl-heidelberg.de/, accessed on 18 June 2021), KEGG (http://www.genome.jp/kegg/, accessed on 18 June 2021).

4.7. Statistical Analysis

All experiments were run in three biological triplicates independently. Data were presented as averages of three triplicates. Significance was analyzed by Student's t-test (GraphPad Prism 6), and the significance were presented as follows, * $p < 0.05$, ** $p < 0.01$, *** $p < 0.001$, and no marked means no significant.

4.8. Data Availability

The transcriptomic data of *S. bingchenggensis* BC04 collected on day 3 can be found in GEO database (GSE147644).

5. Conclusions

In this work, we identified a new and highly conserved regulator of antibiotic biosynthesis, SspH, in *Streptomyces*. SspH is a small and single domain protein with only the HATPase domain. Overexpression of *sspH* repressed the transcription of milbemycin biosynthetic genes and consequently decreased milbemycin production. The *sspH* overexpression also differentially influenced the expression of many other antibiotic biosynthetic core genes present in the genome of *S. bingchenggensis*. Moreover, SspH and its orthologs from *S. coelicolor* and *S. avermitilis* shared the same function, and all could promote the production of actinorhodin and inhibit the production of avermectin. These results extended our understanding of the regulatory network of antibiotic biosynthesis, provided effective targets for antibiotic discovery and overproduction, and also provided a useful method to mine new antibiotic regulators.

Supplementary Materials: The following supporting information can be downloaded at: https://www.mdpi.com/article/10.3390/antibiotics11050538/s1, Figure S1: Secondary and three-dimensional (3D) structure of SspH, Figure S2: Amino acid sequence alignment of SspH and its orthologs, Table S1: Genes whose products contain only the HATPase domain in several *Streptomyces* species, Table S2: Amino acid sequences alignment of homologous proteins from *S. bingchenggensis* and other *Streptomyces* species, Table S3: Number of hits of SspH homologous sequence in different bacteria after PSI-BLAST search, Table S4: Three milbemycin biosynthetic genes (*milR, milF* and *milA1*) and their functions, Table S5: 31 antibiotic biosynthetic core genes and their putative functions, Table S6: Strains used in this work, Table S7: Plasmids used in this work, Table S8: Primers used in this work.

Author Contributions: X.Y. performed experiments, bioinformatics analysis and wrote part of the manuscript; Y.Z. performed bioinformatics analysis, conceived and supervised the whole research work, wrote and revised the manuscript; S.L. revised the manuscript; L.Y. assisted with revised the figures; X.W. revised the manuscript; W.X. supervised the whole research work and also revised the manuscript. All authors have read and agreed to the published version of the manuscript.

Funding: This work received financial support from the National Natural Science Foundation of China (31872936, 31972291 and 31972348).

Institutional Review Board Statement: Not applicable.

Informed Consent Statement: Not applicable.

Data Availability Statement: The datasets supporting the conclusions of this article are included within the article and the Supplementary Materials.

Acknowledgments: We would like to thank Weihong Jiang (Institute of Plant Physiology and Ecology, Chinese Academy of Sciences, Shanghai, China) and Yinhua Lu (Shanghai Normal University, Shanghai, China) for providing pSETddCpf1.

Conflicts of Interest: The authors declare that they have no competing interest.

References

1. Nett, M.; Ikeda, H.; Moore, B.S. Genomic basis for natural product biosynthetic diversity in the actinomycetes. *Nat. Prod. Rep.* **2009**, *26*, 1362–1384. [CrossRef] [PubMed]
2. Zhou, Q.; Ning, S.; Luo, Y. Coordinated regulation for nature products discovery and overproduction in *Streptomyces*. *Synth. Syst. Biotechnol.* **2020**, *5*, 49–58. [CrossRef] [PubMed]
3. Xia, H.; Li, X.; Li, Z.; Zhan, X.; Mao, X.; Li, Y. The Application of Regulatory Cascades in *Streptomyces*: Yield Enhancement and Metabolite Mining. *Front. Microbiol.* **2020**, *11*, 406. [CrossRef] [PubMed]
4. López-Agudelo, V.A.; Gómez-Ríos, D.; Ramirez-Malule, H. Clavulanic Acid Production by *Streptomyces clavuligerus*: Insights from Systems Biology, Strain Engineering, and Downstream Processing. *Antibiotics* **2021**, *10*, 84. [CrossRef] [PubMed]
5. Bu, Q.; Li, Y.; Xie, H.; Li, J.; Lv, Z.; Su, Y.; Li, Y. Rational engineering strategies for achieving high-yield, high-quality and high-stability of natural product production in actinomycetes. *Metab. Eng.* **2021**, *67*, 198–215. [CrossRef]
6. Liu, N.; Guan, H.; Niu, G.; Jiang, L.; Li, Y.; Zhang, J.; Li, J.; Tan, H. Molecular mechanism of mureidomycin biosynthesis activated by introduction of an exogenous regulatory gene ssaA into *Streptomyces roseosporus*. *Sci. China Life Sci.* **2021**, *64*, 1949–1963. [CrossRef]
7. Liu, G.; Chater, K.F.; Chandra, G.; Niu, G.; Tan, H. Molecular regulation of antibiotic biosynthesis in *Streptomyces*. *Microbiol. Mol. Biol. Rev.* **2013**, *77*, 112–143. [CrossRef]
8. Yu, Z.; Zhu, H.; Dang, F.; Zhang, W.; Qin, Z.; Yang, S.; Tan, H.; Lu, Y.; Jiang, W. Differential regulation of antibiotic biosynthesis by DraR-K, a novel two-component system in *Streptomyces coelicolor*. *Mol. Microbiol.* **2012**, *85*, 535–556. [CrossRef]
9. Wang, L.; Tian, X.; Wang, J.; Yang, H.; Fan, K.; Xu, G.; Yang, K.; Tan, H. Autoregulation of antibiotic biosynthesis by binding of the end product to an atypical response regulator. *Proc. Natl. Acad. Sci. USA* **2009**, *106*, 8617–8622. [CrossRef]
10. Dutta, R.; Inouye, M. GHKL, an emergent ATPase/kinase superfamily. *Trends Biochem. Sci.* **2000**, *25*, 24–28. [CrossRef]
11. Robinson, V.L.; Buckler, D.R.; Stock, A.M. A tale of two components: A novel kinase and a regulatory switch. *Nat. Struct. Biol.* **2000**, *7*, 626–633. [CrossRef] [PubMed]
12. Zhang, Y.; He, H.; Liu, H.; Wang, H.; Wang, X.; Xiang, W. Characterization of a pathway-specific activator of milbemycin biosynthesis and improved milbemycin production by its overexpression in *Streptomyces bingchenggensis*. *Microb. Cell Fact.* **2016**, *15*, 152. [CrossRef]
13. Wei, K.; Wu, Y.; Li, L.; Jiang, W.; Hu, J.; Lu, Y.; Chen, S. MilR2, a novel TetR family regulator involved in 5-oxomilbemycin A3/A4 biosynthesis in *Streptomyces hygroscopicus*. *Appl. Microbiol. Biotechnol.* **2018**, *102*, 8841–8853. [CrossRef] [PubMed]
14. He, H.; Ye, L.; Li, C.; Wang, H.; Guo, X.; Wang, X.; Zhang, Y.; Xiang, W. SbbR/SbbA, an Important ArpA/AfsA-Like System, Regulates Milbemycin Production in *Streptomyces bingchenggensis*. *Front. Microbiol.* **2018**, *9*, 1064. [CrossRef] [PubMed]

15. Wang, H.; Liu, Y.; Cheng, X.; Zhang, Y.; Li, S.; Wang, X.; Xiang, W. Titer improvement of milbemycins via coordinating metabolic competition and transcriptional co-activation controlled by *Streptomyces* antibiotic regulatory protein family regulator in *Streptomyces bingchenggensis*. *Biotechnol. Bioeng.* **2022**, *119*, 1252–1263. [CrossRef] [PubMed]
16. Ye, L.; Zhang, Y.; Li, S.; He, H.; Ai, G.; Wang, X.; Xiang, W. Transcriptome-guided identification of a four-component system, SbrH1-R, that modulates milbemycin biosynthesis by influencing gene cluster expression, precursor supply, and antibiotic efflux. *Synth. Syst. Biotechnol.* **2022**, *7*, 705–717. [CrossRef]
17. Schumacher, M.A.; den Hengst, C.D.; Bush, M.J.; Le, T.B.K.; Tran, N.T.; Chandra, G.; Zeng, W.; Travis, B.; Brennan, R.G.; Buttner, M.J. The MerR-like protein BldC binds DNA direct repeats as cooperative multimers to regulate *Streptomyces* development. *Nat. Commun.* **2018**, *9*, 1139. [CrossRef]
18. Liu, X.; Cheng, Y.; Lyu, M.; Wen, Y.; Song, Y.; Chen, Z.; Li, J. Redox-sensing regulator Rex regulates aerobic metabolism, morphological differentiation, and avermectin production in *Streptomyces avermitilis*. *Sci. Rep.* **2017**, *7*, 44567. [CrossRef]
19. Zhang, X.; Andres, S.N.; Elliot, M.A. Interplay between Nucleoid-Associated Proteins and Transcription Factors in Controlling Specialized Metabolism in *Streptomyces*. *mBio* **2021**, *12*, e0107721. [CrossRef]
20. Lu, T.; Zhu, Y.; Zhang, P.; Sheng, D.; Cao, G.; Pang, X. SCO5351 is a pleiotropic factor that impacts secondary metabolism and morphological development in *Streptomyces coelicolor*. *FEMS Microbiol. Lett.* **2018**, *365*, fny15. [CrossRef]
21. Al-Bassam, M.M.; Bibb, M.J.; Bush, M.J.; Chandra, G.; Buttner, M.J. Response regulator heterodimer formation controls a key stage in *Streptomyces* development. *PLoS Genet.* **2014**, *10*, e1004554. [CrossRef] [PubMed]
22. Bush, M.J. The actinobacterial WhiB-like (Wbl) family of transcription factors. *Mol. Microbiol.* **2018**, *110*, 663–676. [CrossRef] [PubMed]
23. Yan, L.; Zhang, Q.; Virolle, M.J.; Xu, D. In conditions of over-expression, WblI, a WhiB-like transcriptional regulator, has a positive impact on the weak antibiotic production of *Streptomyces lividans* TK24. *PLoS ONE* **2017**, *12*, e0174781.
24. Wang, H.; Zhang, J.; Zhang, Y.; Zhang, B.; Liu, C.; He, H.; Wang, X.; Xiang, W. Combined application of plasma mutagenesis and gene engineering leads to 5-oxomilbemycins A3/A4 as main components from *Streptomyces bingchenggensis*. *Appl. Microbiol. Biotechnol.* **2014**, *98*, 9703–9712. [CrossRef]
25. Kim, M.S.; Cho, W.J.; Song, M.C.; Park, S.W.; Kim, K.; Kim, E.; Lee, N.; Nam, S.J.; Oh, K.H.; Yoon, Y.J. Engineered biosynthesis of milbemycins in the avermectin high-producing strain *Streptomyces avermitilis*. *Microb. Cell Fact.* **2017**, *16*, 9. [CrossRef]
26. Nowak, E.; Panjikar, S.; Morth, J.P.; Jordanova, R.; Svergun, D.I.; Tucker, P.A. Structural and functional aspects of the sensor histidine kinase PrrB from *Mycobacterium tuberculosis*. *Structure* **2006**, *14*, 275–285. [CrossRef]
27. Niu, G. Genomics-Driven Natural Product Discovery in Actinomycetes. *Trends Biotechnol.* **2018**, *36*, 238–241. [CrossRef]
28. Xu, J.; Zhang, J.; Zhuo, J.; Li, Y.; Tian, Y.; Tan, H. Activation and mechanism of a cryptic oviedomycin gene cluster via the disruption of a global regulatory gene, adpA, in *Streptomyces ansochromogenes*. *J. Biol. Chem.* **2017**, *292*, 19708–19720. [CrossRef]
29. Chen, L.; Lu, Y.; Chen, J.; Zhang, W.; Shu, D.; Qin, Z.; Yang, S.; Jiang, W. Characterization of a negative regulator AveI for avermectin biosynthesis in *Streptomyces avermitilis* NRRL8165. *Appl. Microbiol. Biotechnol.* **2008**, *80*, 277–286. [CrossRef]
30. Zou, Z.; Du, D.; Zhang, Y.; Zhang, J.; Niu, G.; Tan, H. A γ-butyrolactone-sensing activator/repressor, JadR3, controls a regulatory mini-network for jadomycin biosynthesis. *Mol. Microbiol.* **2014**, *94*, 490–505. [CrossRef]
31. Van Schaik, W.; Tempelaars, M.H.; Zwietering, M.H.; de Vos, W.M.; Abee, T. Analysis of the role of RsbV, RsbW, and RsbY in regulating σ^B activity in *Bacillus cereus*. *J. Bacteriol.* **2005**, *187*, 5846–5851. [CrossRef] [PubMed]
32. Pathak, D.; Jin, K.S.; Tandukar, S.; Kim, J.H.; Kwon, E.; Kim, D.Y. Structural insights into the regulation of SigB activity by RsbV and RsbW. *IUCrJ* **2020**, *7*, 737–747. [CrossRef] [PubMed]
33. Bentley, S.D.; Chater, K.F.; Cerdeño-Tárraga, A.M.; Challis, G.L.; Thomson, N.R.; James, K.D.; Harris, D.E.; Quail, M.A.; Kieser, H.; Harper, D.; et al. Complete genome sequence of the model actinomycete *Streptomyces coelicolor* A3(2). *Nature* **2002**, *417*, 141–147. [CrossRef] [PubMed]
34. Ikeda, H.; Ishikawa, J.; Hanamoto, A.; Shinose, M.; Kikuchi, H.; Shiba, T.; Sakaki, Y.; Hattori, M.; Omura, S. Complete genome sequence and comparative analysis of the industrial microorganism *Streptomyces avermitilis*. *Nat. Biotechnol.* **2003**, *21*, 526–531. [CrossRef] [PubMed]
35. Sambrock, J.; Russel, D.J.I. *Molecular Cloning: A Laboratory Manual*; Cold Spring Harbour Laboratory Press: New York, NY, USA, 2001; Volume 49, pp. 895–909.
36. Keiser, T.; Bipp, M.; Buttner, M.; Chater, K.F.; Hopwood, D.A. *Practical Streptomyces Genetics*; The John Innes Foundation: Norwich, UK, 2000.
37. Gao, H.; Liu, M.; Liu, J.; Dai, H.; Zhou, X.; Liu, X.; Zhuo, Y.; Zhang, W.; Zhang, L. Medium optimization for the production of avermectin B1a by *Streptomyces avermitilis* 14-12A using response surface methodology. *Bioresour. Technol.* **2009**, *100*, 4012–4016. [CrossRef] [PubMed]
38. Bierman, M.; Logan, R.; O'Brien, K.; Seno, E.T.; Rao, R.N.; Schoner, B.E. Plasmid cloning vectors for the conjugal transfer of DNA from *Escherichia coli* to *Streptomyces* spp. *Gene* **1992**, *116*, 43–49. [CrossRef]
39. Li, L.; Wei, K.; Zheng, G.; Liu, X.; Chen, S.; Jiang, W.; Lu, Y. CRISPR-Cpf1-Assisted Multiplex Genome Editing and Transcriptional Repression in *Streptomyces*. *Appl. Environ. Microbiol.* **2018**, *84*, e00827-18. [CrossRef]

40. Strauch, E.; Takano, E.; Baylis, H.A.; Bibb, M.J. The stringent response in *Streptomyces coelicolor* A3(2). *Mol. Microbiol.* **1991**, *5*, 289–298. [CrossRef]
41. Livak, K.J.; Schmittgen, T.D. Analysis of relative gene expression data using real-time quantitative PCR and the 2(-Delta Delta C(T)) Method. *Methods* **2001**, *4*, 402–408. [CrossRef]

antibiotics

MDPI

Article

Comparative Transcriptome-Based Mining of Genes Involved in the Export of Polyether Antibiotics for Titer Improvement

Xian Liu [1,2], Yuanting Wu [1,2], Xiaojie Zhang [1,2], Qianjin Kang [1,2], Yusi Yan [1,3,*] and Linquan Bai [1,2,*]

1 State Key Laboratory of Microbial Metabolism, Shanghai-Islamabad-Belgrade Joint Innovation Center on Antibacterial Resistances, School of Life Sciences & Biotechnology, Shanghai Jiao Tong University, Shanghai 200240, China; liuxian19950811@sjtu.edu.cn (X.L.); yuanting_wu@163.com (Y.W.); xiaojiezhang0831@163.com (X.Z.); qjkang@sjtu.edu.cn (Q.K.)
2 Joint International Research Laboratory of Metabolic & Developmental Sciences, Shanghai Jiao Tong University, Shanghai 200240, China
3 Institute of Biopharmaceuticals, Taizhou University, Taizhou 318000, China
* Correspondence: yanyusi@sjtu.edu.cn (Y.Y.); bailq@sjtu.edu.cn (L.B.)

Abstract: The anti-coccidiosis agent salinomycin is a polyether antibiotic produced by *Streptomyces albus* BK3-25 with a remarkable titer of 18 g/L at flask scale, suggesting a highly efficient export system. It is worth identifying the involved exporter genes for further titer improvement. In this study, a titer gradient was achieved by varying soybean oil concentrations in a fermentation medium, and the corresponding transcriptomes were studied. Comparative transcriptomic analysis identified eight putative transporter genes, whose transcription increased when the oil content was increased and ranked top among up-regulated genes at higher oil concentrations. All eight genes were proved to be positively involved in salinomycin export through gene deletion and trans-complementation in the mutants, and they showed constitutive expression in the early growth stage, whose overexpression in BK3-25 led to a 7.20–69.75% titer increase in salinomycin. Furthermore, the heterologous expression of *SLNHY_0929* or *SLNHY_1893* rendered the host *Streptomyces lividans* with improved resistance to salinomycin. Interestingly, *SLNHY_0929* was found to be a polyether-specific transporter because the titers of monensin, lasalocid, and nigericin were also increased by 124.6%, 60.4%, and 77.5%, respectively, through its overexpression in the corresponding producing strains. In conclusion, a transcriptome-based strategy was developed to mine genes involved in salinomycin export, which may pave the way for further salinomycin titer improvement and the identification of transporter genes involved in the biosynthesis of other antibiotics.

Keywords: *Streptomyces*; comparative transcriptome; polyether antibiotics; salinomycin; exporter genes

Citation: Liu, X.; Wu, Y.; Zhang, X.; Kang, Q.; Yan, Y.; Bai, L. Comparative Transcriptome-Based Mining of Genes Involved in the Export of Polyether Antibiotics for Titer Improvement. *Antibiotics* **2022**, *11*, 600. https://doi.org/10.3390/antibiotics11050600

Academic Editor: Jesús F. Aparicio

Received: 27 March 2022
Accepted: 28 April 2022
Published: 29 April 2022

1. Introduction

Polyether antibiotics, also called polyether ionophores, are a broad class of natural compounds produced by actinomycetes, with the vast majority being derived from the genera *Streptomyces* and *Actinomadura* [1]. In recent years, with the discovery of over 120 novel molecules, these chemicals have received more and more attention. Typical polyether antibiotics, including salinomycin, nigericin, lasalocid, and monesin (Figure 1), feature 2–5 ether oxygen atoms and a carboxyl group [2]. This structure enables them to chelate with metal cations, such as Na^+ and K^+, and protons to form neutral coordination compounds, which cross the cell membrane and subsequently change ion gradients and osmotic pressures, thus resulting in cell death [1]. Salinomycin, a polyether antibiotic produced by *Streptomyces albus* DSM41398 and its derived strains [3], is widely applied in husbandry because it has properties that kill Gram-positive bacteria and coccidia [4]. Recent studies have found that salinomycin also inhibits the growth of leukemia stem cells [5] and epithelial cancer stem cells [6], indicating that it is a potential anti-tumor drug [6].

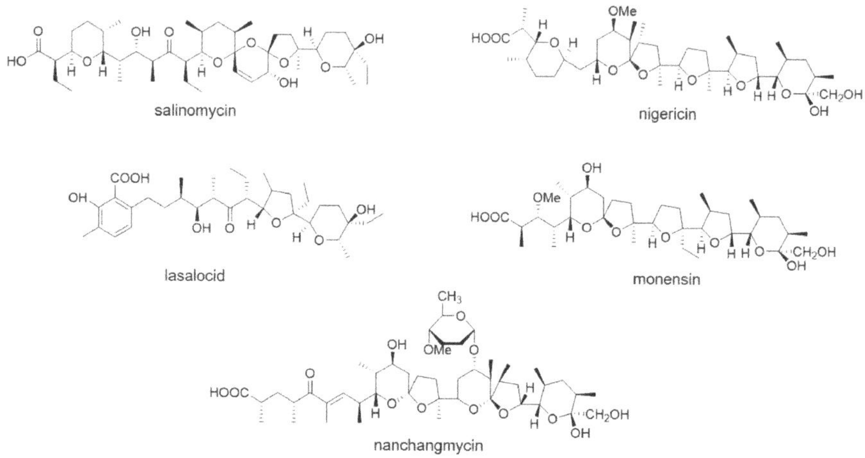

Figure 1. Structures of polyether antibiotics.

The high-titer *Streptomyces albus* strain BK3-25 produces 18 g/L salinomycin under lab conditions [7], but the intracellular accumulation of salinomycin poses a threat to cell growth, which can be released by the strain's resistance ability [8–10]. However, the mechanism underlying this antibiotic resistance remains elusive. According to our previous work, *SLNHY_261* (*slnTII*) and *SLNHY_262* (*slnTI*) in a salinomycin biosynthetic gene cluster (BGC), encoding the ATP-binding subunit and transmembrane subunit, respectively [11], were thought to form an ATP-binding cassette (ABC) complex participating in salinomycin export. When *slnTI* and *slnTII* were deleted in *Streptomyces albus* XM 211, the salinomycin titers of the corresponding mutants declined by only 27.2% and 45.4%, respectively [12], indicating that there are additional genes involved in salinomycin export, which may be located beyond the salinomycin BGC regions.

Actinomycetes have large-capacity transporter protein systems, which participate in cell metabolism, intercellular communication, biosynthesis, and proliferation [13]. The ABC superfamily [14] and major facilitator superfamily (MFS) [15] are two well-studied classes of transporters. Whole-genome sequencing has illustrated that there are numerous ABC and MFS transporter genes in an actinomycetal genome both inside and outside of secondary metabolite BGCs [13,16]. Wang et al. analyzed transcriptome expression differences with expression profile chips and discovered 13 candidate transporter genes outside the natamycin BGC from *Streptomyces chattanoogensis* L10 [17]. Chu et al. built a step-by-step workflow based on the TCDB database BLAST, and they included substrate analysis, transporter classification analysis, and phylogenetic analysis to mine BGC-independent exporters. Together with a tunable plug-and-play exporter module with replaceable promoters and ribosome-binding sites, they realized the titer improvement of macrolide biopesticides in different *Streptomyces* producers [18]. Nevertheless, the current commonly used approaches are mainly sequence-dependent or based on known exporters, and the methods of BGC-independent exporter mining still need development.

Soybean oil serves as the main carbon source in salinomycin fermentation, supplying energy through primary metabolism and precursors, such as malonyl-CoA, methylmalonyl-CoA, and ethylmalonyl-CoA, for salinomycin biosynthesis. Usually, 15% (*w*/*v*) of soybean oil is added to the fermentation medium, which is extremely high compared with other antibiotic fermentations. Our previous work revealed that increased soybean oil addition resulted in higher salinomycin production [19], and, thus, we hypothesized that higher concentrations of soybean oil cause a higher transcription of exporter genes.

Herein, a strategy based on comparative transcriptomic analysis under different concentrations of soybean oil supplementation was developed to identify salinomycin exporter genes (Figure S1, Supporting Information). Our work provided universal exporters for the titer improvement of polyether antibiotics in *Streptomyces*. Furthermore, our method might broaden transporter engineering toolkits for the titer improvement of other valuable products in *Streptomyces*.

2. Materials and Methods

2.1. Strains, Plasmids, and Culture Conditions

The bacterial strains and plasmids used in this study are listed in Table S1, Supporting Information.

S. albus BK3-25 (from Zhejiang Shenghua Biok Biology Co., Ltd., Deqing, China) and its mutants were grown on ISP4 medium (10 g/L soluble starch, 2 g/L $(NH_4)_2SO_4$, 1 g/L K_2HPO_4, 2 g/L $CaCO_3$, 1 g/L NaCl, 1 g/L $MgSO_4$, 100 µL of trace element solution (1% $ZnSO_4$, 1% $MnCl_2$, 1% $FeSO_4$ (*w*/*v*)), 20 g/L agar) for 7 days for sporulation. The conjugation of *Streptomyces* with *Escherichia coli* was carried out on ISP4 plates supplemented with 20 mM $MgCl_2$. The fermentation procedure was as follows: *S. albus* BK3-25 and its mutants were grown in 30 mL of TSBY medium [20] (30 g/L tryptone soya broth, 5 g/L yeast extract, 103 g/L sucrose) at 30 °C and 220 r.p.m. for 48 h. Then, 1 mL of the culture was transferred into 30 mL of the seed medium (30 g/L soybean meal, 10 g/L yeast extract, 2 g/L $CaCO_3$, 80 mL/L 50% glucose) and cultivated at 33 °C and 220 r.p.m. for 16 h. Finally, 5 mL of the seed culture was transferred into 50 mL of the fermentation medium (8 g/L germ powder, 5 g/L soybean meal, 2.2 g/L KCl, 1 g/L NaCl, 1.6 g/L urea, 2 g/L tartaric acid, 0.1 g/L $MgSO_4$, 0.1 g/L K_2HPO_4, 5 g/L $CaCO_3$, pH 6.6–6.9, supplemented with 7.5 g/50 mL soybean oil) and cultured at 33 °C and 220 r.p.m. for 9 days [11].

S. lividans TK24, *S. cinnamonensis* ATCC 15413, *S. lasaliensis* ATCC 31180, *S. hygroscopicus* XM201-*ga32* and their mutants were grown on SFM medium (20 g/L soybean meal, 20 g/L mannitol, 20 g/L agar) for 7 days for sporulation. The conjugation of *Streptomyces* with *E. coli* was carried out on SFM plates supplemented with 20 mM $MgCl_2$.

For fermentation, *S. cinnamonensis* ATCC 15413 and its mutants were grown in 25 mL of the seed medium (20 g/L dextrin, 15 g/L soybean meal, 2.5 g/L yeast extract, 5 g/L glucose, 1 g/L $CaCO_3$, pH 6.7–6.8), and then 2.5 mL of the seed culture was transferred into 25 mL of the fermentation medium (20 g/L soybean oil, 45 g/L glucose, 40 g/L soybean meal, 2.2 g/L $NaNO_3$, 2.2 g/L Na_2SO_4, 0.07 g/L $Al_2(SO_4)_3$, 0.1 g/L $FeSO_4$, 0.33 g/L $MnCl_2$, 0.075 g/L K_2HPO_4, 2.5 g/L $CaCO_3$, pH 6.7–6.8) and cultured at 32 °C and 250 r.p.m. for 10 days.

S. lasaliensis ATCC 31180 and its mutants were grown in 25 mL of the seed medium (20 g/L sucrose, 20 g/L soybean meal, 5 g/L tryptone, 5 g/L malt extract, 2 g/L NaCl, 4 g/L $CaCO_3$, pH 7.0), and then 2.5 mL of the seed culture was transferred into 25 mL of the fermentation medium (5 g/L glucose, 40 g/L dextrin, 35 g/L soybean meal, 7.5 g/L corn starch, 3 g/L NaCl, 4 g/L KH_2PO_4, 2 g/L $MgSO_4 \cdot 7H_2O$, pH 7.0) and cultured at 28 °C and 200 r.p.m. for 6 days [21].

S. hygroscopicus XM201-*ga32* and its mutants were grown in 50 mL of the seed medium (10 g/L glucose, 10 g/L tryptone, 5 g/L yeast extract), and then 7.5 mL of the seed culture was transferred into 50 mL of the fermentation medium (30 g/L corn starch, 70 g/L glucose, 40 g/L soybean meal, 3 g/L $(NH_4)_2SO_4$, 0.01 g/L $CoCl_2$, 10 g/L $CaCO_3$, 1 g/L soybean oil, pH 6.8–7.0) and cultured at 30 °C and 220 r.p.m. for 7 days [22].

E. coli ET12567 (pUZ8002) was used for conjugation. The *E. coli* cells were cultured in Luria–Bertani (LB) broth at 37 °C.

2.2. Transcriptome Sequencing of BK3-25

For transcriptome sequencing, mycelia were harvested on the third day of fermentation. The total RNA was extracted using Redzol according to the manufacturer's instructions. Transcriptome sequencing was performed by the Shanghai Biotechnology Corporation, and the expression level of each gene was calculated as fragments per kilobase of exon per megabase of library size (FPKM).

2.3. Construction of Plasmids for Deletion and Over-Expression of Eight Candidate Genes

For gene deletion through homologous recombination, left and right flanking regions of each gene were obtained using PCR amplification, ligated to *Eco*RV-digested pBluescript SK, and sequenced (Figure S2A). Then, these plasmids were digested with *Xba*I/*Eco*RI or *Eco*RI/*Hind*III and ligated to *Xba*I/*Hind*III-digested plasmid pJTU1278. The primers used for gene deletion are listed in Table S2.

For gene overexpression, eight candidate genes were obtained using PCR amplification, ligated to *Eco*RV-digested pBluescript SK, and sequenced. Then, these plasmids were digested with *Xba*I/*Not*I (*Nde*I/*Eco*RI), and the fragments containing genes were ligated with the *Xba*I/*Not*I-digested plasmid pIB139 or *Nde*I/*Eco*RI-digested plasmid pLQ646 (Figure S2B,C). The primers used for gene over-expression are listed in Table S2.

2.4. Conjugation between Streptomyces and E. coli

All the plasmids were successively introduced into the non-methylating *E. coli* strain ET12567 (pUZ8002) and *S. albus* strains. Spores (~10^9 CFU) were heat-shocked at 50 °C for 10 min, pregerminated for 2.5 h, and then mixed with *E. coli* cells. The suspensions were spread onto non-selective plates containing ISP4 medium supplemented with 20 mM $MgCl_2$. Apramycin was overlaid on the plates after 17 h of incubation at 30 °C, and exconjugants typically appeared after 3 days.

For gene deletion, the exconjugants were assessed as single-crossover mutants using PCR amplification with primers SLNHY_X-YZ-F/R (Table S2). After two rounds of sporulation without antibiotic selection, double-crossover mutants were verified using PCR with primers SLNHY_X-YZ-F/R (Table S2 and Figure S2A). For gene over-expression, the exconjugants were verified using PCR with primers pIB139-over-YZ-F (pLQ648-over-YZ-F) and SLNHY_X-over-YZ-R (Table S2).

2.5. HPLC Analysis of Antibiotics

For the detection of total salinomycin, 1 mL of fermentation broth was mixed with 9 mL of methanol, followed by sonication at 40 kHz for 30 min. Then, 1 mL of the mixture was taken and centrifuged at 12,000 r.p.m. for 1 min, and the supernatant was filtrated and subjected to HPLC analysis. For the detection of intracellular salinomycin, 1 mL of the fermentation broth was centrifuged at 12,000 r.p.m. for 5 min, and the supernatant was discarded. The mycelia were washed twice with water and mixed with 1 mL of methanol, followed by sonication at 40 kHz for 30 min. Then, the mixture was centrifuged at 12,000 r.p.m. for 5 min, and the supernatant was filtrated and subjected to HPLC analysis. HPLC was performed on Agilent series 1260 (Agilent Technologies, Santa Clara, CA, USA) with an Agilent TC-18 column (2.1 × 150 mm, 5 μm). In this process, 8% A (water, 2% acetic acid) and 92% B (acetonitrile) were used as the mobile phase with a flow rate of 1 mL/min, and the detection time was 20 min using UV spectroscopy at 210 nm [11]. The concentrations of salinomycin were calculated according to the standard curve of salinomycin (Figure S3).

For the detection of lasalocid, 1 mL of fermentation broth was mixed with 1 mL of methanol, followed by sonication at 40 kHz for 30 min. Then, the mixture was centrifuged at 12,000 r.p.m. for 1 min, and the supernatant was filtrated and subjected to HPLC analysis. HPLC was performed on Agilent series 1260 (Agilent Technologies, USA) with an Agilent TC-18 column (2.1 × 150 mm, 5 μm) at 40 °C. In this process, 15% A (water, 0.125 mol/L ammonium acetate, pH 4.8) and 85% B (acetonitrile) were used as the mobile phase with a flow rate of 1 mL/min, and the detection time was 20 min using UV spectroscopy at 305 nm [21]. The concentrations of lasalocid were calculated according to the standard curve of lasalocid (Figure S4).

For the detection of monensin, 1 mL of fermentation broth was centrifuged at 12,000 r.p.m. for 5 min, and the supernatant was discarded. The mycelia were washed twice with water and mixed with 1 mL of ethanol, followed by sonication at 40 kHz for 30 min. Then, the mixture was centrifuged at 12,000 r.p.m. for 5 min, and the supernatant was filtrated and subjected to HPLC analysis. HPLC was performed on Agilent series 1260 (Agilent Technologies, Santa Clara, CA, USA) with an Agilent TC-18 column (2.1 × 150 mm, 5 μm). In this process, 20 mM ammonium acetate and methanol were used as the mobile phase with a flow rate of 1 mL/min. For 0–25 min, the ratio of methanol increased from 80% to 100%, and for 25–30 min, it decreased from 100% to 80%. Evaporative light-scattering detection (ELSD) was conducted for 30 min at 85 °C [23]. The concentrations of monensin were calculated according to the standard curve of monensin (Figure S5).

For the detection of nigericin, 1 mL of fermentation broth was mixed with 9 mL of methanol, followed by sonication at 40 kHz for 30 min. Then, 1 mL of the mixture was taken and centrifuged at 12,000 r.p.m. for 1 min, and the supernatant was filtrated and subjected to HPLC analysis. HPLC was performed on Agilent series 1260 (Agilent Technologies, USA) with an Agilent TC-18 column (2.1 × 150 mm, 5 μm). A gradient elution (1 mL/min flow rate) was performed using A (methanol/water, 9:1 ratio, 0.1% TFA from 0 to 25 min) and B (methanol, 100%, 0.1% TFA, from 25 to 50 min). ELSD detection was conducted for 50 min at 85 °C [24]. The concentrations of nigericin were calculated according to the standard curve of nigericin (Figure S6).

2.6. RNA Extraction and RT-qPCR Analysis

Mycelia of *S. albus* BK3-25 were harvested, and the total RNA was extracted using Redzol according to the manufacturer's instructions (SBS Genetech, Shanghai, China) [25]. The quality of the RNA was determined using a NanoDrop 2000 spectrophotometer. For RT-qPCR experiments, total RNA was reversely transcribed into cDNA using RevertAid^{TM} H Minus First Strand cDNA Synthesis Kit (Thermo Fisher, Waltham, MA USA). The RT-qPCR experiments were carried out on a 7500 Fast Real-time RCR system (Applied Biosystems, Waltham, MA USA) using Maxima^{TM} SYBR Green/ROX qPCR Maxter Mix (Thermo Fisher, Waltham, MA USA) according to the manufacturer's procedure. The expression values of the target genes were calculated using $2^{-\Delta\Delta CT}$ methods with the housekeeping gene *hrdB* as internal control [26].

2.7. Biomass Determination under Fermentation Condition

Due to the insoluble residues in the liquid medium, total intracellular nucleic acid rather than dry cell weight was determined to represent the growth of *Streptomyces*. The concentration of intracellular nucleic acid was detected as follows: 1 mL fermentation broth was centrifuged and washed twice to eliminate the interference of the medium. Then, 1 mL of Solution A (1.5 g diphenylamine, 100 mL acetic acid, 1.5 mL concentrated sulfuric acid, 1 mL 1.6% acetaldehyde) was mixed with the mycelia and put in water bath at 60 °C for 1 h. Then, the mixture was centrifuged, and 150 μL of supernatant was transferred into 96-well plates and detected at 595 nm (Infinite M200 PRO, TECAN, Männedorf, Switzerland).

3. Results

3.1. Transcriptome-Based Identification of Candidate Exporter Genes

We initially assumed that the higher the salinomycin titer, the higher the transcription of the involved exporter genes. In order to establish a titer gradient, different concentrations of soybean oil (5%, 10%, and 15%) were supplemented to the fermentation broth, and the corresponding salinomycin titers were 5.30 g/L, 12.50 g/L, and 17.40 g/L, respectively. Using these three samples, transcriptomic data were collected using RNA-seq technology. According to the hypothetic concurrent relationship between salinomycin titers and the transcription of exporter genes, eight exporter genes with increasing expression patterns and that topped the fold-change of transcription at a higher salinomycin titer were selected from 248 ABC transporter genes and 23 MFS genes in the BK3-25 genome. These eight genes include seven ABC transporter genes, *SLNHY_3363*, *SLNHY_4037*, *SLNHY_6316*, *SLNHY_6652*, *SLNHY_0818*, *SLNHY_0199*, and *SLNHY_1893*, and one MFS gene, *SLNHY_0929* (Figure 2 and Table S3). Although the transcriptions of *SLNHY_3363* and *SLNHY_0199* at 15% oil supplementation were lower than those at 10%, both of their transcriptions dramatically rose when the oil contents were shifted from 5% to 10%.

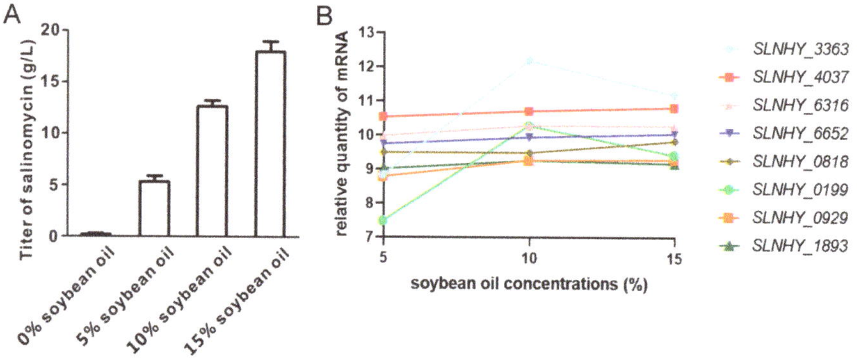

Figure 2. (**A**) Salinomycin titers under 0, 5, 10, and 15% soybean oil supplementation. (**B**) Transcription profiles of eight candidate transporter genes under 5, 10, and 15% (w/v) soybean oil supplementation.

In order to verify the transcriptomic data, the transcription levels of the eight candidate genes were measured using RT-qPCR with cultures collected on the third day of fermentation supplemented with 5% or 15% soybean oil, and all genes showed higher expressions with 15% oil supplementation, which was consistent with the transcriptomic data. Among them, *SLNHY_929* demonstrated the highest expression, followed by *SLNHY_3363* and *SLNHY_1893* (Table S4).

3.2. All Eight Candidate Genes Were Positively Involved in Salinomycin Export

To investigate whether these eight genes are involved in salinomycin production, they were knocked out through homologous recombination. As shown in Figure 3A,B, the total salinomycin titers of all mutants decreased to 11.62–27.36% of that of BK3-25, and the intracellular concentrations of salinomycin increased to 143.61–237.89% of that of BK3-25, indicating that these genes were positively related to salinomycin biosynthesis. Moreover, the deletion of Δ*SLNHY_0199* was the most pronounced, with a dramatic decrease in the salinomycin titer from 13.34 g/L to 1.55 g/L.

Further verification of the above conclusion was conducted through trans-complementation of each mutant with the corresponding gene cloned under the control of *PermE**. All individually complemented strains returned to 78.77–88.77% of the original titer of salinomycin (Figure 3C), and the intracellular accumulations of salinomycin synchronically returned to 62.36–105.19% of the level of BK3-25 (Figure 3D), providing more proof of the involvement of these eight genes in salinomycin biosynthesis and, most likely, in its export.

In addition, these eight genes were individually over-expressed in BK3-25 to see whether they played vital roles in salinomycin titer improvement. As expected, compared with the control strain bearing the empty vector pIB139, the excessive expression of these genes all increased salinomycin titers by 7.20–69.75%, especially BK3-25::*SLNHY_3363* and BK3-25::*SLNHY_0929*, which had improved titers of 24.60 g/L and 22.85 g/L, respectively (Figure 3E). Accordingly, the intracellular salinomycin accumulations of all mutants showed a marked fall to 24.35–46.23% of the same level of BK3-25 (Figure 3F).

3.3. These Eight Exporter Genes Were Constitutively Expressed

In order to determine whether the expressions of these eight genes were constitutive or induced by salinomycin, the transcription profiles of each gene were obtained using RT-qPCR with samples collected each day during the whole fermentation period (Figure 4). Compared with the house-keeping gene *hrdB*, these exporter genes were actively transcribed at the very beginning and then gradually decreased along with the fermentation process. Since salinomycin obviously accumulated after the first day, as shown in Figure S7, we can safely draw the conclusion that these genes were constitutively expressed rather than being induced by salinomycin.

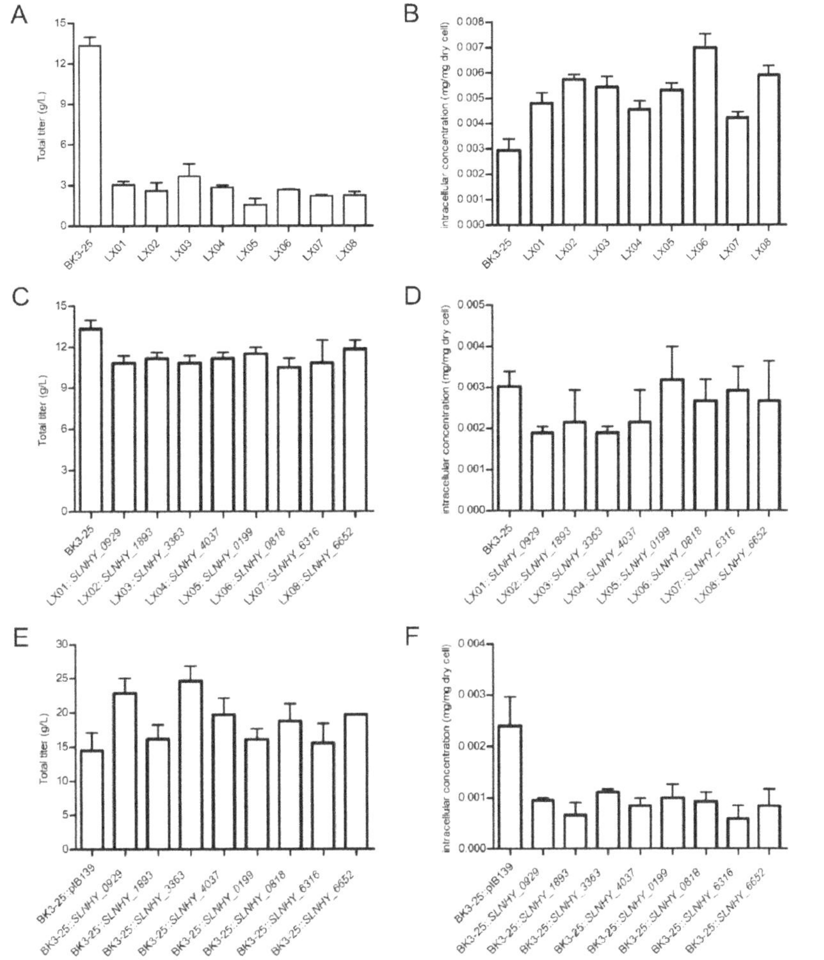

Figure 3. Salinomycin production of exporter gene mutants. (**A**) Total salinomycin titers of gene deletion mutants of BK3-25. (**B**) Intracellular salinomycin concentrations of gene deletion mutants. (**C**) Total salinomycin titers of gene complement mutants. (**D**) Intracellular salinomycin concentrations of gene complement mutants. (**E**) Total salinomycin titers of gene over-expression mutants. (**F**) Intracellular salinomycin concentrations of gene over-expression mutants. The total salinomycin titers were calculated based on the fermentation broth volume (**A**,**C**,**E**), and the intracellular salinomycin concentrations were calculated based on the dry cell weight (**B**,**D**,**F**). LX01, BK3-25Δ*SLNHY_0929*; LX02, BK3-25Δ*SLNHY_1893*; LX03, BK3-25Δ*SLNHY_3363*; LX04, BK3-25Δ*SLNHY_4037*; LX05, BK3-25Δ*SLNHY_0199*; LX06, BK3-25Δ*SLNHY_0818*; LX07, BK3-25Δ*SLNHY_6316*; LX08, BK3-25Δ*SLNHY_6652*.

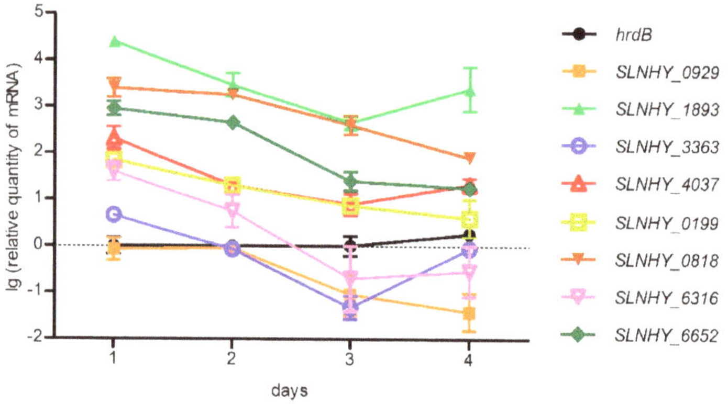

Figure 4. Transcription profiles of eight exporter genes. Line with black circle, *hrdB*; orange square, *SLNHY_0929*; green equilateral solid triangle, *SLNHY_1893*; blue circle, *SLNHY_3363*; red equilateral hollow triangle, *SLNHY_4037*; earthy yellow hollow square, *SLNHY_0199*; brown inverted solid triangle, *SLNHY_0818*; pink inverted hollow triangle, *SLNHY_6316*; dark green rhombus, *SLNHY_6652*. Logarithm of transcription data were taken as ordinate; if not, it would be difficult to show their wide fluctuation scales.

3.4. SLNHY_0929 and SLNHY_1893 Improved Resistance to Salinomycin in Streptomyces lividans

Even though these eight exporter genes were proved to be involved in salinomycin export, we still needed to determine why they functioned in this way. *Streptomyces lividans* TK24 was found to be susceptible to high concentrations of salinomycin, and the minimal inhibition concentration (MIC) was 0.5 mmol/L for the control strain TK24::pIB139. These eight genes were individually introduced into *S. lividans* TK24. Although most mutants carrying the introduced exporter genes maintained similar susceptibility to salinomycin, TK24::*SLNHY_0929* and TK24::*SLNHY_1893* rendered the host with an improved resistance as high as 1.0 mmol/L (Figure 5A,B). These data strongly suggest the salinomycin export ability of *SLNHY_0929* and SLNHY_1893, which exported the assimilated exogenous salinomycin out of *S. lividans* TK24.

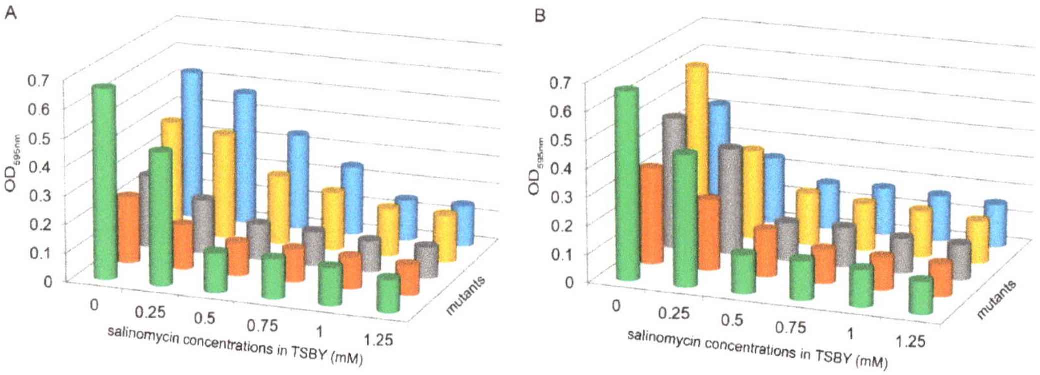

Figure 5. Biomass profiles of *S. lividans* TK24 mutants under different concentrations of salinomycin. Total intracellular nucleic acid was determined to represent the growth of *S. lividans*. (**A**) Green column, TK24::pIB139; orange column, TK24::*SLNHY_3363*; gray column, TK24::*SLNHY_4037*; yellow column, TK24::*SLNHY_0929*; blue column, TK24::*SLNHY_1893*. (**B**) Green column, TK24::pIB139; orange column, TK24::*SLNHY_0199*; gray column, TK24::*SLNHY_0818*; yellow column, TK24::*SLNHY_6316*; blue column, TK24::*SLNHY_6652*.

3.5. SLNHY_0929 Was a Universal Exporter for Polyether Antibiotics with Similar Structure with Salinomycin

Since most polyether antibiotics shared similar hydrophobic structures, we wondered whether the exporter genes were universal in pumping them out and whether the genes played roles in improving their titers. Therefore, the three exporter genes with the most substantial effects on salinomycin titers when over-expressed in BK3-25, i.e., *SLNHY_0929*, *SLNHY_3363* and, *SLNHY_4037*, were heterologously expressed in *Streptomyces lasaliensis* ATCC 31180 (a lasalocid producer), *Streptomyces cinnamonensis* ATCC 15413 (a monensin producer), and *Streptomyces hygroscopicus* XM201-*ga*32 (a nigericin producer). Herein, the previously used *PermE** promoter was replaced by a stronger promoter *kasO*p*, since the latter was reported to work better in XM201 [22,27].

Interestingly, the heterologous expression of *SLNHY_0929* resulted in a significant improvement in all three antibiotics, with lasalocid titers from 163.60 mg/L to 262.50 mg/L in *S. lasaliensis* (Figure 6A), monensin titers from 572.40 mg/L to 1,286.00 mg/L in *S. cinnamonensis* (Figure 6B), and nigericin titers from 116.77 mg/L to 207.27 mg/L in *S. hygroscopicus* (Figure 6C). Surprisingly, the heterologous expressions of *SLNHY_3363* and *SLNHY_4037* had no positive effects on the production of these three polyether antibiotics, even with unexpected, dropped titers, and the reason needed further exploration. Overall, these results clearly show that *SLNHY_0929* is a universal exporter for salinomycin, lasalocid, monensin, and nigericin, which shared similar molecular structures.

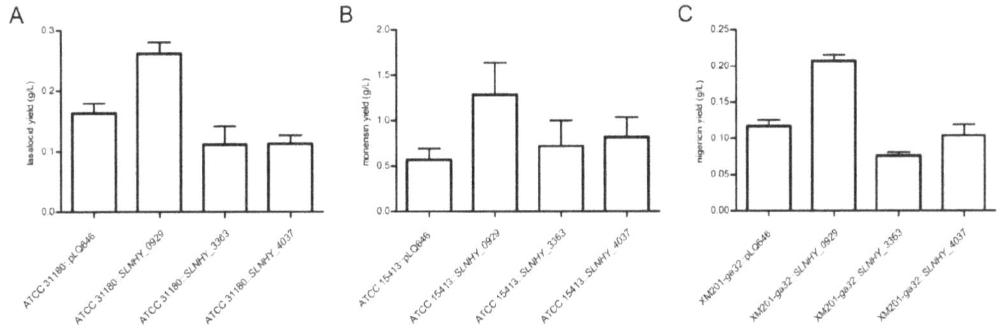

Figure 6. Production of polyether antibiotics with heterologous expressions of exporters in other Streptomyces hosts. (**A**) Lasalocid production of *S. lasaliensis* mutants. (**B**) Monensin production of *S. cinnamonensis* mutants. (**C**) Nigericin production of *S. hygroscopicus* mutants. pLQ646, integrative vector with *kasO*p* promoter.

Transporter engineering has been considered as a promising strategy to maximize secondary metabolite production in bacterial hosts, such as *Streptomyces* spp. and *Aspergillus* spp. [28,29]. Except for the exporters located in BGCs, BGC-independent exporters caused by horizontal gene transfer may also contribute to metabolite export [30,31]. To discover transporter genes located far from BGCs, expression profiling, genome-wide knockout studies, stress-based selection, and the inhibitor strategy have often been used [32].

Herein, due to *Streptomyces albus'* highly efficient utilization of soybean oil, we found that salinomycin production rose as oil addition rose, and the transcription levels of the genes involved in salinomycin PKS, β-oxidation, and precursor biosynthesis also increased [7]. Since transporters are the essential channels of both precursor import and salinomycin export, they are more necessary with the rise in ionophore product synthesis, so we decided to focus on the study of transporters. Next, we managed to mine eight transporter genes outside salinomycin BGC based on comparative transcriptome data under different salinomycin titers. Furthermore, all eight selected genes proved to be correlated with salinomycin synthesis by gene deletion and over-expression. Among the eight genes, only *SLNHY_0929* and *SLNHY_1893* encoded proteins that could pump salinomycin out of the cell and, thus, provide self-resistance to the host, according to the heterologous expression in the model strain *S. lividans* and the salinomycin supplement experiments. Our work constructed a novel method for antibiotic transporter gene mining. The fermentation characteristics of *Streptomyces albus* with a soybean oil preference were focused and combined with transcriptome sequencing technology, which showed a very successful titer improvement of salinomycin through exporter engineering and might apply to other antibiotics with oil-derived precursors. Therefore, this study also provides an application with great potential in promoting the more cost-effective production of salinomycin and other chemicals.

MFS-type transporters are channels of multiple substrates, such as monose, polysaccharides, amino acids, polypeptides, vitamins, cofactors, secondary metabolites, chromophores, and bases. They can either function independently or cooperate with ABC transporters in metabolite efflux [17]. In our work, the MFS family gene, *SLNHY_0929*, was heterologously expressed in the producers of other polyether antibiotics, and it showed a broad spectrum of substrate identification, whose protein pumped out intracellular salinomycin, lasalocid, monensin, and nigericin, as well as increasing the titers of each. As a matter of fact, *S. cinnamonensis* ATCC 15413 possesses a homolog of the exporter gene *SLNHY_0929*, namely, *orf6552*, with 93% coverage and 75.81% identity, while none of the homolog is present in *S. lasaliensis* ATCC 31180 or *S. hygroscopicus* XM201-*ga32*. Thus, ORF6552 is considered to be an endogenous MFS transporter, which exports monensin from the host. Whether or not it served as another universal ionophore pump remains to be explored. Compared to specific exporters slnTI and slnTII in the gene cluster, *SLNHY_0929* contributed more to salinomycin biosynthesis [12]. Meanwhile, endogenous exporters in lasalocid, monensin, and nigericin BGCs, which provide self-resistance to producing strains, have been reported. Lsd5 in *S. lasaliensis* showed 53% identity with MonT in *S. cinnamonensis*, and in *S. hygroscopicus*, R14/R15 formed an ABC transporter with 71% and 49% identities with the TMD and NBD of the transporter in *S. avermitilis*'s BGC, respectively [33–36]. Until now, there have been no data about these self-exporters' functions on antibiotic production, so we could not compare them with the non-specific exporter *SLNHY_0929*. However, introducing this MFS transporter improved the polyether antibiotics' titers by over 60% and almost doubled monensin's titer, which demonstrates the efficacy of our method for transporter mining. Due to the constitutive expression of these exporter genes, we can safely draw the conclusion that *SLNHY_0929* served as a universal and stable pump by flexibly identifying polyether compounds. Whether or not it could recognize other types of chemicals remains unknown. SLNHY_1893, SLNHY_3363, SLNHY_4037, SLNHY_0199, SLNHY_0818, SLNHY_6316, and SLNHY_6652 all belong to ABC transporters; however, they cannot export salinomycin according to the MIC results implemented in *Streptomyces lividans*. Their functions were speculated to be importers of soybean oil or other small nutrient molecules, or exporters of other secondary metabolites, such as actinopyranone and elaiophylin, whose BGCs were detected in the *S. albus* genome. Hence, on the one hand, in a future study, intracellular acyl-CoA concentrations will be detected to study whether these proteins pump fatty acids and glycerol derived from soybean oil [37]. On the other hand, we would like to strengthen or weaken the expression of these pumps and observe the production of possible metabolites to identify the pumps' function.

Additionally, since each transporter exported salinomycin and improved its biosynthesis, it would be interesting to examine whether these pumps are competitive or cooperative. Thus, tandem over-expression of two or more genes will be employed to study their relationship, which may push the salinomycin titer to a higher level. By means of electronic microscope observation and molecular dynamics simulation, the conformation change of transporters during the pumping of salinomycin will be analyzed. Besides export, other resistance strategies, including the inactivation of antibiotics and the modification of function targets, are worthy of study with regard to salinomycin. Further bioinformatic and functional analyses are likely to provide answers to these questions.

4. Conclusions

In summary, we constructed a novel method for salinomycin exporter mining by combining soybean oil preference and transcriptome analysis. We identified eight BGC-independent transporters and verified their functions. Finally, one of them was proved to be efficient in multiple polyether antibiotic-producing hosts. Our work contributes new strategies to further the improvement of salinomycin titers, and it paves the way for increasing the titers of other antibiotics through transporter engineering.

Supplementary Materials: The following supporting information can be downloaded at https://www.mdpi.com/article/10.3390/antibiotics11050600/s1, Table S1: Strains and plasmids used in this study; Table S2: Primers used in this study; Table S3: Transcription data and annotation of eight candidate genes; Table S4: RT-qPCR analysis of candidate genes; Figure S1: Flowchart of the strategy for exporter gene identification; Figure S2: The process of gene deletion and over-expression of *SLNHY_0929*; Figure S3: HPLC analysis of salinomycin in the fermentation cultures of BK3-25 and mutant LX01 with *SLNHY_0929* deleted; Figure S4: HPLC analysis of lasalocid in the fermentation culture of *S. lasaliensis* and its mutants; Figure S5: HPLC analysis of monensin in the fermentation culture of *S. cinnamonensis* and its mutants; Figure S6: HPLC analysis of nigericin in the fermentation culture of *S. hygroscopicus* and its mutants; Figure S7: Growth (A) and salinomycin titer (B) curves of *Streptomyces albus* BK3-25. References [38–43] are cited in the supplementary materials.

Author Contributions: Conceptualization, L.B. and Y.Y.; methodology, X.L., Y.W. and X.Z.; validation, X.L.; formal analysis, L.B. and Q.K.; investigation, X.L.; data curation, L.B.; writing—original draft preparation, X.L.; writing—review and editing, L.B. and Y.Y.; supervision, L.B. and Y.Y.; project administration, L.B.; funding acquisition, L.B. and Y.Y. All authors have read and agreed to the published version of the manuscript.

Funding: This research was funded by the National Key R&D Program of China (2021YFC2100600, 2019YFA0905400), the Science and Technology Commission of Shanghai Municipality (19430750600, 19JC1413000), and the National Natural Science Foundation of China (31830104, 31801036).

Institutional Review Board Statement: Not applicable.

Informed Consent Statement: Not applicable.

Data Availability Statement: Not applicable.

Acknowledgments: We thank Weishan Wang for providing the *kas*Op* promoter and Yongquan Li for providing the BK3-25 strain.

Conflicts of Interest: The authors declare no conflict of interest. The funders had no role in the design of the study; in the collection, analyses, or interpretation of data; in the writing of the manuscript; or in the decision to publish the results.

References

1. Wang, H.; Liu, N.; Xi, L.; Rong, X.; Ruan, J.; Huang, Y. Genetic screening strategy for rapid access to polyether ionophore producers and products in Actinomycetes. *Appl. Environ. Microbiol.* **2011**, *77*, 3433–3442. [CrossRef] [PubMed]
2. Cane, D.E. Unified stereochemical model of polyether antibiotic structure and biogenesis. *J. Am. Chem. Soc.* **1983**, *2*, 3594–3600. [CrossRef]
3. Yurkovich, M.E.; Tyrakis, P.A.; Hong, H.; Sun, Y.; Samborskyy, M.; Kamiya, K.; Leadlay, P.F. A late-stage intermediate in salinomycin biosynthesis is revealed by specific mutation in the biosynthetic gene cluster. *ChemBioChem* **2012**, *13*, 66–71. [CrossRef] [PubMed]
4. Wietrzyk, J.; Antoszczak, M.; Popiel, K.; Stefa, J. Synthesis, cytotoxicity and antibacterial activity of new esters of polyether antibiotic salinomycin. *Eur. J. Med. Chem.* **2014**, *76*, 435–444.
5. Fuchs, D.; Daniel, V.; Sadeghi, M.; Opelz, G.; Naujokat, C. Salinomycin overcomes ABC transporter-mediated multidrug and apoptosis resistance in human leukemia stem cell-like KG-1a cells. *Biochem. Biophys. Res. Commun.* **2010**, *394*, 1098–1104. [CrossRef]
6. Gupta, P.B.; Onder, T.T.; Jiang, G.; Tao, K.; Kuperwasser, C.; Weinberg, R.A.; Lander, E.S. Identification of selective inhibitors of cancer stem cells by high-throughput screening. *Cell* **2009**, *138*, 645–659. [CrossRef]
7. Lu, C.; Zhang, X.; Jiang, M.; Bai, L. Enhanced salinomycin production by adjusting the supply of polyketide extender units in *Streptomyces albus*. *Metab. Eng.* **2016**, *35*, 129–137. [CrossRef]

8. Martín, J.F.; Liras, P. Engineering of regulatory cascades and networks controlling antibiotic biosynthesis in *Streptomyces*. *Curr. Opin. Microbiol.* **2010**, *13*, 263–273. [CrossRef]
9. Hopwood, D.A. MicroCommentary How do antibiotic-producing bacteria ensure their self-resistance before antibiotic biosynthesis incapacitates them? *Mol. Microbiol.* **2007**, *63*, 937–940. [CrossRef]
10. Mak, S.; Xu, Y.; Nodwell, J.R. Micro Review The expression of antibiotic resistance genes in antibiotic-producing bacteria. *Mol. Microbiol.* **2014**, *93*, 391–402. [CrossRef]
11. Jiang, C.; Wang, H.; Kang, Q.; Liu, J.; Bai, L. Cloning and characterization of the polyether salinomycin biosynthesis gene cluster of *Streptomyces albus* XM211. *Appl. Environ. Microbiol.* **2012**, *78*, 994–1003. [CrossRef] [PubMed]
12. Liu, J.; Jiang, C.; Zhang, B.; Bai, L. Involvement of ABC transporter genes *slnTI* and *slnTII* in salinomycin biosynthesis. *Microbiol. China* **2014**, *41*, 58–66.
13. Bentley, S.D.; Chater, K.F.; Cerdeño-Tárraga, A.-M.; Challis, G.L.; Thomson, N.R.; James, K.D.; Harris, D.E.; Quail, M.A.; Kieser, H.; Harper, D.; et al. Complete genome sequence of the model actinomycete *Streptomyces coelicolor* A3(2). *Nature* **2002**, *417*, 141–147. [CrossRef]
14. Méndez, C.; Salas, J.A. The role of ABC transporters in antibiotic-producing organisms: Drug secretion and resistance mechanisms. *Nucleic Acids Res.* **2001**, *152*, 341–350. [CrossRef]
15. Saier, M.H.; Reddy, V.S.; Tamang, D.G.; Västermark, Å. The transporter classification database. *Nucleic Acids Res.* **2014**, *42*, 251–258. [CrossRef]
16. Omura, S.; Ikeda, H.; Ishikawa, J.; Hanamoto, A.; Takahashi, C.; Shinose, M.; Takahashi, Y.; Horikawa, H.; Nakazawa, H.; Osonoe, T.; et al. Genome sequence of an industrial microorganism *Streptomyces avermitilis*: Deducing the ability of producing secondary metabolites. *Proc. Natl. Acad. Sci. USA* **2001**, *98*, 12215–12220. [CrossRef]
17. Wang, T.; Shan, Y.; Li, H. Multiple transporters are involved in natamycin efflux in *Streptomyces chattanoogensis* L10. *Mol. Microbiol.* **2017**, *103*, 713–728. [CrossRef]
18. Chu, L.; Li, S.; Dong, Z.; Zhang, Y.; Jin, P.; Ye, L.; Wang, X.; Xiang, W. Mining and engineering exporters for titer improvement of macrolide biopesticides in *Streptomyces*. *Microb. Biotechnol.* **2021**, *15*, 1120–1132. [CrossRef]
19. Zhang, X.; Lu, C.; Bai, L. Conversion of the high-yield salinomycin producer *Streptomyces albus* BK3-25 into a surrogate host for polyketide production. *Sci. China Life Sci.* **2017**, *60*, 1000–1009. [CrossRef]
20. Hopwood, D.A. Highlights of *Streptomyces* genetics. *Heredity* **2019**, *123*, 23–32. [CrossRef]
21. Ling, Y.; Zeng, Z. Optimization of fermentation conditions of polyether antibiotic lasalocid by response surface method. *Chin. J. Antibiot.* **2017**, *42*, 28–32.
22. Wang, X.; Ning, X.; Zhao, Q.; Kang, Q.; Bai, L. Improved PKS gene expression with strong endogenous promoter resulted in geldanamycin yield increase. *Biotechnol. J.* **2017**, *12*, 1700321. [CrossRef] [PubMed]
23. Tang, Z.; Li, X.; Pang, A.; Lin, C.; Zhang, Y.; Zhang, J.; Qiao, J.; Zhao, G. Characterization of three pathway-specific regulators for high production of monensin in *Streptomyces cinnamonensis*. *Appl. Microbiol. Biotechnol.* **2017**, *101*, 6083–6097. [CrossRef] [PubMed]
24. Leulmi, N.; Sighel, D.; Defant, A.; Khenaka, K.; Boulahrouf, A.; Mancini, I. Enhanced production and quantitative evaluation of nigericin from the Algerian soil-living *Streptomyces youssoufiensis* SF10 strain. *Fermentation* **2019**, *5*, 13. [CrossRef]
25. Qu, S.; Kang, Q.; Wu, H.; Wang, L.; Bai, L. Positive and negative regulation of GlnR in validamycin A biosynthesis by binding to different loci in promoter region. *Appl. Microbiol. Biotechnol.* **2015**, *99*, 4771–4783. [CrossRef]
26. Livak, K.J.; Schmittgen, T.D. Analysis of relative gene expression data using real-time quantitative PCR and the $2^{-\Delta\Delta CT}$ method. *Methods* **2001**, *25*, 402–408. [CrossRef]
27. Wang, W.; Li, X.; Wang, J.; Xiang, S.; Feng, X.; Yang, K. An engineered strong promoter for streptomycetes. *Appl. Environ. Microbiol.* **2013**, *79*, 4484–4492. [CrossRef]
28. Du, D.; Zhu, Y.; Wei, J.; Tian, Y.; Niu, G.; Tan, H. Improvement of gougerotin and nikkomycin production by engineering their biosynthetic gene clusters. *Appl. Microbiol. Biotechnol.* **2013**, *97*, 6383–6396. [CrossRef]
29. Qiu, J.; Zhuo, Y.; Zhu, D.; Zhou, X.; Zhang, L.; Bai, L.; Deng, Z. Overexpression of the ABC transporter AvtAB increases avermectin production in *Streptomyces avermitilis*. *Appl. Microbiol. Biotechnol.* **2011**, *92*, 337–345. [CrossRef]
30. Nigam, S.K. What do drug transporters really do? *Nat. Rev. Drug Discov.* **2014**, *14*, 29–44. [CrossRef]
31. Jiang, X.; Ellabaan, M.M.H.; Charusanti, P.; Munck, C.; Blin, K.; Tong, Y.; Weber, T.; Sommer, M.O.A.; Lee, S.Y. Dissemination of antibiotic resistance genes from antibiotic producers to pathogens. *Nat. Commun.* **2017**, *8*, 15784. [CrossRef] [PubMed]
32. Kell, D.B.; Swainston, N.; Pir, P.; Oliver, S.G. Membrane transporter engineering in industrial biotechnology and whole cell biocatalysis. *Trends Biotechnol.* **2015**, *33*, 237–246. [CrossRef] [PubMed]
33. Migita, A.; Watanabe, M.; Hirose, Y.; Watanabe, K.; Tokiwano, T.; Kinashi, H.; Oikawa, H. Identification of a gene cluster of polyether antibiotic lasalocid from *Streptomyces lasaliensis*. *Biosci. Biotechnol. Biochem.* **2009**, *73*, 169–176. [CrossRef] [PubMed]
34. Oliynyk, M.; Stark, C.B.W.; Bhatt, A.; Jones, M.A.; Hughes-Thomas, Z.A.; Wilkinson, C.; Oliynyk, Z.; Demydchuk, Y.; Staunton, J.; Leadlay, P.F. Analysis of the biosynthetic gene cluster for the polyether antibiotic monensin in *Streptomyces cinnamonensis* and evidence for the role of *monB* and *monC* genes in oxidative cyclization. *Mol. Microbiol.* **2003**, *49*, 1179–1190. [CrossRef] [PubMed]
35. Harvey, B.M.; Mironenko, T.; Sun, Y.; Hong, H.; Deng, Z.; Leadlay, P.F.; Weissman, K.J.; Haydock, S.F. Insights into polyether biosynthesis from analysis of the nigericin biosynthetic gene cluster in *Streptomyces* sp. DSM4137. *Chem. Biol.* **2007**, *14*, 703–714. [CrossRef]

36. Zhang, Y.; Lin, C.Y.; Li, X.M.; Tang, Z.K.; Qiao, J.; Zhao, G.R. DasR positively controls monensin production at two-level regulation in *Streptomyces cinnamonensis*. *J. Ind. Microbiol. Biotechnol.* **2016**, *43*, 1681–1692. [CrossRef]
37. Armando, J.W.; Boghigian, B.A.; Pfeifer, B.A. LC-MS/MS quantification of short-chain acyl-CoA's in *Escherichia coli* demonstrates versatile propionyl-CoA synthetase substrate specificity. *Lett. Appl. Microbiol.* **2012**, *54*, 140–148. [CrossRef]
38. Zhang, X.; Lu, C.; Bai, L. Mechanism of salinomycin overproduction in *Streptomyces albus* as revealed by comparative functional genomics. *Appl. Microbiol. Biotechnol.* **2017**, *101*, 4635–4644. [CrossRef] [PubMed]
39. Hopwood, D.A.; Hintermann, G.; Kieser, T.; Wright, H.M. Integrated DNA sequences in three streptomycetes form related autonomous plasmids after transfer to *Streptomyces lividans*. *Plasmid* **1984**, *11*, 1–16. [CrossRef]
40. Day, L.E.; Chamberlin, J.W.; Gordee, E.Z.; Chen, S.; Gorman, M.; Hamill, R.L.; Ness, T.; Weeks, R.E.; Stroshane, R. Biosynthesis of monensin. *Antimicrob. Agents Chemother.* **1973**, *4*, 410–414. [CrossRef]
41. Smith, P. High efficiency intergeneric conjugal transfer of plasmid DNA from *Esc herichia coli* to methyl DNA-restricting streptomycetes. *FEMS Microbiol. Lett.* **1997**, *155*, 223–229.
42. Wang, X.; Wang, R.; Kang, Q.; Bai, L. The antitumor agent ansamitocin P-3 binds to cell division protein FtsZ in *Actinosynnema pretiosum*. *Biomolecules* **2020**, *10*, 699. [CrossRef] [PubMed]
43. Wang, Z.; Bai, L.; Liang, J.; Zhou, X.; Deng, Z. Two pHZ1358 derivative vectors for efficient gene knockout in *Streptomyces*. *J. Microbiol. Biotechnol.* **2010**, *20*, 678–682.

antibiotics

MDPI

Communication

Genome Mining Discovery of a New Benzazepine Alkaloid Pseudofisnin A from the Marine Fungus *Neosartorya pseudofischeri* F27-1

Xiao-Xin Xue [1], Lin Chen [1] and Man-Cheng Tang [1,2,*]

1 State Key Laboratory of Microbial Metabolism, School of Life Sciences and Biotechnology, Shanghai Jiao Tong University, Shanghai 200240, China
2 Zhangjiang Institute for Advanced Study, Shanghai Jiao Tong University, Shanghai 200240, China
* Correspondence: tangmc@sjtu.edu.cn

Abstract: L-Kynurenine (Kyn) is an intermediate in the kynurenine pathway and is also found to be a building block or biosynthetic precursor to bioactive natural products. Recent studies revealed that L-Kyn can be incorporated via nonribosomal peptide synthetase (NRPS) biosynthetic routes to generate 1-benzazepine-containing compounds, while 1-benzazepine is a pharmaceutically important scaffold that is rarely found in natural products. Using a core biosynthetic enzyme-guided genome-mining approach, we discovered a biosynthetic gene cluster from *Neosartorya pseudofischeri* and identified that it encodes for the biosynthesis of pseudofisnins, novel 1-benzazepine-containing compounds. The biosynthetic pathway of pseudofisnins was elucidated through in vivo and in vitro experiments. The methyltransferase PseC from the pathway was biochemically characterized to be an iterative methyltransferase that catalyzes off-NRPS line di-methylation on an amine group.

Keywords: 1-benzazepine; L-kynurenine; genome mining; biosynthesis; iterative methyltransferase

Citation: Xue, X.-X.; Chen, L.; Tang, M.-C. Genome Mining Discovery of a New Benzazepine Alkaloid Pseudofisnin A from the Marine Fungus *Neosartorya pseudofischeri* F27-1. *Antibiotics* **2022**, *11*, 1444. https://doi.org/10.3390/antibiotics11101444

Academic Editor: Madan K. Kharel

Received: 1 October 2022
Accepted: 19 October 2022
Published: 20 October 2022

1. Introduction

L-Kynurenine (Kyn) (Figure 1A) is a metabolite of the L-tryptophan (Trp) metabolism and is known as the biosynthetic precursor to synthesize nicotinamide adenine dinucleotide in the well-known kynurenine pathway [1]. The conversion of L-Trp to *N*-formyl-L-Kyn, which is rapidly hydrolyzed to L-Kyn, is catalyzed by indoleamine-2,3-dioxygenase (IDO) or tryptophan-2,3-dioxygenase (TDO) [2,3]. L-Kyn is present in milk proteins, lens crystallins, and human Cu^{2+}/Zn^{2+} superoxide dismutase in vivo [1]. In addition, L-Kyn is a non-proteinogenic amino acid with roles in a number of biochemical signaling pathways [4,5]. However, L-Kyn is rarely found as a building block or biosynthetic precursor to bioactive natural products. In the past few decades, a limited number of examples have been reported. For example, L-Kyn is a building block of daptomycin, which is a lipodepsipeptide isolated from *Streptomyces roseoporus* that is used in the treatment of Gram-positive pathogen skin infections [6]. Recently, several research groups reported the genome mining discovery of novel natural products, such as nanangelenin A and aspcandine, from fungi with new chemical scaffolds derived from L-Kyn, which is incorporated by nonribosomal peptide synthetase (NRPS) [7–9]. An interesting finding in these studies is that all the identified gene clusters contain genes encoding for NRPS and IDO. Considering the accumulation of fungal genome sequence data, the discovery of novel L-Kyn-derived bioactive compounds could be achieved through the mining of the available fungal genomes.

Benzazepines are a class of nitrogenous heterocyclic compounds with diverse chemical structures and broad bioactivities, exemplified by the heart-rate-lowering agent ivabradine and the angiotensin-converting enzyme inhibitor, benazepril [10,11]. Natural products containing a benzazepine structural unit are found to be widespread in nature, especially the 2- and 3-benzazepines, such as communesin [12] and rhoeadine [13]. However, 1-benzazepine-containing natural products are relatively rare. In recent studies, novel 1-benzazepine-containing compounds have been discovered from different fungal species via genome mining [7–9]. Biosynthetic studies have revealed that the 1-benzazepine unit within these compounds was directly derived from L-Kyn by NRPS pathways (Figure 1B). Since there are many cryptic NRPS biosynthetic gene clusters that exist in the fungal genomes, we reasoned that other novel 1-benzazepine-containing natural products could be discovered through genome mining.

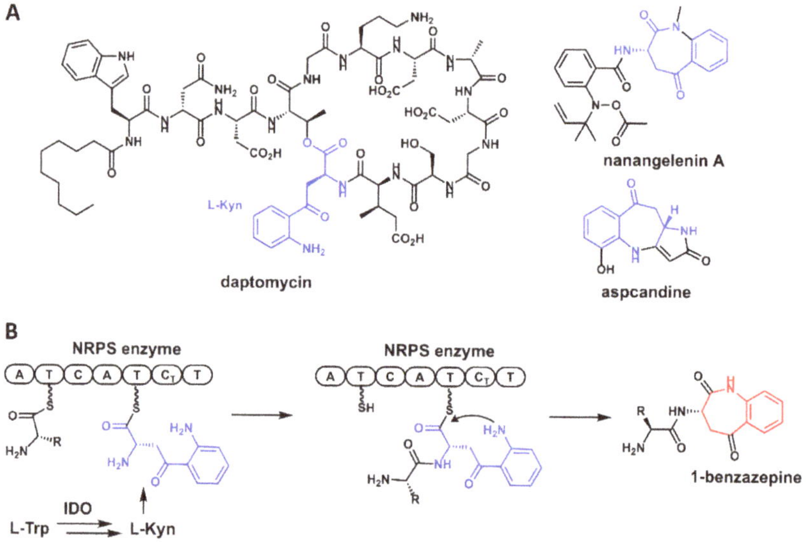

Figure 1. The structures of L-kynurenine (L-Kyn)-derived natural products (**A**) and the representative biosynthetic route to generating 1-benzazepine-containing compounds from L-Kyn (**B**).

Here, we reported the discovery of pseudofisnins, 1-benzazepine-containing compounds from the marine fungus *Neosartorya pseudofischeri* F27-1, via a core biosynthetic enzyme-guided genome mining approach. A biosynthetic pathway to generate these compounds was established by both in vivo and in vitro experiments. In addition, a methyltransferase PseC from the pathway was characterized to be an iterative methyltransferase catalyzing off-NRPS line methylation modification.

2. Results and Discussion

2.1. Identification of the Pse Cluster through Core Biosynthetic Enzyme-Guided Genome Mining

As reported in the biosynthesis of nanangelenins, the core biosynthetic enzymes involved in the nanangelenin scaffold assembly are the NRPS NanA and the IDO NanC [7]. A number of cryptic biosynthetic gene clusters (BGCs) can also be found in the sequenced fungal genomes, using NanA and NanC as the search query [7]. Therefore, we reasoned that the combination of NanA and NanC homolog pairs with different tailoring enzymes could lead to the generation of novel compounds belonging to the benzazepine alkaloid family.

Using the core biosynthetic enzymes, NanA and NanC, as the search query, we identified a compact BGC *pse* cluster from the genome of the marine fungus, *N. pseudofischeri* F27-1 (Figure 2A). Apart from the genes encoding NanA and NanC homologs, *pseA* and *pseB*, respectively, the *pse* cluster also encodes a methyltransferase (*pseC*), a flavine-dependent monooxygenase (*pseD*), a hypothetical protein (*pseE*), a deaminase (*pseF*), and a UbiH type hydroxylase (*pseG*). Since the other genes (*pseC-G*) of the *pse* cluster are different from the tailoring genes in the nanangelenin biosynthetic pathway, we hypothesized that the *pse* cluster might encode for novel benzazepine-containing natural products.

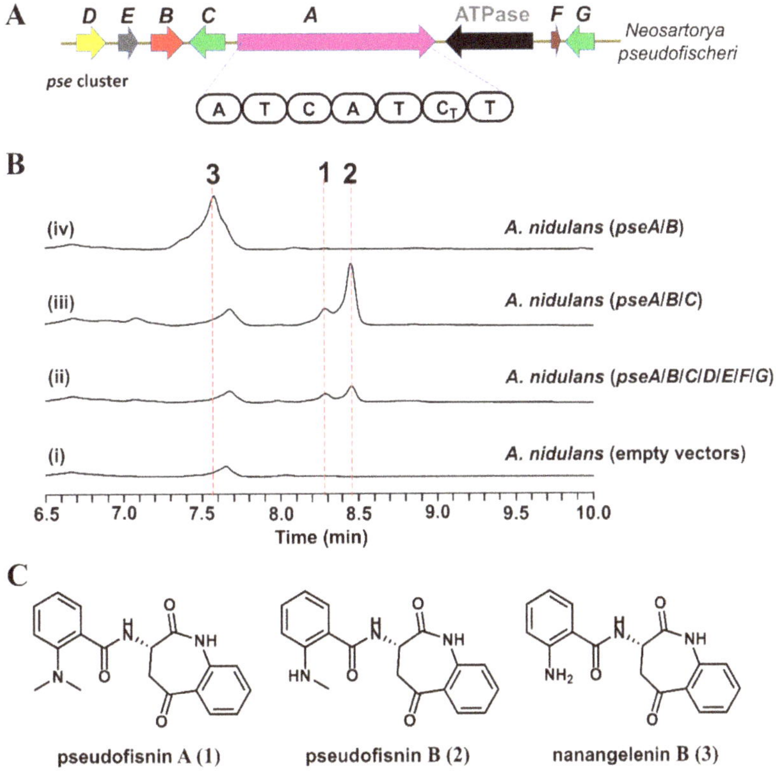

Figure 2. The *pse* cluster is responsible for the biosynthesis of pseudofisnins. (**A**) The *pse* cluster identified from *Neosartorya pseudofischeri*. *pseA* encodes NRPS, *pseB* encodes IDO, *pseC* encodes methyltransferase, *pseD* encodes flavine-dependent monooxygenase, *pseE* encodes a hypothetical protein, *pseF* encodes deaminase, and *pseG* encodes UbiH type hydroxylase. (**B**) Product profiles of *A. nidulans*, transformed with combinations of *pse* genes. The traces are HPLC with λ = 312 nm. (**C**) The structures of the characterized compounds.

2.2. Heterologous Expression of the Pse Cluster and Characterization of the Products

To explore the natural product encoded by the *pse* cluster, all seven putative biosynthetic genes (*pseA-G*) were introduced into an engineered *Aspergillus nidulans* expression host on three episomal vectors (Tables S2 and S3 in the Supplementary Materials) [14]. Compared to the negative control, two new metabolites, **1** (with a molecular weight of 337) and **2** (with a molecular weight of 323), were identified from the extract of *A. nidulans* expressing *pseA-G* (Figure 2B, trace (ii), and Figure S3 in the Supplementary Materials). For the purposes of structural elucidation, these two compounds were isolated and purified. Based on 1D and 2D Nuclear Magnetic Resonance (NMR) spectroscopy data (Tables S5 and S6 and Figures S7–S18 in the Supplementary Materials), **1** (named pseudofisnin A) and **2** (named pseudofisnin B) were characterized to be 1-benzazepine-containing compounds (the structures of which are shown in Figure 2C). The only difference between **1** and **2** is that the amine group attached to the benzene ring is di-methylated in **1** but mono-methylated in **2**.

The characterization of **1** and **2** validated our hypothesis that the *pse* cluster encodes the enzymes for the biosynthesis of natural products belonging to the benzazepine alkaloid family. In the previously reported studies of nanangelelins, nanangelenin B (**3**, the structure shown in Figure 2C) was identified to be the product of the NRPS assembly line and the biosynthetic intermediate to other nanangelelin compounds [7]. As **1** and **2** are structural analogs to **3**, we reasoned that **3** could also be the biosynthetic precursor to **1** and **2**. Combined with the bioinformatics analysis results, we proposed that the biosynthesis of pseudofisnins would involve the IDO encoding gene, *pseB*, the NRPS encoding gene, *pseA*, and one or two methyltransferase encoding genes, such as *pseC*.

2.3. Functional Identification of the Genes Involved in the Biosynthesis of Pseudofisnins

To test our hypothesis for the biosynthesis of pseudofisnins, different combinations of other genes from the *pse* cluster were co-expressed together with *pseA* and *pseB* in *A. nidulans*. Interestingly, when the putative methyltransferase PseC was co-expressed with PseA and PseB, the production of **1** and **2** was detected (Figure 2B, trace (iii)). When *pseC* was removed from the co-expression strain, the production of **1** and **2** was abolished (Figure 2B, trace (iv)). Instead, another new metabolite (**3**) with an MW of 309 emerged, which is identical to the MW of nanangelenin B. Structural elucidation of this new metabolite, based on NMR data (Table S4 and Figures S4–S6 in the Supplementary Materials), confirmed it to be nanangelenin B.

These results confirmed that only three genes, encoding for the NRPS PseA, the IDO PseB, and the methyltransferase PseC, are required for the biosynthesis of pseudofisnins, which is consistent with our hypothesis. Similar to the biosynthesis of nanangelenins, PseA and PseB are responsible for the biosynthesis of nanangelenin B. In addition, the results from the heterologous expression studies suggested that the methyltransferase PseC might catalyze the methylation of **3** to afford **1** and **2**, acting as an iterative methyltransferase.

2.4. In Vitro Characterization of PseC as an Iterative Methyltransferase

To further validate the function of PseC, the intron-free *pseC* gene was cloned from the extracted cDNA of *A. nidulans*, expressing the *pse* cluster and being overexpressed in *Escherichia coli* BL21 (DE3) as an *N*-terminal His_6-tagged fusion protein. In the presence of PseC and the cofactor, *S*-adenosyl methionine (SAM), in PIPES (1,4-piperazinediethanesulfonic acid) buffer (pH 7.4), the conversion of **3** to **1** and **2** was clearly detected after 1 h of incubation (Figure 3, trace (iv)). To further investigate, **2** was tested as the substrate under the same reaction conditions. As shown in Figure 3, **2** could be totally converted into **1** by PseC (Figure 3, trace (ii)). These results confirmed that PseC is an iterative methyltransferase catalyzing di-methylation on the amine group of **3** to **1** by a stepwise mechanism.

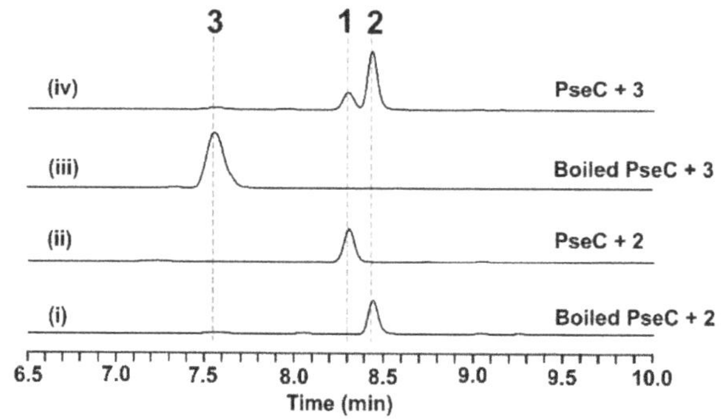

Figure 3. Biochemical characterization of the methyltransferase PseC in vitro. The traces are of HPLC with λ = 312 nm.

To explore the catalytic efficiency of PseC in the first methylation step and the second methylation step, kinetic studies of PseC on different substrates were carried out under optimized reaction conditions. When **3** was tested as the substrate, to minimize the formation of **1** and accumulate **2** as the major product, the enzymatic reaction mixtures were only incubated for 10 min at 30 °C. As shown in Figure S2 in the Supplementary Materials, the kinetic measurements showed PseC to have a K_M of 72 ± 16 µM and k_{cat} of 9.8 ± 0.8 min^{-1} toward **3** for the first methylation step, and a K_M of 14 ± 5 µM and k_{cat} of 0.29 ± 0.04 min^{-1} toward **2** for the second methylation step. The results indicated that PseC has a stronger binding affinity toward **2** over **3**, which could suggest that PseC will efficiently catalyze the second methylation step to complete the biosynthesis, to form **1**.

2.5. The Proposed Biosynthetic Pathway of **1**

By combining the results from both in vivo and in vitro experiments, a plausible biosynthetic pathway to **1** was proposed. As shown in Figure 4, the IDO homolog PseB could catalyze the ring-opening oxidation of L-Trp to form *N*-formyl-L-Kyn, which is then hydrolyzed to form L-Kyn, either spontaneously or via the unknown endogenous kynurenine formamidase. Then, the NRPS PseA incorporates one molecule of anthranilic acid and one molecule of L-Kyn to assemble the dipeptide scaffold, which is further released by the C_T domain through regioselective lactamization to form **3**, with a benzazepine core structure. Finally, the iterative methyltransferase PseC catalyzes the stepwise methylation of **3**, to complete the biosynthesis of **1**.

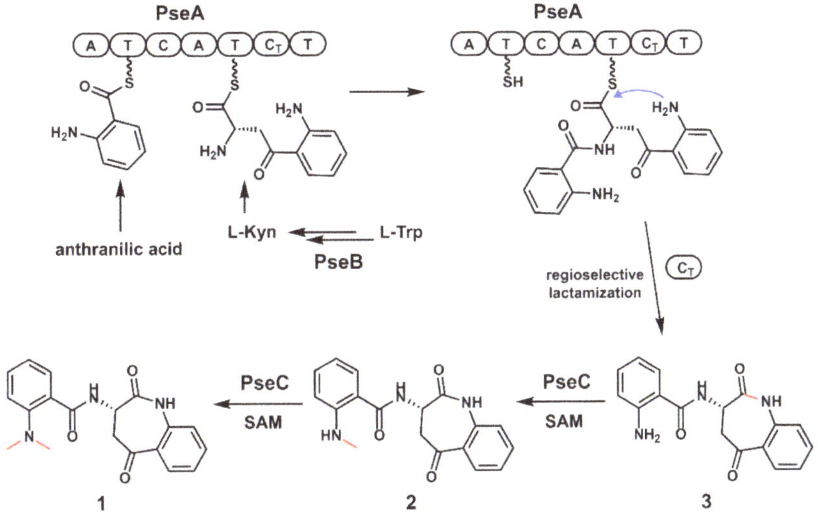

Figure 4. The proposed biosynthetic pathway of pseudofisnins.

3. Conclusions

In summary, by combining genome mining and heterologous expression, a biosynthetic gene cluster, including IDO and NRPS encoding genes, was mined from the genome of the marine fungus *N. pseudofischeri* and was characterized as being responsible for the biosynthesis of the novel benzazepine alkaloid, pseudofisnin A. The biosynthetic pathway of pseudofisnin A was proposed and established based on in vivo and in vitro studies. An iterative methyltransferase was biochemically characterized to catalyze the off-line post-di-methylation on the amine group of nanangelenin B, to complete the biosynthesis of pseudofisnin A. Our results provide a successful example of the discovery of novel natural products through core biosynthetic enzyme-guided genome mining.

4. Materials and Methods

4.1. Strains and Culture Conditions

The *Neosartorya pseudofischeri* strain F27-1 was grown at 28 °C over 4 days on PDA medium (Potato Dextrose Water, 2% agar) for genome extraction. The *Saccharomyces cerevisiae* strain BJ5464-NpgA (*MATα ura3-52 his3-Δ200 leu2-Δ1 trp1 pep4::HIS3 prb1 Δ1.6R can1 GAL*) was used for in vivo homologous recombination and was cultured in YPD (yeast extract 1%, peptone 2%, glucose 2%) media or SDCt (uracil dropout) medium at 28 °C. *Aspergillus nidulans* A1145 Δ*EM*Δ*ST* [15] was used as the host for heterologous expression, cultured at 37 °C on CD medium (1% glucose, 50 mL/L of 20× nitrate salts, 1 mL/L of trace elements, pH 6.5) for 3 days for sporulation, or fermented at 28 °C in CD-ST (starch 2%, tryptone 2%, 50 mL/L of 20× nitrate salts, 1 mL/L of trace elements, pH 6.5). The 20× nitrate salts solution was prepared by dissolving 120 g $NaNO_3$, 10.4 g KCl, 10.4 g $MgSO_4 \cdot 7H_2O$, and 30.4 g KH_2PO_4 in 1 L double-distilled water. The trace element solution (100 mL) consisted of 2.20 g of $ZnSO_4 \cdot 7H_2O$, 1.10 g of H_3BO_3, 0.50 g of $MnCl_2 \cdot 4H_2O$, 0.16 g of $FeSO_4 \cdot H_2O$, 0.16 g of $CoCl_2 \cdot 5H_2O$, 0.16 g of $CuSO_4 \cdot 5H_2O$, and 0.11 g of $(NH_4)_6Mo_7O_{24} \cdot 4H_2O$, and the pH was adjusted to 6.5. All *Escherichia coli* were cultured in LB medium at 37 °C. *E. coli* BL21 (DE3) (Tsingke, Beijing, China) was used for protein expression.

4.2. General DNA Manipulation Techniques

The genomic DNA of *N. pseudofischeri* F27-1 was extracted using the CTAB method [16]. *E. coli* HB101 and DH5α were used for cloning, following the standard recombinant DNA techniques. DNA restriction enzymes were used as recommended by the manufacturer (New England Biolabs, NEB, Ipswich, MA, USA). PCR reactions were performed using Phanta® Max Super-Fidelity DNA Polymerase (Vazyme, Nanjing, China). The gene-specific primers are listed in Table S2 in the Supplementary Materials. The plasmids (Table S3 in the Supplementary Materials) that are used for heterologous expression in *A. nidulans* were constructed by yeast homologous recombination using the Frozen-EZ Yeast Transformation II Kit™ (Zymo Research, Irvine, CA, USA). The yeast plasmid extraction was performed with Zymoprep™ Yeast Plasmid Miniprep I (Zymo Inc., Irvine, CA, USA).

4.3. Plasmids Construction

The vectors pYTP, pYTU, and pYTR, with auxotrophic markers for pyridoxine (*pyroA*), uracil (*pyrG*), and riboflavin (*riboB*), respectively, were used for insertion into target genes. All three plasmids were digested with *Pac*I and *Swa*I. The target genes were amplified from the genomic DNA of *N. pseudofischeri* F27-1, with their native terminators added to the ends, and the amplified fragments were flanked by homologous arms for recombination. The prepared fragments and vectors were co-transformed into *S. cerevisiae* BJ5464-NpgA yeast-competent cells to yield the plasmids for gene expression in *A. nidulans*, which were verified by sequencing.

To construct the protein expression plasmid of PseC for *E. coli* BL21 (DE3), the intron-free *pseC* gene was amplified from the cDNA of the *A. nidulans* strain expressing the *pse* gene cluster, using the primers listed in Table S2 in the Supplementary Materials. The expression vector pET28a was digested with *Nde*I and *Bam*H I. The digested vector and the target fragment with homologous arms were recombined using the ClonExpress® II One-Step Cloning Kit (Vazyme). The resulting plasmid pXXX1-7 (Table S3 in the Supplementary Materials) was verified by enzyme digestion and sequencing.

4.4. Protoplast Preparation and Transformation of A. nidulans

The spores of *A. nidulans* A1145 Δ*EM*Δ*ST* were inoculated into 50 mL of liquid CD, containing 10 mM uridine, 5 mM uracil, 0.5 µg/mL pyridoxine and 0.125 µg/mL riboflavin, and then germinated at 37 °C at 220 rpm for approximately 9 h. The mycelia were harvested by centrifugation at 5000 rpm for 5 min at 4 °C and washed with 15 mL osmotic buffer (1.2 mol/L $MgSO_4 \cdot 7H_2O$, 10 mM sodium phosphate, pH 5.8). The mycelium was then transferred to 10 mL osmotic buffer containing 30 mg lysing enzymes and 20 mg Yatalase and was then shaken at 28 °C and 80 rpm for 14 h. After the protoplasts were observed under the microscope, the liquid was poured into a sterile 50 mL centrifuge tube and gently covered with an equal volume of trapping buffer (0.6 M sorbitol, 0.1 M Tris-HCl, pH 7.0). Centrifugation at 3750 rpm for 15 min at 4 °C layered the protoplasts at the interface of the two buffers. Then the protoplasts were washed and resuspended with STC buffer (1.2 M sorbitol, 10 mM $CaCl_2$, 10 mM Tris-HCl, pH 7.5).

For transformation, recombinant plasmids were added to the protoplasts of *A. nidulans* and placed on ice for 30 min. First, 60% PEG buffer (60% PEG 4000, 0.56% $CaCl_2$, 50 mM Tris, pH 7.5) was added to the mixture and then induced at room temperature for 1 h. The mixture was grown on the regeneration dropout solid medium (CD medium with 1.2 mM sorbitol and appropriate supplements) at 37 °C for about 2 days. The transformants were inoculated on fresh dropout solid medium at 37 °C for about 2 days.

4.5. Chemical Analysis and Compound Isolation and Characterization

In order to perform small-scale metabolite analysis in *A. nidulans*, different transformants were grown on liquid CD-ST media at 28 °C for 3–5 days, then extracted with ethyl acetate (EtOAc). UPLC–MS analysis was carried out for 15 min on a Shimadzu LC-30A system connected to a single quadrupole mass spectrometer MS2020 (ESI) (Shimadzu, Kyoto, Japan), using a C18 reverse-phase column (shim pack XR-ODS III, 2.0 mm × 75 mm, 1.6 μm) with a linear gradient of 5−95% MeCN-H_2O and a flow rate of 0.2 mL/min.

For the isolation of nanangelenin B (**1**), pseudofisnin A (**2**), and pseudofisnin B (**3**), the corresponding transformants of *A. nidulans* strains were grown on 4 L of liquid CD media for 4 days at 28 °C and then extracted with EtOAc. After concentration, the crude extract was chromatographed on a silica gel column using EtOAc and *n*-hexane as eluent. The fractions containing the target compounds were combined and further purified by semi-preparative HPLC, using a Pntulips C18 reverse-phase column (C18, 5 μm, 250 × 10 mm) (Puningtech, Shanghai, China).

The 1D and 2D NMR spectra were recorded in $CDCl_3$ or DMSO-d_6 using Bruker 600 MHz spectrometers and tetramethylsilane (TMS) as an internal standard. HR-ESIMS (high-resolution electrospray ionization mass) data were measured on an Acquity 2D-UPLC/Acquity UPC2/Xevo G2-XS QTOF (Waters, Milford, MA, USA).

4.6. Protein Expression and Purification of PseC

The correct plasmid was transformed into BL21 (DE3) to express the *N*-terminal His_6-tagged fusion protein. The protein purification of PseC was performed as follows. The transformed strains were first cultured overnight in 5 mL LB containing 50 μg/mL kanamycin and were then transferred to 2 L LB medium containing antibiotics at a ratio of 1:100. The culture was shaken at 37 °C until the optical density at 600 nm (OD_{600}) reached 0.6~0.8, and then 0.2 mM isopropylthio-β-D-galactoside (IPTG) was added for induction. After induction at 16 °C for 22 h, the cells were collected via centrifugation at 5000 rpm. The collected cells were resuspended in lysis buffer (20 mM Tris-HCl, 300 mM NaCl, 5 mM β-mercaptoethanol, 10 mM imidazole, 1 mM PMSF, 10% glycerol, pH 8.0) and sonicated on ice. After centrifugation (12,000 rpm, 30 min, 4 °C), the supernatant was collected and added to Ni-NTA agarose resin (Smart-Lifesciences, Changzhou, China). Then, the resin was loaded into a gravity flow column and eluted in a gradient with elution buffer (20 mM Tris-HCl, 300 mM NaCl, 10% glycerol, pH 8.0) containing progressively increasing concentrations of imidazole (20 mM to 300 mM). After analysis by 12.5% (*w*/*v*) acrylamide sodium dodecyl sulfate polyacrylamide gel electrophoresis (SDS-PAGE) (Figure S1 in the Supplementary Materials), the protein fractions containing PseC were combined and concentrated using Amicon-Ultra Centrifugal Filters (Millipore, Burlington, MA, USA). The protein solution was concentrated to 2.5 mL and further desalted using a PD-10 column (GE Healthcare, Chicago, IL, USA) with elution buffer (50 mM HEPES, 100 mM NaCl, 10% glycerol, pH 7.5). The concentration of the protein was measured with a Bradford protein assay (Bio-Rad, Hercules, CA, USA). Finally, the proteins were flash-frozen in liquid nitrogen and stored at −80 °C.

4.7. Enzymatic Assays of PseC In Vitro

To verify the function of PseC in vitro, 10 μM PseC was incubated with 120 μM nanangelenin B (**1**) or 30 μM pseudofisnin A (**2**), with 2 mM SAM in 50 mM PIPES (pH 7.4) buffer. The reaction mixture was incubated at 30 °C for 1 h and then quenched with 2 times the volume of acetonitrile. Boiled PseC was used in the negative control assays. After centrifugation, the supernatant was analyzed via LC-MS, using the method mentioned above.

Kinetic studies of the protein were performed using the following conditions: 1 μM PseC was incubated with 4–480 μM nanangelenin B (**1**) or 2.5–115 μM pseudofisnin A (2), with 2 mM SAM in 50 mM PIPES (pH 7.4) buffer for 10 min at 30 °C. After quenching with acetonitrile, the reaction mixtures were analyzed by LC-MS to determine the product formation. Each data point represents a minimum of three replicate endpoint assays; kinetic constants were obtained by non-linear regression analysis using GraphPad Prism 9 (GraphPad Software, Inc., San Diego, CA, USA).

Supplementary Materials: The following supporting information can be downloaded at https://www.mdpi.com/article/10.3390/antibiotics11101444/s1, Figure S1: SDS-PAGE analysis of purified PseC; Figure S2: Kinetic analysis of the methylation reaction catalyzed by PseC; Figure S3: HR-ESIMS of (a) pseudofisnin B and (b) pseudofisnin A; Figure S4: ^{1}H NMR spectrum of nanangelenin B (**3**) in DMSO-d_6. Figure S5: ^{13}C NMR spectrum of nanangelenin B (**3**) in DMSO-d_6; Figure S6: DEPT 135 spectrum of nanangelenin B (**3**) in DMSO-d_6; Figure S7: ^{1}H NMR spectrum of pseudofisnin B (**2**) in $CDCl_3$; Figure S8: ^{13}C NMR spectrum of pseudofisnin B (**2**) in $CDCl_3$; Figure S9: DEPT 135 spectrum of pseudofisnin B (**2**) in $CDCl_3$; Figure S10: COSY spectrum of pseudofisnin B (**2**) in $CDCl_3$; Figure S11: HSQC spectrum of pseudofisnin B (**2**) in $CDCl_3$; Figure S12: HMBC spectrum of pseudofisnin B (**2**) in $CDCl_3$; Figure S13: ^{1}H NMR spectrum of pseudofisnin A (**1**) in $CDCl_3$; Figure S14: ^{13}C NMR spectrum of pseudofisnin A (**1**) in $CDCl_3$; Figure S15: DEPT 135 spectrum of pseudofisnin A (**1**) in $CDCl_3$; Figure S16: COSY spectrum of pseudofisnin A (**1**) in $CDCl_3$; Figure S17: HSQC spectrum of pseudofisnin A (**1**) in $CDCl_3$; Figure S18: HMBC spectrum of pseudofisnin A (**1**) in $CDCl_3$; Table S1: Bioinformatics analysis of the *pse* gene cluster; Table S2: Primers used in this study; Table S3: Plasmids used in this study; Table S4: NMR data of nanangelenin B (**3**) in DMSO-d_6; Table S5: NMR data of pseudofisnin B (**2**) in $CDCl_3$; Table S6: NMR data of pseudofisnin A (**1**) in $CDCl_3$.

Author Contributions: Conceptualization, X.-X.X., L.C. and M.-C.T.; methodology, X.-X.X., L.C. and M.-C.T.; software, X.-X.X., and L.C.; validation, X.-X.X., L.C. and M.-C.T.; formal analysis, X.-X.X., and L.C.; investigation, X.-X.X.; resources, M.-C.T.; data curation, X.-X.X., L.C. and M.-C.T.; writing—original draft preparation, X.-X.X., L.C. and M.-C.T.; writing—review and editing, X.-X.X., L.C. and M.-C.T.; visualization, X.-X.X., and L.C.; supervision, M.-C.T.; project administration, M.-C.T.; funding acquisition, M.-C.T. All authors have read and agreed to the published version of the manuscript.

Funding: This research was funded by the National Key Research and Development Program of China (2018YFA0901900) and the National Natural Science Foundation of China (31971345).

Institutional Review Board Statement: Not applicable.

Informed Consent Statement: Not applicable.

Data Availability Statement: Data are available in publicly accessible repository and within the article or Supplementary Materials.

Acknowledgments: We thank Wen-Jian Lan from Sun Yat-sen University for providing the marine fungus *Neosartorya pseudofischeri* F27-1.

Conflicts of Interest: The authors declare no conflict of interest.

References

1. Cervenka, I.; Agudelo, L.Z.; Ruas, J.L. Kynurenines: Tryptophan's metabolites in exercise, inflammation, and mental health. *Science* **2017**, *357*, eaaf9794. [CrossRef] [PubMed]
2. Heyes, M.P.; Achim, C.L.; Wiley, C.A.; Major, E.O.; Saito, K.; Markey, S.P. Human microglia convert L-tryptophan into the neurotoxin quinolinic acid. *Biochem. J.* **1996**, *320*, 595–597. [CrossRef] [PubMed]
3. Batabyal, D.; Yeh, S.R. Human tryptophan dioxygenase: A comparison to indoleamine 2,3-dioxygenase. *J. Am. Chem. Soc.* **2007**, *129*, 15690–15701. [CrossRef] [PubMed]
4. Stone, T.W.; Darlington, L.G. Endogenous kynurenines as targets for drug discovery and development. *Nat. Rev. Drug Disc.* **2002**, *1*, 609–620. [CrossRef] [PubMed]
5. Malpass, K. Neurodegenerative disease: The kynurenine pathway–promising new targets and therapies for neurodegenerative disease. *Nat. Rev. Neurol.* **2011**, *7*, 417. [CrossRef] [PubMed]
6. Karas, J.A.; Carter, G.P.; Howden, B.P.; Turner, A.M.; Paulin, O.K.A.; Swarbrick, J.D.; Baker, M.A.; Li, J.; Velkov, T. Structure-Activity Relationships of Daptomycin Lipopeptides. *J. Med. Chem.* **2020**, *63*, 13266–13290. [CrossRef] [PubMed]
7. Li, H.; Gilchrist, C.L.M.; Phan, C.S.; Lacey, H.J.; Vuong, D.; Moggach, S.A.; Lacey, E.; Piggott, A.M.; Chooi, Y.H. Biosynthesis of a New Benzazepine Alkaloid Nanangelenin A from *Aspergillus nanangensis* Involves an Unusual L-Kynurenine-Incorporating NRPS Catalyzing Regioselective Lactamization. *J. Am. Chem. Soc.* **2020**, *142*, 7145–7152. [CrossRef] [PubMed]
8. Caesar, L.K.; Robey, M.T.; Swyers, M.; Islam, M.N.; Ye, R.; Vagadia, P.P.; Schiltz, G.E.; Thomas, P.M.; Wu, C.C.; Kelleher, N.L.; et al. Heterologous Expression of the Unusual Terreazepine Biosynthetic Gene Cluster Reveals a Promising Approach for Identifying New Chemical Scaffolds. *mBio* **2020**, *11*, e01691-20. [CrossRef] [PubMed]
9. Chen, L.; Tang, J.W.; Liu, Y.Y.; Matsuda, Y. Aspcandine: A Pyrrolobenzazepine Alkaloid Synthesized by a Fungal Nonribosomal Peptide Synthetase-Polyketide Synthase Hybrid. *Org. Lett.* **2022**, *24*, 4816–4819. [CrossRef] [PubMed]
10. Jigar, H.S.; Rama, M.H.; Hari, N.P. Pharmacological and biological activities of benzazepines: An overview. *Curr. Bioact. Compd.* **2015**, *11*, 170–188.
11. Webb, R.L.; Miller, D.; Traina, V.; Gomez, H.J. Benazepril. *Cardiovasc. Drug Rev.* **1990**, *8*, 89–104. [CrossRef]
12. Lin, H.-C.; Chiou, G.; Chooi, Y.-H.; McMahon, T.C.; Xu, W.; Garg, N.K.; Tang, Y. Elucidation of the concise biosynthetic pathway of the communesin indole alkaloids. *Angew. Chem. Int. Ed.* **2015**, *54*, 3004–3007. [CrossRef] [PubMed]
13. Rönsch, H. Chapter 1 Rhoeadine alkaloids. In *The Alkaloids: Chemistry and Pharmacology*; Brossi, A., Ed.; Academic Press: Orlando, FL, USA, 1986; Volume 28, pp. 1–93.
14. Li, L.; Yu, P.; Tang, M.C.; Zou, Y.; Gao, S.S.; Hung, Y.S.; Zhao, M.; Watanabe, K.; Houk, K.N.; Tang, Y. Biochemical Characterization of a Eukaryotic Decalin-Forming Diels-Alderase. *J. Am. Chem. Soc.* **2016**, *138*, 15837–15840. [CrossRef] [PubMed]
15. Liu, N.; Hung, Y.S.; Gao, S.S.; Hang, L.; Zou, Y.; Chooi, Y.H.; Tang, Y. Identification and Heterologous Production of a Benzoyl-Primed Tricarboxylic Acid Polyketide Intermediate from the Zaragozic Acid A Biosynthetic Pathway. *Org. Lett.* **2017**, *19*, 3560–3563. [CrossRef] [PubMed]
16. Cubero, O.F.; Crespo, A.; Fatehi, J.; Bridge, P.D. DNA extraction and PCR amplification method suitable for fresh, herbarium-stored, lichenized, and other fungi. *Plant Syst. Evol.* **1999**, *216*, 243–249. [CrossRef]

antibiotics

MDPI

Article

Genome Mining and Metabolic Profiling Reveal Cytotoxic Cyclodipeptides in *Streptomyces hygrospinosus* var. Beijingensis

Dashan Zhang, Junbo Wang, Yongjian Qiao, Baixin Lin, Zixin Deng, Lingxin Kong * and Delin You *

State Key Laboratory of Microbial Metabolism, Joint International Research Laboratory of Metabolic and Developmental Sciences, School of Life Sciences and Biotechnology, Shanghai Jiao Tong University, Shanghai 200240, China
* Correspondence: konglingxin7@sjtu.edu.cn (L.K.); dlyou@sjtu.edu.cn (D.Y.)

Abstract: Two new cyclodipeptide (CDP) derivatives (**1–2**) and another seven known cyclodipeptides (**3–9**) were isolated from *Streptomyces* 26D9-414 by the genome mining approach combined with genetic dereplication and the "one strain many compounds" (OSMAC) strategy. The structures of the new CDPs were established on the basis of 1D- and 2D-NMR and comparative electronic circular dichroism (ECD) spectra analysis. The biosynthetic gene clusters (BGCs) for these CDPs were identified through antiSMASH analysis. The relevance between this *cdp* cluster and the identified nine CDPs was established by genetic interruption manipulation. The newly discovered natural compound **2** displayed comparable cytotoxicity against MDA-MB-231 and SW480 with that of cisplatin, a widely used chemotherapeutic agent for the treatment of various cancers.

Keywords: genome mining; genetic dereplication; cyclodipeptide; OSMAC; cytotoxicity; *Streptomyces*

Citation: Zhang, D.; Wang, J.; Qiao, Y.; Lin, B.; Deng, Z.; Kong, L.; You, D. Genome Mining and Metabolic Profiling Reveal Cytotoxic Cyclodipeptides in *Streptomyces hygrospinosus* var. Beijingensis. *Antibiotics* **2022**, *11*, 1463. https://doi.org/10.3390/antibiotics11111463

Academic Editor: Jesús F. Aparicio

Received: 13 October 2022
Accepted: 21 October 2022
Published: 24 October 2022

1. Introduction

Actinomyces provide a rich source of natural products (NPs) with potential therapeutic applications, and modern "omics"-based technologies have revealed their potent potential for encoding diverse natural products [1]. Genome-guided discovery of clostrubin A [2], closthioamide [3] and cytotoxic benzolactones [4] has reinvigorated NP research, making it a more targeted and systematic research endeavor. To avoid the re-isolation of known NPs, the "genetic dereplication" strategy [5] and the "one strain many compounds" (OSMAC) approach [6] have been successfully used during large-scale culture for discovering NPs with novel skeletons (such as alterbrassinoids A-D [7] and waikikiamides [8]) and novel NPs derived from post-modifications (such as the branched cyclic peptide lyciumin [9] and highly modified polytheonamide-like peptides [10]).

Cyclodipeptides (CDPs), also called 2,5-diketopiperazines (DKPs), are the smallest cyclic peptides formed via the condensation of two α-amino acids. CDPs are mainly produced by *Streptomyces* [11]. CDPs exhibit important and diverse biological properties, such as antibacterial, antifungal, antiviral, antitumor, immunosuppressive and anti-inflammatory activities [12]. Owing to the great potential for activation of specific binding sites in enzymes or proteins, CDPs have become important pharmacophores in pharmaceutical chemistry [13]. Natural CDPs can be biosynthesized through two different machineries; one is catalyzed by the large multi-modular nonribosomal peptide synthetases (NRPSs), and the other is mediated by cyclodipeptide synthases (CDPSs) [14]. The former utilizes free amino acids, and the latter hijacks aminoacyl-tRNAs (AA-tRNAs) from primary metabolism [15]. Generally, CDPSs catalyze the production of representative 2,5-DKPs, which then will be modified by cyclodipeptide-tailoring enzymes (such as methyltransferases, prenyltransferases, oxidoreductases and cytochrome P450 enzymes) to form their intriguing molecular character [14,15].

Aiming at mining more natural products with structural diversity and bioactivity, the OSMAC strategy was carried out using *Streptomyces hygrospinosus* var. beijingensis, which is rich in secondary metabolites including tetramycin [16], anisomycin [17], nystatin and toyocamycin [18]. Among them, tetramycin, anisomycin and structurally related derivatives were high-yield products in the wild-type strain. To avoid the rediscovery of already characterized compounds and to reduce the interference effect of tetramycin and anisomycin, the BGCs of those compounds were genetically deleted and the resultant mutant strain (named *S. hygrospinosus* 26D9-414) was used as the starting strain in this study. When *S. hygrospinosus* 26D9-414 was incubated in a new medium different from the one used for tetramycin and anisomycin production, nine CDPs of two types, diketopiperazines with phenylalanine (**1** and **4–9**) and pyrazinones with arginine (**2** and **3**), were successfully identified. Among the identified CDPs, compounds **1** and **2** (argilein) were new compounds, and compound **4** was reported as a natural product for the first time here. Compound **4** has been used as an important substrate for antitumor spirotryprostatin B synthesis [19]. The other six CDPs have been reported before and were known as argvalin (**3**) [20], albonoursin (**5**) [21], 3,6-Dibenzylidene-2,5-dioxopiperazine (**6**) [22], 3-benzylidenepiperazine-2,5-dione (**7**) [23], 3-benzylidene-6-methylpiperazine-2,5-dione (**8**) [24] and 3-Benzyl-6-benzylidene-2,5-dioxopiperazine (**9**) [25], respectively. antiSMASH analysis of the genome sequence revealed a possible *cdp* cluster for the nine CDPs, and genetic deletion of *cdpA-C* confirmed the correlation of genes with compounds. Finally, the antibacterial and cytotoxic properties of **1–5** were evaluated.

2. Results and Discussion

The "genetic dereplication" strain *Streptomyces* 26D9-414 [18], in which BGCs of tetramycin and anisomycin were deleted, was selected for OSMAC screening of new natural products. The original medium for anisomycin production and the other ten liquid media PYJ1-J10 were selected for the mining of new compounds. Comparative HPLC analysis of the secondary metabolites was conducted, and the metabolic profile of the PYJ1 medium gave many new peaks characteristic of absorption at 224 nm and 296 nm (Figure S1). Repeat rounds of fractionation alternating between silica gel chromatography and Sephadex LH-20 column chromatography followed by semi-preparative reversed-phase HPLC afforded compounds **1–9** (Figure 1). Their structures were elucidated by spectroscopic methods and HR-ESI-MS data.

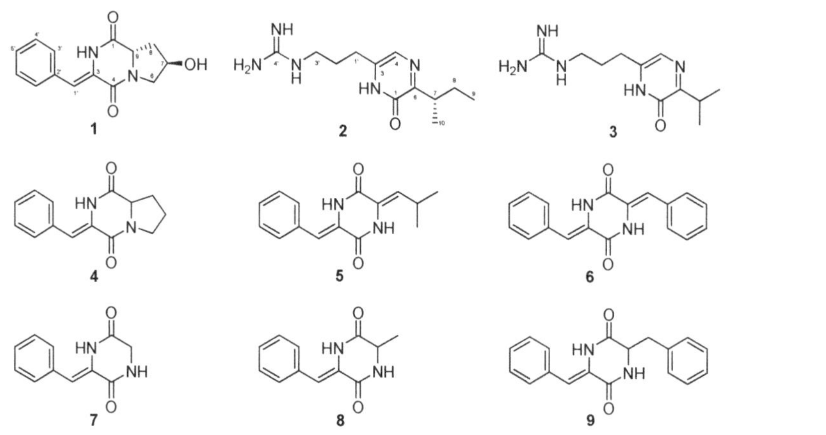

Figure 1. Structures of compounds **1–9**.

Compound **1** was isolated as a white amorphous powder. The UV characteristic absorptions (224 nm and 296 nm) were similar to those of the already-known compound **4**. The similarity suggested a diketopiperazine moiety within **1**. Based on HR-ESI-MS ions at *m/z* 259.1079 [M + H]$^+$, the molecular formula of **1** was determined as $C_{14}H_{14}N_2O_3$. According to the NMR data, it contained nine degrees of unsaturation. The diketopiperazine nature of **1** was confirmed through the ^{1}H and ^{13}C spectroscopic data analysis (Table 1), in which two amidic carbonyls (C-1 and C-4) were observed (Figure 1). The ^{1}H NMR combined with H-H COSY spectrum showed characteristic 4-hydroxyproline residues at δ_H 3.71 (1H, dd, *J* = 12.7, 4.7 Hz), 3.32 (1H, d, *J* = 12.7 Hz), 4.34 (1H, m), 2.12 (1H, dd, *J* = 12.6, 6.4 Hz), 2.01 (1H, td, *J* = 12.6, 4.4 Hz), 4.58 (1H, dd, *J* = 12.6, 6.4 Hz), 5.17 (1H, d, *J* = 2.9 Hz). Meanwhile, the ^{13}C and 2D NMR spectrum of **1** revealed an α,β-unsaturated phenylalanine residue (δ_C 158.7, 133.5, 129.3, 128.6, 127.9, 114.7) (Table 1). Finally, according to the HMBC correlations of H-2/C-9 and H-6/C-4, two fragments concatenated to form diketopiperazine (Figure 2). Therefore, the planar structure of compound **1** was a new pyrrolidine-containing and hydroxylated analog of compound **4**.

Table 1. ^{1}H (600 MHz) and ^{13}C (150 MHz) NMR data for **1** and **2**.

No.	**1**		**2**	
	δ_H, **Mult (*J* in Hz)**	δ_C, **Type**	δ_H, **Mult (*J* in Hz)**	δ_C, **Type**
1		167.5, C		158.8, C
2	10.03, s		12.10, s	
3		128.6, C		137.5, C
4		158.7, C	7.84, s	120.1, C
5				
6	3.71, dd (12.7, 4.7) 3.32, d (12.7)	54.5, CH_2		156.0, C
7	4.34, m	66.6, CH	3.08, m	35.7, CH
8	2.12, dd (12.6, 6.4) 2.01, td (12.6, 4.4)	37.3, CH_2	1.68, m 1.41, m	27.1, CH_2
9	4.58, dd (12.6, 6.4)	56.8, CH	0.79, t (7.4)	11.9, CH_3
10			1.07, d (6.9)	17.8, CH_3
1′	6.67, s	114.8, CH	2.43, t (7.5)	26.7, CH_2
2′		133.6, C	1.78, m	27.2, CH_2
3′	7.54, d (7.6)	129.4, CH	3.10, t (6.4)	40.1, CH_2
4′	7.40, t (7.6)	128.6, CH		156.9, C
5′	7.30, t (7.6)	128.0, CH		
7-OH	5.17, d (2.9)			

See Supplementary Materials for NMR spectra. Spectra were recorded in DMSO-d_6.

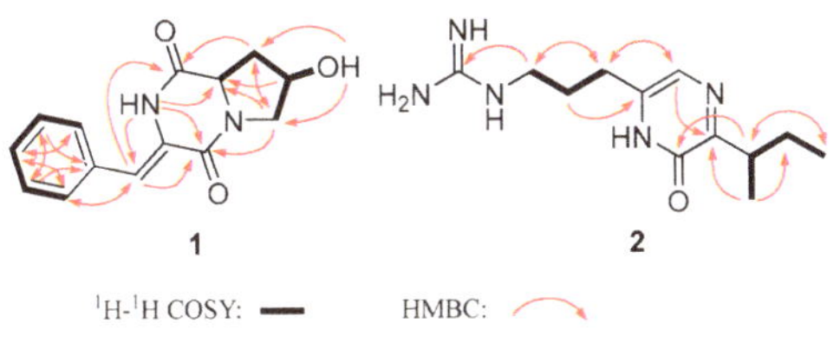

Figure 2. COSY and key HMBC correlations of compounds **1** and **2**.

The relative configuration of compound **1** was determined based on the NOESY spectrum (Figure 3A). The correlations between H-9 and 7-OH instead of H-9 and H-7 indicated that H-7 is located on the opposite side of H-9. Based on the NOE cross-peaks between H-2 and H-3′, the configurations of the β,γ-unsaturated bond in **1** were assigned as (*Z*). The absolute configuration was established by electronic circular dichroism (ECD), and the experimental ECD spectra matched well with the calculated ECD curves of 7*R*, 9*S* (Figure 3B). Considering the biosynthetic origins, the *S* configuration at C-9 was consistent with natural L-proline. Taken together, compound **1** was identified as (7*R*,9*S*)-3-((*Z*)-benzylidene)-7-hydroxy-hexahydropyrrolo [1,2-a] pyrazine-1,4-dione.

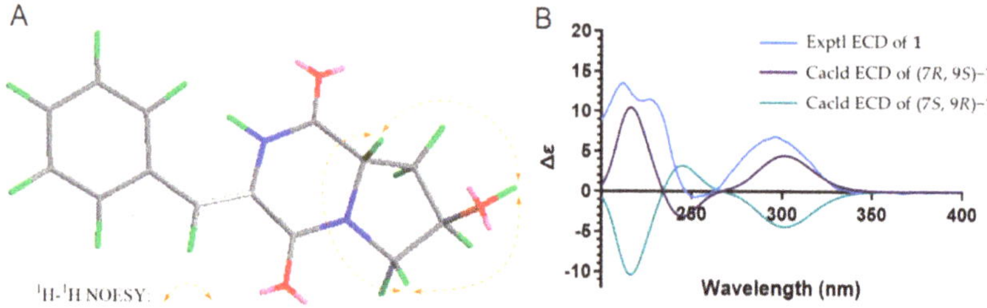

Figure 3. Key NOESY correlations and ECD curves of compound **1**.

Compound **2** was detected as an intracellular product with different UV spectra from compound **1**. The UV characteristic absorptions of **2** (228 nm and 322 nm) were more consistent with the known argvalin (**3**) [20]. Its molecular formula $C_{12}H_{21}N_5O$ was established on the basis of HR-ESI-MS data *m/z* 252.1823 $[M + H]^+$ (Cal. 252.1819), which increased by 14 Da compared to **3**. The 1D NMR data of **2** were similar to those of **3** except for the two methyl groups with different chemical shifts and the presence of an extra methylene at δ_C 27.1 (Table 1). These data suggest that isoleucine, rather than valine, was condensed with arginine, leading to the formation of compound **2** as a new pyrazine derivative. Considering the arginine origin, compound **2** was named argilein. This is the third arginine-containing pyrazine derivative found in natural products. Similarly, the absolute configuration of **2** was 7*S*, which was consistent with the L-isoleucine (Figure 4).

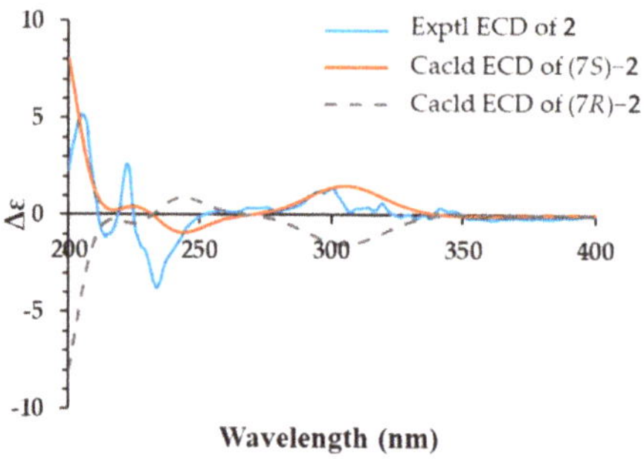

Figure 4. Experimental and calculated ECD curves of compound **2**.

The compounds **1–5** were evaluated for cytotoxicity against human lung carcinoma (A549), human leukemia (HL60), human hepatocellular carcinoma (SMMC-7721), human colon cancer (SW480) and human breast carcinoma (MDA-MB-231) cell lines by MTS assay. Among the five compounds, only compound **2** exhibited selective inhibitory activity against MDA-MB-231 and SW480, the IC_{50} values of which were 18.26 μM and 13.42 μM, respectively (Table S3). The cytotoxicity of compound **2** was comparable with that of cisplatin, which has been widely used as a chemotherapeutic agent for the treatment of various cancers [26]. As to the antibiotic activity, all compounds showed no obvious inhibitory activity against all tested bacteria and fungi, except for compound **2**, which exhibited weak activity against *Xanthomonas albilineans, Candida albicans* and *Candida sake* with MIC values of 0.25 mg/mL, 1.0 mg/mL and 1.0 mg/mL, respectively (Table S4). This finding was consistent with the reported weak antibacterial activities of CDPs (MICs of 0.5–10 mg/mL) [27,28].

To correlate BGCs with the isolated nine diketopiperazines, antiSMASH analysis of the genome sequence of *S. hygrospinosus* var. beijingensis was conducted. The arrangement and sequence of genes within the *cdp* cluster showed high similarity with those of the *alb* cluster, which was reported to be responsible for albonoursin (**5**) production (Figure S2) [21]. *albC* encodes cyclodipeptide synthase (CDPS), which catalyzes the cyclic dipeptide precursor formation. The heterologous expression of *albC* led to the synthesis of various cyclodipeptides, including cyclo(Phe-Pro) [29,30], the possible precursor of compounds **1** and **4**. The deletion of *cdpA-cdpC* abolished the production of nine cyclodipeptides (Figures 5 and S4). To the best of our knowledge, this is the first finding of a cyclodipeptide synthesized by CDPS using arginine and also the first report of a proline-derived cyclodipeptide (compounds **1** and **4**) from the original producing strain.

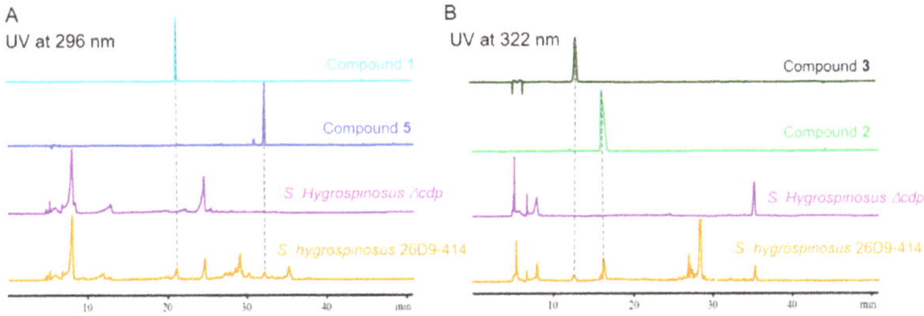

Figure 5. HPLC analysis of metabolites in *S. hygrospinosus* 26D9-414 and in mutant *S. hygrospinosus* Δ*alb*. (**A**) The UV detection of piperazines were performed at 296 nm; (**B**) The UV detection of pyrazinones at 322 nm.

Cyclodipeptide oxidases (CDOs) AlbA and AlbB usually catalyze the dehydrogenation of cyclodipeptides to form dehydrogenated cyclodipeptide derivatives [31]. Whether the hydroxyl group in **1** and pyrazinone in **2** are catalyzed by CDO candidates CdpA and CdpB still awaits discovery. CDPSs and CDOs both possess broad substrate selectivity and can be used to synthesize various dehydrogenated cyclodipeptide derivatives, which serve as important precursors for the development of pharmaceutical intermediates [30,32]. Gene c-blast analysis revealed that *cdp* gene analogs were mainly distributed in *Streptomyces* and *Nocardiopsis*, and a few were also found in *Nonomuraea*, *Goodfellowiella*, *Bailinhaonella*, *Saccharopolyspora* and *Actinomadura* (Figure S5).

3. Conclusions

Modern "omics"-based technologies have revealed the potent potential of *Actinobacteria* for encoding natural products with diverse structures and biologically active compounds. To reveal the diversity of NPs encoded by *Streptomyces hygrospinosus* var. beijingensis, the "genetic dereplication" strategy and OSMAC approach were used in this study. Nine CDP derivatives of two types were identified from *S. hygrospinosus* 26D9-414 through the genome mining strategy. The relevance between the *cdp* cluster and all the isolated CDPs was confirmed by genetic manipulation. These findings increase the repertoire of natural DKPs and reveal a CDPS with a broad range of substrates that could be developed as a biocatalyst for the future development of therapeutic agents.

4. Material and Methods

4.1. General Experimental Procedures

Optical rotations were recorded with a JASCO P-2000 digital polarimeter. UV spectra were recorded on a Thermofisher Evolution 300 UV-vis spectrophotometer. The 1D-NMR and 2D-NMR spectra were obtained on a Bruker AVANCE III 600 MHz spectrometer with TMS as an internal standard. HR-ESI-MS spectra were recorded on an Agilent 1290 HPLC system coupled to a 6230 TOF system mass spectrometer. ECD spectra were recorded using a JASCO J-1500-150ST. HPLC analysis and semi-preparative HPLC were performed with Agilent 1260 HPLC system using an Agilent ZORBAX SB-C18 column (5 μm, 4.6 × 250 mm) and an Agilent ZORBAX SB-C18 column (5 μm, 9.4 × 250 mm), respectively. All comparative studies of crude extracts obtained based on the OSMAC strategy were based on HPLC analysis, the mobile phases were CH_3OH-H_2O and 1‰ formic acid or trifluoroacetic acid in the water. The gradient was chosen as CH_3OH-H_2O: 5% 0–5 min, 5–50% 5–30 min, 50–95% 30–45 min, 95% 45–50 min, 95–5% 50–51 min, 5% 51–60 min, 0.5 mL/min. The HPLC methods used for the separation of compounds **1–9** are described in detail in Section 4.5. Silica gel (100–200, 200–300 mesh, Qingdao Haiyang Chemical Co., Ltd., Qingdao, China) and Sephadex LH-20 gel (Uppsala, Sweden) were used for column chromatography (CC). Precoated silica gel GF254 plates (Qingdao Marine Chemical Ltd., Qingdao, China) were used for TLC monitoring combined with UV light and 10% H_2SO_4 in EtOH. Taq DNA polymerase and KOD-plus high-fidelity polymerase were obtained from Takara. All restriction enzymes were purchased from Thermo Scientific or Vazyme Biotech Co., Ltd. E.Z.N.A. Gel Extraction Kit and Plasmid Mini Kit were purchased from OMEGA. PCR primers were synthesized by GENEWIZ. All solvents used for CC were of analytical grade (Shanghai Chemical Reagents Co., Ltd., Shanghai, China), and solvents used for HPLC were of HPLC grade (Sigma-Aldrich, St. Louis, MO, USA).

4.2. Bacterial Strains, Plasmids, Primers and Culture Conditions

The strains, plasmids and primers used in this study were listed in Tables S1 and S2. *Streptomyces* and its derivatives were grown at 30 °C on solid SFM medium (2% mannitol, 2% soya flour and 1.5% agar) for sporulation and conjugation, and in TSBY liquid medium (3% tryptone soy broth, 10.3% sucrose and 0.5% yeast extract) for the isolation of chromosomal DNA [33]. All *E. coli* strains including DH10B and ET12567/pUZ8002 were grown in liquid Luria–Bertani (LB) medium or on LB agar at 37 °C. Apramycin (50 μg/mL) and trimethoprim (50 μg/mL) were used when necessary. All plasmid subcloning experiments were performed in *E. coli* DH10B following standard protocols. General procedures for *E. coli* or *Streptomyces* manipulation were carried out according to the published procedures [34].

4.3. Construction of *S. hygrospinosus* Δ*cdp* Mutant

To construct the *cdpA-cdpC* deletion mutant, the 1624 bp DNA fragment covering total *cdpA-cdpC* was substituted by *aac(3)IV* + *oriT* cassette (Apr^R gene), amplified from pIJ773 primers cdp-apr-P1/P2 and cdp-apr-P1/P2 (Table S2). Two homologous arms of 2004 bp and 1964 bp containing the upstream and downstream regions flanking *cdpA-cdpC* were amplified by PCR with primers cdp-L-P1/P2 and cdp-R-P1/P2, respectively (Table S2). The entire PCR product was cloned into the *BamHI/EcoRI*-digested pJTU1278, generating the recombinant plasmid vector pZDS-1, using the Vazyme one-step cloning kit (Vazyme Biotech Co., Ltd., Nanjing, China). The resultant plasmid was firstly transferred into *E. coli* ET12567/pUZ8002 and then introduced into *S. hygrospinosus* 26D9-414 strain for the construction of *cdpA-C*-deleted strain *S. hygrospinosus* Δ*cdp*. According to the previously described procedure [35], the double-crossover strains were obtained through antibiotic selection and confirmed by PCR verification using primers cdp1-P1/P2 and apr-P1/P2 (Table S2).

4.4. Strain Fermentation and Chemical Analysis

The mutant *S. hygrospinosus* 26D9-414 was cultivated in TSBY liquid medium at 30 °C for 2 days to afford seed broth. The seed broth was next inoculated into a fermentation medium (5% (*v*/*v*)) and incubated at 30 °C with shaking for a further 6 days. Ten different fermentation liquid media (Table S4) including the original medium (containing 1% corn starch, 2% soluble starch, 1% soya flour, 0.02% KH_2PO_4, 0.3% NaCl, 0.3% NH_4Cl and 0.4% $CaCO_3$ per liter) were selected for tetramycin and anisomycin production [36]. The EtOAc extracts of all fermentation broths were analyzed by high-performance liquid chromatography (HPLC). For the accumulation of nine CDPs, medium PYJ1 (containing 3% soluble starch, 4% glucose, 1% glycerin, 1.5% tryptone soy broth, 1% beef extract, 1% peptone, 0.65% yeast extract, 0.05% $MgSO_4$, 0.1% NaCl and 0.2% $CaCO_3$ per liter) was used.

4.5. Fermentation and Isolation

Large-scale fermentation for the isolation and purification of DKPs was conducted according to the standard method described before. A total of 10 L fermentation broth was centrifuged to afford the mycelia and the liquid phase. The liquid phase was extracted with an equal volume of EtOAc three times at room temperature. The EtOAc crude extract was concentrated under reduced pressure to yield dark brown matter (2.6 g). The crude extract was separated using a column of silica gel eluted with $CHCl_3$/MeOH (from 100:1 to 1:1, v/v) to obtain eight fractions (Fr.A-Fr.H). Then Fr.B was directly separated by semi-preparative HPLC (CH_3OH-H_2O: 10% 0–5 min, 10–90% 5–40 min, 1.5 mL/min, 296 nm) to afford compounds **1** (22.4 mg, t_R = 23 min) and **7** (16.8 mg, t_R = 22 min); Fr.C was subjected to Sephadex LH-20 column chromatography and elution with $CHCl_3$/MeOH (1:1, v/v) to give four fractions (Fr.C1-C4); Fr.C3 was separated by semi-preparative HPLC (CH_3OH-H_2O: 5% 0–5 min, 5–80% 5–40 min, 1.5 mL/min, 296 nm) to afford compounds **4** (16.6 mg, t_R = 27 min), **5** (6.3 mg, t_R = 34 min), **6** (2.6 mg, t_R = 36 min), **8** (12.1 mg, t_R = 26 min) and **9** (3.2 mg, t_R = 30 min). The mycelia were extracted with acetone (1 L) and ultrasound for 2 h, and the organic solvents were dried under vacuum to yield a dark brown crude extract (1.1 g). The extract was separated into six fractions (Fr.a-Fr.f) by silica gel column chromatography using $CHCl_3$/MeOH mixtures of increasing polarities (100:1 to 5:1, v/v). Fr.e was separated by semi-preparative HPLC (CH_3OH-H_2O: 5–50% 30 min, 1.5 mL/min, 322 nm, 1‰ TFA in water) to afford compounds **2** (7.4 mg, t_R = 22 min) and **3** (5.5 mg, t_R = 20 min).

Compound 1: white powder; $[\alpha]_D^{25}$ +13.5 (c 0.2 MeOH); UV (MeOH) λ_{max} (log ε) 214 (3.31), 299 (3.32) nm; ^{1}H NMR (600 MHz, DMSO-d_6) and ^{13}C NMR (150 MHz, DMSO-d_6) data in Table 1; HR-ESI-MS *m/z* 259.1079 $[M + H]^+$ (calcd for $C_{14}H_{15}N_2O_3^+$, 259.1077).

Compound 2: pale-yellow powder; $[\alpha]_D^{25}$ −12.1 (c 0.2 MeOH); UV (MeOH) λ_{max} (log ε) 230 (3.12), 323 (3.17) nm; ^{1}H NMR (600 MHz, DMSO-d_6) and ^{13}C NMR (150 MHz, DMSO-d_6) data in Table 1; HR-ESI-MS *m/z* 252.1823 $[M + H]^+$ (calcd for $C_{12}H_{22}N_5O^+$, 252.1819).

4.6. ECD Calculations

Conformational analyses for compounds **1–2** were performed via Spartan'14 software using the MMFF94 molecular mechanics force field calculation. Conformers within a 10 kcal/mol energy window were generated and optimized using DFT calculations at the B3LYP/6-31G(d) level. Conformers with a Boltzmann distribution over 1% were chosen for the ECD calculations in MeOH at the B3LYP/6-311 + G (2d, p) level. The IEF-PCM solvent model for MeOH was used. The calculated ECD spectra were obtained by DFT and time-dependent DFT (TD-DFT) using Gaussian 09 and analyzed using SpecDis v1.71.

4.7. Cytotoxicity Assays

To determine the cytotoxicity of compounds **1–5**, five human cancer cell lines (HL60, A549, SMMC-7721, SW480, MDA-MB-23) were evaluated by MTS assay. Each cell line was exposed to the tested compounds at concentrations of 40, 8, 1.6, 0.32 and 0.064 μM in triplicate. Cell viability was determined using MTS Kit according to the manufacturer's instructions [37].

Supplementary Materials: The following supporting information is free of charge and can be downloaded at: https://www.mdpi.com/article/10.3390/antibiotics11111463/s1. Figure S1: HPLC analysis of fermentation products of *S. hygrospinosus* 26D9-414 from eleven different mediums. Figure S2: Biosynthetic analysis of CDPs. Figure S3: Schematic construction and PCR verification of S. hygrospinosus Δ*cdp* mutant. Figure S4: HPLC analysis of metabolites in *S. hygrospinosus* 26D9-414 and in *S. hygrospinosus* Δ*cdp* mutant. Figure S5: Distribution statistics of putative *cdp* gene analogues. Figures S6–S12: 1D and 2D NMR, HR-ESI-MS, and UV spectra of compound **1**. Figures S13–S20: 1D and 2D NMR, HR-ESI-MS, and UV spectra of compound **2**. Table S1: Strains and plasmids used in this study. Table S2: Primers used in this study. Table S3: Cytotoxicity assay of compounds **1–5**. Table S4: Antibacteria activity assay of compounds **1** and **2**. Table S5: Ten mediums in this study. References [18,21,38–46] are cited in Supplementary Materials file.

Author Contributions: Conceptualization, D.Y.; methodology, D.Y. and D.Z.; investigation, D.Z., J.W., Y.Q. and B.L.; data curation, D.Z. and L.K.; writing—original draft preparation, L.K. and D.Z.; writing—review and editing, L.K. and D.Y.; supervision, D.Y. and Z.D.; project administration, D.Y.; funding acquisition, L.K. and D.Y. All authors have read and agreed to the published version of the manuscript.

Funding: This work was supported by grants from the National Key Research and Development Program of China (2021YFC2100600, 2018YFA0900400, 2021YFA0909500, 2021YFC2100100), the National Natural Science Foundation of China (32170077, 32170075) and the Startup Fund for Young Faculty at Shanghai Jiao Tong University (SFYF at SJTU) (21X010500720).

Institutional Review Board Statement: Not applicable.

Informed Consent Statement: Not applicable.

Data Availability Statement: All data generated or analyzed during this study are included in this published article.

Conflicts of Interest: The authors declare no conflict of interest.

References

1. Scherlach, K.; Hertweck, C. Mining and unearthing hidden biosynthetic potential. *Nat. Commun.* **2021**, *12*, 3864. [CrossRef] [PubMed]
2. Shabuer, G.; Ishida, K.; Pidot, S.J.; Roth, M.; Dahse, H.M.; Hertweck, C. Plant pathogenic anaerobic bacteria use aromatic polyketides to access aerobic territory. *Science* **2015**, *350*, 670–674. [CrossRef] [PubMed]
3. Dunbar, K.L.; Dell, M.; Gude, F.; Hertweck, C. Reconstitution of polythioamide antibiotic backbone formation reveals unusual thiotemplated assembly strategy. *Proc. Natl. Acad. Sci. USA* **2020**, *117*, 8850–8858. [CrossRef] [PubMed]
4. Niehs, S.P.; Dose, B.; Richter, S.; Pidot, S.J.; Dahse, H.M.; Stinear, T.P.; Hertweck, C. Mining symbionts of a spider-transmitted fungus lluminates uncharted biosynthetic pathways to cytotoxic benzolactones. *Angew. Chem. Int. Ed. Engl.* **2020**, *59*, 7766–7771. [CrossRef] [PubMed]
5. Chiang, Y.M.; Ahuja, M.; Oakley, C.E.; Entwistle, R.; Asokan, A.; Zutz, C.; Wang, C.C.; Oakley, B.R. Development of genetic dereplication strains in *Aspergillus nidulans* results in the discovery of aspercryptin. *Angew. Chem. Int. Ed. Engl.* **2016**, *55*, 1662–1665. [CrossRef]
6. Hu, Z.; Ye, Y.; Zhang, Y. Large-scale culture as a complementary and practical method for discovering natural products with novel skeletons. *Nat. Prod. Rep.* **2021**, *38*, 1775–1793. [CrossRef]
7. Li, F.; Lin, S.; Zhang, S.; Hao, X.; Li, X.N.; Yang, B.; Liu, J.; Wang, J.; Hu, Z.; Zhang, Y. Alterbrassinoids A-D: Fusicoccane-derived diterpenoid dimers featuring different carbon skeletons from *Alternaria brassicicola*. *Org. Lett.* **2019**, *21*, 8353–8357. [CrossRef]
8. Wang, F.; Sarotti, A.M.; Jiang, G.; Huguet-Tapia, J.C.; Zheng, S.L.; Wu, X.; Li, C.; Ding, Y.; Cao, S. Waikikiamides A-C: Complex diketopiperazine dimer and diketopiperazine-polyketide hybrids from a Hawaiian marine fungal strain *Aspergillus* sp. FM242. *Org. Lett.* **2020**, *22*, 4408–4412. [CrossRef]
9. Kersten, R.D.; Weng, J.K. Gene-guided discovery and engineering of branched cyclic peptides in plants. *Proc. Natl. Acad. Sci. USA* **2018**, *115*, e10961–e10969. [CrossRef]
10. Bhushan, A.; Egli, P.J.; Peters, E.E.; Freeman, M.F.; Piel, J. Genome mining- and synthetic biology-enabled production of hypermodified peptides. *Nat. Chem.* **2019**, *11*, 931–939. [CrossRef]
11. Gondry, M.; Jacques, I.B.; Thai, R.; Babin, M.; Canu, N.; Seguin, J.; Belin, P.; Pernodet, J.L.; Moutiez, M. A comprehensive overview of the cyclodipeptide synthase family enriched with the characterization of 32 new enzymes. *Front. Microbiol.* **2018**, *9*, 46. [CrossRef] [PubMed]
12. Belin, P.; Moutiez, M.; Lautru, S.; Seguin, J.; Pernodet, J.L.; Gondry, M. The nonribosomal synthesis of diketopiperazines in tRNA-dependent cyclodipeptide synthase pathways. *Nat. Prod. Rep.* **2012**, *29*, 961–979. [CrossRef]
13. Stark, T.; Hofmann, T. Structures, sensory activity, and dose/response functions of 2,5-diketopiperazines in roasted cocoa nibs (*Theobroma cacao*). *J. Agric. Food. Chem.* **2005**, *53*, 7222–7231. [CrossRef] [PubMed]
14. Harken, L.; Li, S.M. Modifications of diketopiperazines assembled by cyclodipeptide synthases with cytochrome P(450) enzymes. *Appl. Microbiol. Biotechnol.* **2021**, *105*, 2277–2285. [CrossRef]
15. Canu, N.; Moutiez, M.; Belin, P.; Gondry, M. Cyclodipeptide synthases: A promising biotechnological tool for the synthesis of diverse 2,5-diketopiperazines. *Nat. Prod. Rep.* **2020**, *37*, 312–321. [CrossRef] [PubMed]
16. Cao, B.; Yao, F.; Zheng, X.; Cui, D.; Shao, Y.; Zhu, C.; Deng, Z.; You, D. Genome mining of the biosynthetic gene cluster of the polyene macrolide antibiotic tetramycin and characterization of a P450 monooxygenase involved in the hydroxylation of the tetramycin B polyol segment. *Chembiochem* **2012**, *13*, 2234–2242. [CrossRef] [PubMed]
17. Zheng, X.; Cheng, Q.; Yao, F.; Wang, X.; Kong, L.; Cao, B.; Xu, M.; Lin, S.; Deng, Z.; Chooi, Y.H.; et al. Biosynthesis of the pyrrolidine protein synthesis inhibitor anisomycin involves novel gene ensemble and cryptic biosynthetic steps. *Proc. Natl. Acad. Sci. USA* **2017**, *114*, 4135–4140. [CrossRef]

18. Wang, J.; Kong, L.; Qiao, Y.; Zhang, D.; Deng, Z.; You, D. Genome mining guided discovery and identification of nystatin and toyocamycin in *Streptomyces hygrospinosus* var. beijingensis. *Acta Microbiol. Sin.* **2021**, *61*, 3076–3086.
19. Xi, Y.-K.; Zhang, H.; Li, R.-X.; Kang, S.-Y.; Li, J.; Li, Y. Total synthesis of spirotryprostatins through organomediated intramolecular umpolung cyclization. *Chemistry* **2019**, *25*, 3005–3009. [CrossRef] [PubMed]
20. Tatsuta, K.; Fujimoto, K.; Yamashita, M.; Tsuchiya, T.; Umezawa, S. Argvalin, a new microbial metabolite: Isolation and structure. *J. Antibiot.* **1973**, *26*, 606–608. [CrossRef]
21. Lautru, S.; Gondry, M.; Genet, R.; Pernodet, J.L. The albonoursin gene Cluster of *S. noursei* biosynthesis of diketopiperazine metabolites independent of nonribosomal peptide synthetases. *Chem. Biol.* **2002**, *9*, 1355–1364. [CrossRef]
22. Brown, R.; Kelley, C.; Wiberley, S.E. The production of 3-benzylidene-6-isobutylidene-2, 5-dioxopiperazine, 3, 6-dibenzylidene-2, 5-dioxopiperazine, 3-benzyl-6-benzylidene-2, 5-dioxopiperazine, and 3, 6-dibenzyl-2, 5-dioxopiperazine by a variant of *Streptomyces noursei*. *J. Org. Chem.* **1965**, *30*, 277–280. [CrossRef]
23. Yang, L.M.; Wu, R.Y.; McPhail, A.T.; Yokoi, T.; Lee, K.H. Neihumicin, a new cytotoxic antibiotic from *Micromonospora neihuensis*. II. structural determination and total synthesis. *J. Antibiot.* **1988**, *41*, 488–493. [CrossRef] [PubMed]
24. Zhang, Q.; Li, S.; Chen, Y.; Tian, X.; Zhang, H.; Zhang, G.; Zhu, Y.; Zhang, S.; Zhang, W.; Zhang, C. New diketopiperazine derivatives from a deep-sea-derived *Nocardiopsis alba* SCSIO 03039. *J. Antibiot.* **2013**, *66*, 31–36. [CrossRef]
25. Kawahara, N.; Nozawa, K.; Nakajima, S.; Yamazaki, M.; Kawai, K. Sulfur-containing dioxopiperazine derivatives from *Emericella heterothallica*. *Heterocycles* **1989**, *29*, 397–402.
26. Achkar, I.W.; Abdulrahman, N.; Al-Sulaiti, H.; Joseph, J.M.; Uddin, S.; Mraiche, F. Cisplatin based therapy: The role of the mitogen activated protein kinase signaling pathway. *J. Transl. Med.* **2018**, *16*, 96. [CrossRef]
27. Cain, C.C.; Lee, D.; Waldo, R.H., 3rd; Henry, A.T.; Casida, E.J., Jr.; Wani, M.C.; Wall, M.E.; Oberlies, N.H.; Falkinham, J.O., 3rd. Synergistic antimicrobial activity of metabolites produced by a nonobligate bacterial predator. *Antimicrob. Agents Chemother.* **2003**, *47*, 2113–2117. [CrossRef]
28. Ström, K.; Sjögren, J.; Broberg, A.; Schnürer, J. Lactobacillus plantarum MiLAB 393 produces the antifungal cyclic dipeptides cyclo(L-Phe-L-Pro) and cyclo(L-Phe-trans-4-OH-L-Pro) and 3-phenyllactic acid. *Appl. Env. Microbiol.* **2002**, *68*, 4322–4327. [CrossRef]
29. Gondry, M.; Sauguet, L.; Belin, P.; Thai, R.; Amouroux, R.; Tellier, C.; Tuphile, K.; Jacquet, M.; Braud, S.; Courçon, M.; et al. Cyclodipeptide synthases are a family of tRNA-dependent peptide bond-forming enzymes. *Nat. Chem. Biol.* **2009**, *5*, 414–420. [CrossRef]
30. Jacques, I.B.; Moutiez, M.; Witwinowski, J.; Darbon, E.; Martel, C.; Seguin, J.; Favry, E.; Thai, R.; Lecoq, A.; Dubois, S.; et al. Analysis of 51 cyclodipeptide synthases reveals the basis for substrate specificity. *Nat. Chem. Biol.* **2015**, *11*, 721–727. [CrossRef]
31. Le Chevalier, F.; Correia, I.; Matheron, L.; Babin, M.; Moutiez, M.; Canu, N.; Gondry, M.; Lequin, O.; Belin, P. In vivo characterization of the activities of novel cyclodipeptide oxidases: New tools for increasing chemical diversity of bioproduced 2,5-diketopiperazines in *Escherichia coli*. *Microb. Cell Fact.* **2020**, *19*, 178. [CrossRef] [PubMed]
32. Mikulski, L.; Schäfer, J.; Brockmeyer, K.; Kraut, R.; Li, S.M. Comparative studies on similarities and differences of cyclodipeptide oxidases for installation of C-C double bonds at the diketopiperazine ring. *Appl. Microbiol. Biotechnol.* **2020**, *104*, 2523–2536. [CrossRef] [PubMed]
33. Li, Y.; Kong, L.; Shen, J.; Wang, Q.; Liu, Q.; Yang, W.; Deng, Z.; You, D. Characterization of the positive SARP family regulator PieR for improving piericidin A1 production in *Streptomyces piomogeues* var. Hangzhouwanensis. *Synth. Syst. Biotechnol.* **2019**, *4*, 16–24. [CrossRef] [PubMed]
34. Kong, L.; Zhang, W.; Chooi, Y.H.; Wang, L.; Cao, B.; Deng, Z.; Chu, Y.; You, D. A multifunctional monooxygenase XanO4 catalyzes xanthone formation in xantholipin biosynthesis via a cryptic demethoxylation. *Cell Chem. Biol.* **2016**, *23*, 508–516. [CrossRef]
35. Wang, Q.; Kong, L.; Zheng, X.; Shen, J.; Wang, J.; Zhang, D.; Qiao, Y.; Wang, J.; Deng, Z.; You, D. Acyltransferase AniI, a tailoring enzyme with broad substrate tolerance for high-level production of anisomycin. *Appl. Env. Microbiol.* **2021**, *87*, e0017221. [CrossRef]
36. Shen, J.; Kong, L.; Li, Y.; Zheng, X.; Wang, Q.; Yang, W.; Deng, Z.; You, D. A LuxR family transcriptional regulator AniF promotes the production of anisomycin and its derivatives in *Streptomyces hygrospinosus* var. beijingensis. *Synth. Syst. Biotechnol.* **2019**, *4*, 40–48. [CrossRef]
37. Wu, Z.; Zhang, X.; Al Anbari, W.H.; Zhang, M.; Chen, X.; Luo, Z.; Li, X.N.; Chen, C.; Liu, J.; Wang, J.; et al. Amiaspochalasins A-H, undescribed aspochalasins with a C-21 ester carbonyl from *Aspergillus micronesiensis*. *J. Org. Chem.* **2019**, *84*, 5483–5491. [CrossRef]
38. Paget, M.S.; Chamberlin, L.; Atrih, A.; Foster, S.J.; Buttner, M.J. Evidence that the extracytoplasmic function sigma factor sigmaE is required for normal cell wall structure in *Streptomyces coelicolor* A3(2). *J. Bacteriol.* **1999**, *181*, 204–211. [CrossRef]
39. He, Y.; Wang, Z.; Bai, L.; Liang, J.; Zhou, X.; Deng, Z. Two pHZ1358-derivative vectors for efficient gene knockout in *Streptomyces*. *J. Microbiol. Biotechnol.* **2010**, *20*, 678–682. [CrossRef]
40. Gust, B.; Challis, G.L.; Fowler, K.; Kieser, T.; Chater, K.F. PCR-targeted *Streptomyces* gene replacement identifies a protein domain needed for biosynthesis of the sesquiterpene soil odor geosmin. *Proc. Natl. Acad. Sci. USA* **2003**, *100*, 1541–1546. [CrossRef]
41. Tan, G.Y.; Deng, K.; Liu, X.; Tao, H.; Chang, Y.; Chen, J.; Chen, K.; Sheng, Z.; Deng, Z.; Liu, T. Heterologous biosynthesis of spinosad: An Omics-guided large polyketide synthase gene cluster reconstitution in *Streptomyces*. *ACS Synth. Biol.* **2017**, *6*, 995–1005. [CrossRef] [PubMed]

42. Sun, X.; Wang, G.; Xiao, H.; Jiang, J.; Xiao, D.; Xing, B.; Li, A.; Zhang, Y.; Sun, K.; Xu, Y.; et al. Strepimidazoles A-G from the plant endophytic *Streptomyces* sp. PKU-EA00015 with inhibitory activities against a plant pathogenic fungus. *J. Nat. Prod.* **2020**, *83*, 2246–2254. [CrossRef] [PubMed]
43. Sun, C.; Liu, Z.; Zhu, X.; Fan, Z.; Huang, X.; Wu, Q.; Zheng, X.; Qin, X.; Zhang, T.; Zhang, H.; et al. Antitubercular ilamycins from marine-derived *Streptomyces atratus* SCSIO ZH16 ΔilaR. *J. Nat. Prod.* **2020**, *83*, 1646–1657. [CrossRef] [PubMed]
44. Perlatti, B.; Lan, N.; Earp, C.E.; AghaAmiri, S.; Vargas, S.H.; Azhdarinia, A.; Bills, G.F.; Gloer, J.B. Arenicolins: C-glycosylated depsides from *Penicillium arenicola*. *J. Nat. Prod.* **2020**, *83*, 668–674. [CrossRef] [PubMed]
45. Huo, C.; Zheng, Z.; Xu, Y.; Ding, Y.; Zheng, H.; Mu, Y.; Niu, Y.; Gao, J.; Lu, X. Naphthacemycins from a *Streptomyces* sp. as protein-tyrosine phosphatase inhibitors. *J. Nat. Prod.* **2020**, *83*, 1394–1399. [CrossRef]
46. Zhang, W.; Wang, L.; Kong, L.; Wang, T.; Chu, Y.; Deng, Z.; You, D. Unveiling the post-PKS redox tailoring steps in biosynthesis of the type II polyketide antitumor antibiotic xantholipin. *Chem. Biol.* **2012**, *19*, 422–432. [CrossRef]

antibiotics

MDPI

Article

Synthesis, Spectroscopic, Chemical Characterizations, Anticancer Capacities against HepG-2, Antibacterial and Antioxidant Activities of Cefotaxime Metal Complexes with Ca(II), Cr(III), Zn(II), Cu(II) and Se(IV)

Eman H. Al-Thubaiti [1], Samy M. El-Megharbel [2], Bander Albogami [3] and Reham Z. Hamza [3,*]

1 Biotechnology Department, College of Sciences, Taif University, Taif-P.O. Box 11099, Taif 21944, Saudi Arabia; i.althubaiti@tu.edu.sa
2 Chemistry Department, College of Sciences, Taif University, Taif-P.O. Box 11099, Taif 21944, Saudi Arabia; s.megherbel@tu.edu.sa
3 Biology Department, College of Sciences, Taif University, Taif-P.O. Box 11099, Taif 21944, Saudi Arabia; b.boqami@tu.edu.sa
* Correspondence: reham.z@tu.edu.sa or dr_reham_z@yahoo.com; Tel.: +966-531355470

Abstract: In this study, metal cefotaxime complexes of Ca(II), Cr(III), Cu(II), Zn(II), and Se(VI) were synthesized and characterized by elemental analysis, conductance measurements, IR, electronic spectra, magnetic measurements, ^{1}HNMR, and XRD, as well as by scanning electron microscopy (SEM) and transmission electron microscopy (TEM). The lower values for molar conductance refer to the nonelectrolyte nature of the complexes. The FTIR and ^{1}H-NMR spectra for the metal complexes of cefotaxime proved that the free cefotaxime antibiotic ligand acted as a monoanionic tridentate ligand through the oxygen atoms of lactam carbonyl, the carboxylate group, and the nitrogen atoms of the amino group. From the magnetic measurements and electronic spectral data, octahedral structures were proposed for the Cr(III) and Se(VI) complexes, while the Cu(II) complex had tetragonal geometry. This study aimed to investigate the effects of cefotaxime and cefotaxime metal complexes on oxidative stress using antioxidant assays including DPPH, ORAC, FARAB, and ABTS, a metal chelation assay, as well as the inhibition of the viability of cancer cells (HepG-2). Regarding the antibacterial activity, the cefotaxime metal complexes were highly effective against both *Bacillus subtilis* and *Escherichia coli*. In conclusion, the cefotaxime metal complexes exhibited highly antioxidant activities. The cefotaxime metal complexes with Zn and Se inhibited HepG-2 cellular viability. Thus, the cefotaxime metal complexes elicited promising results as potent antioxidant and anticancer agents against HepG-2, with potent antibacterial activities at a much lower concentration.

Keywords: metal complexes; spectroscopic studies; antioxidant capacities; hepatic functions; cancer cells; cefotaxime

Citation: Al-Thubaiti, E.H.; El-Megharbel, S.M.; Albogami, B.; Hamza, R.Z. Synthesis, Spectroscopic, Chemical Characterizations, Anticancer Capacities against HepG-2, Antibacterial and Antioxidant Activities of Cefotaxime Metal Complexes with Ca(II), Cr(III), Zn(II), Cu(II) and Se(IV). *Antibiotics* **2022**, *11*, 967. https://doi.org/10.3390/antibiotics11070967

Academic Editors: Yaojun Tong, Zixin Deng and Linquan Bai

Received: 8 June 2022
Accepted: 14 July 2022
Published: 19 July 2022

1. Introduction

Infectious diseases resulting from bacteria are considered a major health problem worldwide, part of which can be attributed to the rapidly increasing resistance to existing antimicrobial drugs. For treating pathogenic multidrug-resistant bacterial strains, developing new antimicrobial compounds is vital. The antibiotic drug, cefotaxime, shown in Figure 1, is considered a third-generation cephalosporin; the resistance to it is due to the inability of cefotaxime to reach its target sites, to alter the binding of penicillin with proteins, which is the main target of the cephalosporins; in addition, β-lactamases (bacterial enzymes) deactivate cephalosporin [1].

The hydrolysis of third-generation cephalosporins has a higher resistance to β-lactamases formed by some Gram-negative bacterial strains than that of the first generation of cephalosporins.

Cefotaxime is used in the treatment of infections of the respiratory tract, meningitis, and septicemia [2].

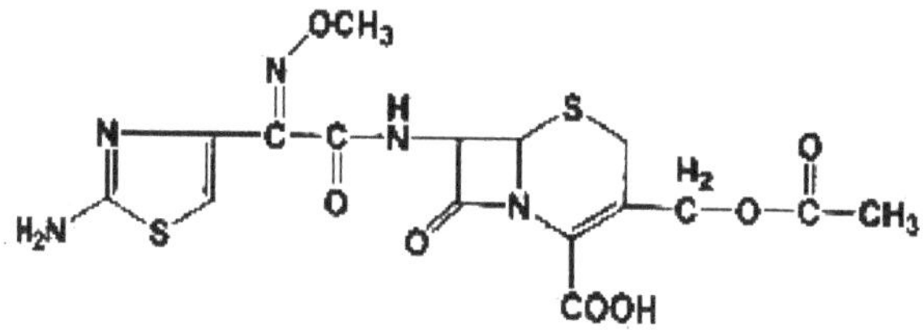

Figure 1. The structure of cefotaxime.

For the action of synthetic and natural metalloantibiotics, metal ions play an important role, as they are involved in specific interactions with some biomolecules, such as nucleic acids, proteins, and membranes [3].

The mode of chelation of cefotaxime occurs through the nitrogen of the NH_2 group, the lactam carbonyl group, and carboxylate; thus, it acts as a tridentate ligand [4]. The reactions of cefotaxime with some transition metal ions, such as Cr(III), Mn(II), Fe(III), Co(II), Ni(II), Cu(II), and Zn(II), have been studied [5] and elucidated using different spectroscopic tools, including IR, electronic, magnetic susceptibility, and ESR spectra. The obtained data confirmed that the cefotaxime may act as a mono, di, tri, and tetra-dentate via the lactam carbonyl, carboxylic, or amide carbonyl groups and the N atom of the thiazole ring. New cefotaxime complexes of the general formula $[ML2(H_2O)_2]$, where M = Co(II), Ni(II), Cu(II), and Zn(II), while L represents the Schiff base ligand, were prepared by reaction of cefotaxime with salicylaldehyde in an ethanolic medium.

The characterizations of the complexes' structures were elucidated by conductance measurements, analyses of C, H, and N, magnetic investigations, and IR and UV spectroscopy [6]. The low conductance values referred to nonelectrolyte types of complexes. According to the electronic spectra and magnetic data, the Co(II), Ni(II), and Zn(II) complexes had an octahedral geometry, while the Cu(II) complex had tetragonal geometry. Many antibiotic drugs have modified toxicological and pharmacological properties with the presence of these drugs in form of metal complexes. Copper(II) is considered to be the most common metal in this field, which plays a role in the treatment of diseases such as gastric ulcers and rheumatoid arthritis [7–10].

These data encourage the study of the biological activity and the coordination chemistry of antibiotic drugs with transition metal ions to investigate the binding modes in the solid state. In continuation of the metal interactions examined in experimental work with derivatives of β-lactam [11–14], we studied the preparation and spectroscopic characterization of metal cefotaxime complexes. The major goal of the present work was to assess the antimicrobial activity of complexes of cefotaxime prepared with metal ions of calcium, chromium, copper, zinc, and selenium.

2. Experimental Methods

2.1. Chemical Reagents

All reagents used in this work, salts of metals, chemical solvents, and the cefotaxime antibiotic drug purchased from Sigma Aldrich were of reagent grade.

2.2. Synthesis of Cefotaxime Metal Complexes

The complexes of: $[M (cefotax)Cl]\cdot nH_2O$ (M = Ca(II), n = 1, Zn(II), n = 2 and Cu(II), n = 3), $[Cr (cefotax)(H_2O)Cl_2]\cdot H_2O$, and $[Se (cefotax) Cl_3]\cdot 2H_2O$ were synthesized by mixing of (1 mmol) from cefotaxime dissolved in 25 mL MeOH and (1 mmol) from metal chlorides of Ca(II), Cr(III), Zn(II), Cu(II), and Se(IV) dissolved in 20 mL distilled H_2O, then reaction mixture was stirred at room temperature for ca. 6 h, and left to stand overnight. The produced precipitates were filtered, washed using mixture of (MeOH distilled H_2O), and finally dried. The analytical data for isolated solid complexes, Table 1, depict the formation of complexes with stoichiometry of 1:1.

Table 1. Physical characterization and micro-analytical and molar conductance data of cefotaxime complexes.

Complexes	M. Wt	Elemental Analysis (%)					Λm ($\Omega^{-1}cm^2mol^{-1}$)
		C	H	N	S	M^{+n}	
[Ca(Cefotax) Cl]·H_2O	565.50	(33.95) 33.88	(3.53) 3.46	(12.37) 12.64	(11.31) 11.74	(7.08) 6.89	18
[Cr(Cefotax) (H_2O) $(Cl)_2$]·H_2O	612	(31.37) 31.84	(3.26) 3.85	(11.43) 11.23	(10.45) 10.92	(8.49) 8.35	23
[Cu(Cefotax) Cl]·$3H_2O$	607	(31.63) 31.63	(3.62) 3.64	(11.53) 11.83	(10.54) 10.71	(10.46) 11.03	16
[Zn(Cefotax) Cl]·$2H_2O$	589	(32.59) 32.72	(3.39) 3.89	(11.88) 11.29	(10.86) 10.45	(11.10) 11.51	17
[Se(Cefotax) $(Cl)_3$]·$2H_2O$	675.50	(28.42) 28.59	(2.96) 3.31	(10.36) 10.83	(9.47) 9.93	(11.69) 11.32	24

2.3. *Instruments*

For free cefotaxime ligand and its metal complexity, IR spectra were recorded as potassium bromide disks using infrared Bruker spectrophotometer ranging from 400–4000 cm^{-1}. UV–vis. spectra were recorded using DMSO solvent within range 800–200 nm using a UV2 Unicam UV/Vis Spectrophotometer fitted with a quartz cell of 1.0 cm path length. C, H, N, and S were performed by the microanalysis unit at Cairo University, Egypt, using a Perkin Elmer CHN 2400 instrument. With concentration of 10^{-3} M in DMSO solvent, conductivity measurements for cefotaxime metal complexity were measured using HACH conductivity meter model. Magnetic measurements were determined at room temperature using "Johnson Matthey" susceptibility balance by using Hg (II) tetrathiocyanato-cobaltate(II) as a calibrant. Spectra of ^1H NMR were investigated using "Bruker AM-500 NMR spectrometer", using TMS and deuterated DMSO as solvent. Scanning electron microscopy (SEM) images were taken with Joel JSM-6390 equipment, with an accelerating voltage of 20 KV. The X-ray diffraction patterns were recorded on X 'Pert PRO PAN analytical X-ray powder diffraction, target copper with secondary monochromate. The images of transmission electron microscopy (TEM) were performed using JEOL 100 s microscopy. The concentrations of metals in all experiments were performed using Flame atomic absorption spectroscopy instrument (Perkin Elmer Analyst 400, Waltham, MA, USA).

2.4. *Antioxidant Activity*

2.4.1. ORAC Assay

For cefotaxime and its metal complexes, the antioxidant activity was assessed based on Liang et al. [15]. For each, 10 μL of the cefotaxime and/or cefotaxime metal complexes were incubated for about 10 min with fluorescein at 37 °C. Subsequently, 70 μL of freshly prepared 2,2′-Azobis (2-amidinopropane) dihydrochloride (AAPH) (300 mM) were added immediately to each well. Fluorescein measurement (485 EX, 520 EM, nm) was continued for 1 h (40 cycles, each one 90 s). Data are represented as means (n = 3) ± SD and the antioxidant effect of the compound/extract was calculated as μM Trolox equivalents by substitution in the linear regression equation y = 4275.8x + 262,311.

2.4.2. The Test of Metal Chelation

The test of metal chelation was performed based on Santos et al. [16]. In total, 20 μL of freshly prepared $FeSO_4$ (0.3 mM) were mixed with about 50 μL of either cefotaxime and/or metal complexes in well plates before adding 30 μL of ferrozine. Next, all the plates were incubated for about 10 min at 37 °C. Any decrease in the intensity of color was measured at 562 nm. Data are represented as means ± SD according to the following equation:

$$\text{percentage inhibition} = ((\text{average absorbance of blank} - \text{average absorbance of the test})/(\text{average absorbance of blank})) \times 100. \quad (1)$$

2.4.3. The Test of DPPH

The antioxidant capacities of cefotaxime and its metal complexes were measured by 100 μL of freshly prepared DPPH reagent (0.1% in methanol) (2,2-diphenyl-1-picryl-hydrazyl-hydrate) free-radical assay carried out according to Boly et al. [17]. In total, 100 μL of the freshly prepared reagent were dissolved in methanol and then were added to cefotaxime and its metal complexes. The chemical reactions were then incubated for about 1/2 h in dark at about 37 °C. The resulting reduction in DPPH color intensity was measured at 540 nm. Data are represented as means ± SD according to the following equation:

$$\text{percentage inhibition} = ((\text{average absorbance of blank} - \text{average absorbance of the test})/(\text{average absorbance of blank})) \times 100.$$

2.4.4. ABTS Assay

The ABTS assay of cefotaxime and its metal complexes was carried out according to Arnao et al. [18]. In total, 192 mg of ABTS reagent were immediately dissolved in 1 mL dist. H_2O and transferred to 50 mL volumetric flask; the volume was then completed with distilled water. One mL of the previous solution was added to 17 μL of $K_2S_2O_8$ (140 mM) and the mixture was left in the dark for 24 h. Next, 1 mL of the reaction mixture was completed to 50 mL with methanol to obtain the final ABTS dilution used in the assay; the chemical reaction was incubated at 37 °C for 1/2 h in dark. At the end of incubation time, the decrease in ABTS color intensity was measured at 734 nm. Data are represented as means ± SD according to the following:

$$\text{percentage inhibition} = ((\text{average absorbance of blank} - \text{average absorbance of the test})/(\text{average absorbance of blank})) \times 100.$$

2.4.5. FARAB Assay

FARAB test was carried out based on Benzie et al. [19], on microplates. We used a freshly prepared TPTZ reagent (300 mM acetate buffer (PH = 3.6), 10 mM TPTZ in 40 mM HCL, and 20 mM $FeCl_3$, in a ratio of 10:1:1 *v/v/v*, respectively. In total, 190 μL from the freshly prepared TPTZ reagent were mixed with 10 μL of the sample on 96-well plates (n = 3). The reaction was incubated for 1/2 h in dark at 37 °C. The resulting blue color was measured at 593 nm. Data are presented as mean ± SD.

2.5. Cell Culture

HepG2 was freshly obtained from Scientific Inc., Cairo, in Egypt. The cells were kept in DMEM medium with streptomycin, penicillin, and fetal bovine serum in a 5% CO_2 atmosphere at about 37 °C. The cell lines were authenticated by STR analysis using the Gene Print 10 system (Promega corporation, Madison, WI, USA) [20].

2.5.1. Cytotoxicity Assay

HepG2 cells were maintained in a DMEM medium supplemented with streptomycin, penicillin, and fetal bovine serum in 5% CO_2 atmosphere at about 37 °C. The cell viability was assessed by SRB assay. Aliquots of about 100 μL of the cell suspension were seeded in well plates and immediately incubated in a complete medium for 24 h. The cells were then treated with another medium containing the drugs at various concentrations. After 3 days of exposure to the cefotaxime and cefotaxime metal complexes, the cells were fixed by replacing the medium with 10% TCA and incubated for 1 h at about 4 °C. The TCA solution was then removed, and the cells were washed 5 times with distilled H_2O. Aliquots of the SRB solution were added and incubated at 37 °C in the dark for about 10 min. The well plates were then washed 3 times with acetic acid (1%) and air-dried. Next, 150 μL of TRIS were added to dissolve the protein-bound SRB stain. The absorbance was measured at 540 nm using a "BMGLABTECH®-FLUOstar" Omega microplate reader (Ortenberg, Germany) [21].

2.5.2. Cytotoxic Activity (IC50 Determination)

The inhibitory concentration (IC50) of cell viability was determined by using Master-Plex 2010 software. The percentage of viable cells was plotted as a function of concentration to obtain the IC50 values. Three independent experiments were performed for all assays. The mean value from the triplicate experiments was calculated, and the results were reported as mean (±standard deviation). Control percentage was considered as 100%.

Inhibition percentage (%) was measured using the following equation:

$$\text{Inhibition percentage (\%)} = (\text{OD of control}) - (\text{OD of sample})/(\text{OD of control}) \times 100$$

where OD = optical density.

IC50 is the tested compound concentration that inhibits or kills 50% of cells and is obtained by plotting the inhibition percentage versus the test-compound concentration.

2.6. Antibacterial Activity against Bacillus Subtilis (ATCC 6633) and Escherichia coli (ATCC 8739)

Discs of *Bacillus subtilis* ATCC 6633 and *Escherichia coli* ATCC 8739 were inoculated into tryptic soy-broth medium and immediately incubated at about 30 °C for (18–24 h). Using a fresh-culture agar plate, a scoop from each broth was gently streaked into agar medium and then incubated at 30 °C. A sterile normal physiological saline (0.9%) was freshly prepared by inoculating about 3 to 4 colonies; the suspension was adjusted to achieve a turbidity equivalent to a 0.5 McFarland standard of each organism using a DensiCHEK© optical device (Anglum Road, Hazelwood, St. Louis, MO, USA). After convenient growth, 1.0 mL Muller Hinton Broth was inoculated into all testing wells (except for the first well). Next, 2 mL from cefotaxime, cefotaxime/Se, cefotaxime/Zn, cefotaxime/Cr, and cefotaxime/Cu were directly inoculated into the first well (without dilution); next, 1 mL was aspirated and transferred to the next well, previously filled with 1.0 mL Muller Hinton Broth. Subsequently, 1 mL was aspirated using a new tip and added to the next 1 mL broth. Next, 1 mL of the prepared inoculum was added to each well (1:2 dilution). This resulted in a final concentration of 5.0×10^5 CFU/well. A further 1 mL from the *Bacillus subtilis* and *Escherichia coli* suspensions was cultured. The growth-control well containing broth, without cefotaxime or its complexes, was added to each sample/plate. A negative-control well containing only the broth without the tested cefotaxime and its novel metal complexes and without bacteria was added to each sample plate [22]. All well plates were then incubated at about 30 °C for 24 h. After incubation, the well plates were removed from the incubator and placed on a dark surface to check for growth. All growth-control wells yielded turbid growth (T), and all negative control wells were found to be clear, showing the validity of the test [23].

2.7. Statistical Analysis

The results are presented as mean ± SE using Statistical Package for Statistical Sciences (SPSS) software version 27 (IBM Corp: Armonk, NY, USA, 2020) and Open Epi version 2.3.1. One-way ANOVA and post hoc power were used to analyze the data. A value of $p < 0.05$ was considered significant (using three replicates).

3. Results

3.1. Microanalytical and Physical Data

The analyses of C, H, N, S, and M^{+n} were convenient due to the stoichiometry of metal: the ligand was 1:1 for all cefotaxime metal complexes (Table 1). All complexes are solids stable in air, soluble in DMF and DMSO. The values of the molar conductivity, measured at room temperature in DMSO, were in agreement with the range of non-electrolytic character [24], suggesting that Cl^- ion was chelated with the metal ions inside the complex sphere.

3.2. Infrared Spectra

The infrared spectral data for the cefotaxime and its metal complexes were similar. The main IR bands are in Table 2 and Figure 2. The obtained spectral data had the same absorption pattern, in the range of 3500–2800 cm^{-1}, which can be attributed to the stretching vibrations of (O-H), (N-H), aromatic (C-H), and aliphatic (C-H) (Guzler and Germlich, 2002) that were observed in the spectra of the free cefotaxime and its metal complexes, with some shifts in the band frequencies owing to changes in the electronic density distribution for the aromatic rings and the main attached functional groups that occurred after the chelation of the metals. A band appeared in the IR spectrum of the cefotaxime at 1775 cm^{-1}, which was characterized as (C=O) lactam, while the (C=O) amide and ester bands overlapped at 1642 cm^{-1}; for the cefotaxime metal complexes, bands ranging from 1715–1740 and 1630–1650 cm^{-1} appeared.

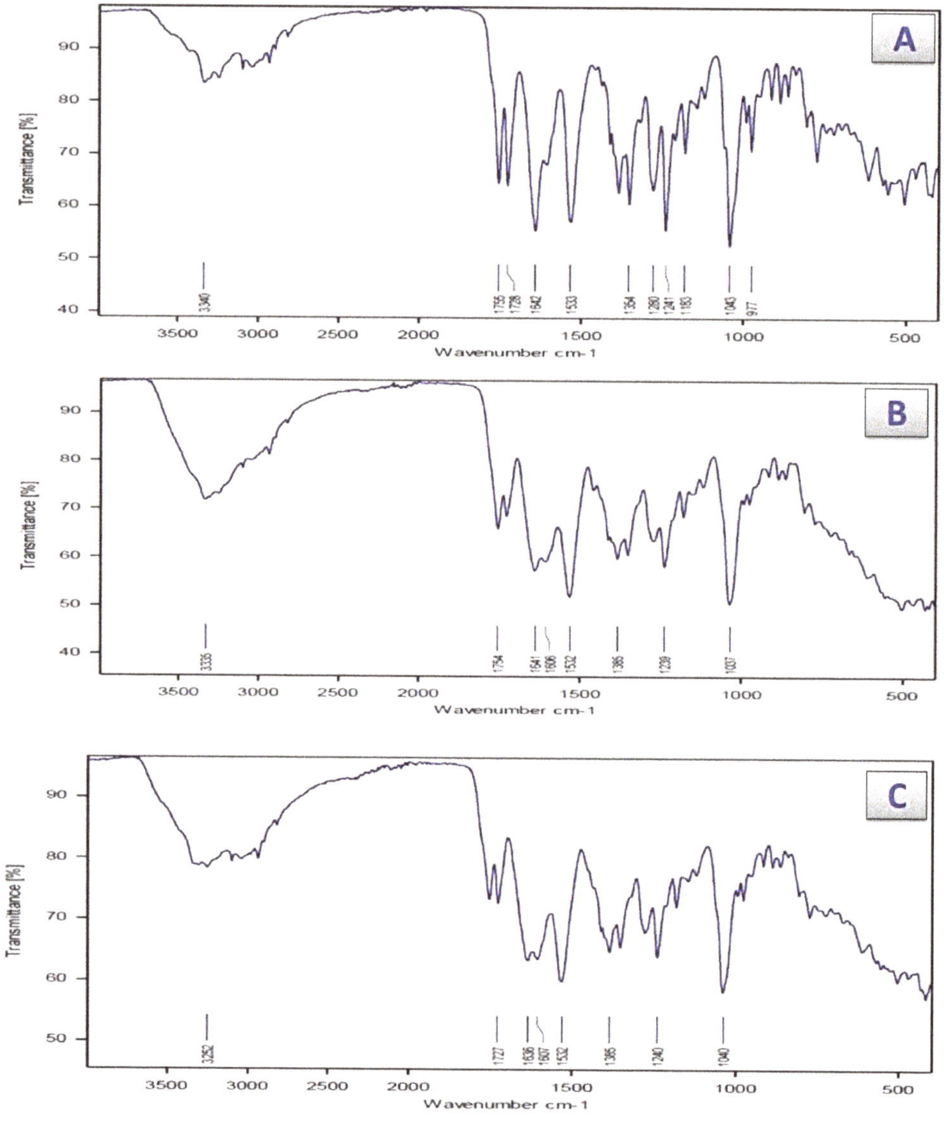

Figure 2. *Cont.*

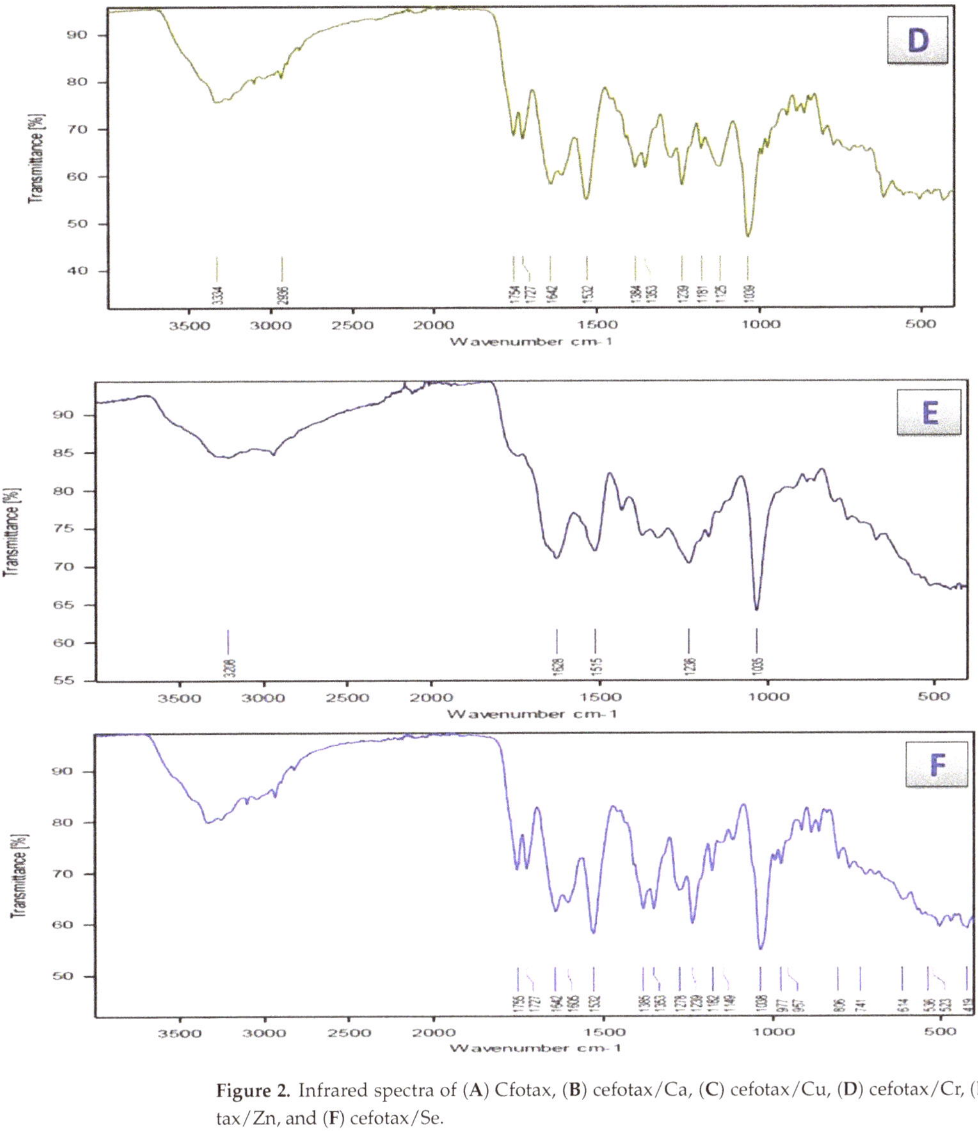

Figure 2. Infrared spectra of (**A**) Cfotax, (**B**) cefotax/Ca, (**C**) cefotax/Cu, (**D**) cefotax/Cr, (**E**) cefotax/Zn, and (**F**) cefotax/Se.

Based on these data, it can be concluded that the cefotaxime free ligand coordinated through the oxygen atom of the (C=O) lactam group rather than that of the (C=O) amide and ester group, since the bands of the (C=O) lactam group were shifted toward fewer frequencies (35–60 cm^{-1}) related to the value of the cefotaxime free ligand, while there was no significant shift for the overlapping bands of the ester and amide carbonyl.

Table 2. Infrared assignments of cefotaxime and its metal complexes.

Assignments		Complexes				
	Cefotax	Ca(II)	Cr(III)	Cu(II)	Zn(II)	Se(IV)
ν(O-H), COOH	3360	–	–	–	–	–
ν(O-H), H_2O	–	3335	3360	3334	3345	3365
ν(N-H), NH_2	3330	3310	3252	3316	3208	3230
ν(C-H); Aromatic	3075	3064	3070	3082	3071	3067
ν(C-H); Aliphatic	2938	2945	2941	2925	2936	2929
ν(C=O), lactam	1775	1720	1728	1715	1733	1740
ν(C=O)ester + ν(C=O)amide	1642	1650	1645	1630	1640	1646
ν_{as} (COO^-)	1598	1606	1607	1600	1612	1604
ν_s (COO^-)	1380	1385	1385	1384	1390	1375
$\Delta\nu = \nu_{as}$ (COO^-)– ν_s (COO^-)	218	221	222	216	222	229
ν (M-O)	--	640 530	618 518	624 523	618 525	614 536
ν (M-N)	--	470	476	495	480	475

The mode of chelation of the carboxylate group of cefotaxime ligand can bind as a monodentate or a bidentate, which leads to shifts in the stretching vibrations of antisymmetric and symmetric motions [25]. The main difference between νas (COO^-) and νs (COO^-): Δν = νas (COO^-)–νs (COO^-) produced values > 200 cm^{-1}, suggesting monodentate chelation for the (COO^-) group. The IR spectra for the cefotaxime complexes presented bands in the ranges of 640–614 cm^{-1} and 536–523 cm^{-1}, which may have been due to the ν(M-O). These bands are not observed in the cefotaxime spectrum. The presence of new bands in the range of 470–495 cm^{-1} in the spectra of the cefotaxime complexes may be attributed to the ν(M-N) (from the NH_2 group). These bands are absent from the cefotaxime free ligand, suggesting cefotaxime's coordination character as a chelating agent with a tridentate monoanionic character [26]. The chelation of the NH_2 group with metal ions was not the only explanation for these bands. When the nitrogen atom of the C=N-OCH_3 group could not chelate with the metal ions in solid complexes due to steric strains, the coordination of this nitrogen atom along with the carboxylate and lactamic carbonyl groups could not occur. This can be attributed to the contribution of the -NH_2 group to the coordination process. Finally, based on the above data from the IR spectra for the cefotaxime complexes, the proposed mode of chelation of cefotaxime with metal ions in nature through the lactam carbonyl, carboxylate group and amino group is tridentate in nature.

3.3. UV–Vis Spectra

The UV–vis spectra and its assignments for the free-ligand cefotaxime and its metal complexes in the DMSO solvent are presented in Table 3. By comparing the spectra for the cefotaxime base ligand and its metal complexes we deduced that cefotaxime has three absorption maxima, at 273, 320, and 350 nm. The band that appeared at 273 nm was assigned to $\pi \rightarrow \pi^*$ transitions due to the organic moiety, while the band that appeared at 320 nm was assigned to intraligand of the $\pi \rightarrow \pi^*$ transitions within the heterocyclic moieties [27,28]. The band at 350 nm was assigned to transition of the intraligand of sulphur atoms of the $n \rightarrow \pi^*$ type, which agreed with the literature data [27,29]. Based on the obtained data for the cefotaxime complexes, the bands due to the S atoms did not shift, suggesting that the S atom did not participate in the chelation process. The calcium(II), chromium(III), zinc(II), and selenium(IV) complexes showed weak bands at (285,322 and 391 nm), (280,327 and 405 nm), (288,331 and 407 nm), and (286,330 and 410 nm), which can be assigned to the $\pi \rightarrow \pi^*$ and $n \rightarrow \pi^*$ transitions.

Table 3. UV–vis spectra of cefotaxime and its metal complexes.

Compound	$\pi \rightarrow \pi^*$	$n \rightarrow \pi^*$
Cefotax	273,320	350
Ca(II)	285,322	391
Cr(III)	280,327	405
Zn(II)	288,331	407
Se(IV)	286,340	410

3.4. H-NMR Spectra

For the cefotaxime free ligand, the spectrum of the ^{1}H NMR (Table 4 and Figure 3) had two single peaks due to the CH_3 groups that appeared at 1.12 and 1.91 ppm. There were three double peaks for the CO-CH and N-CH on the β-lactam ring and the NH at 4.956, 5.770, and 9.534 ppm, respectively. The S-CH_2 and CH_2-O groups presented peaks at 3.127 and 4.174 ppm, respectively. The ^{1}H NMR spectra for the calcium (II), zinc(II), and Se(IV) cefotaxime complexes changed slightly compared with the cefotaxime, where, as expected, signals appeared downfield due to the occurrence of chelation between the cefotaxime ligand and the metal ions, which resulted in increasing conjugation. On the other hand, the signals of the NH_2 group observed at 7.459 ppm shifted to 7.249, 7.255, and 7.229 ppm for the Ca(II), Zn(II), and Se(IV), which means that the NH_2 participated in the coordination process. For the Se (IV) cefotaxime complexity, the band that appeared at 3.831, 3.836, and 3.808 ppm for the calcium, zinc, and selenium complexes was assigned to water molecules, which were not present in the ^{1}H NMR spectrum of the cefotaxime free ligand. Upon comparison with the free-ligand cefotaxime, the signal that appeared at 12 ppm was attributed to the protons in the COOH group. This signal peak disappeared in the spectra of the calcium, zinc, and selenium cefotaxime complexes, confirming the chelation of the cefotaxime ligand with the calcium(II), zinc(II), and Se(IV) ions through the deprotonated COOH group. Based on these data, we can conclude that cefotaxime behaves as a monoanionic tridentate ligand through the carboxylic, amino, and carbonyl beta lactam ring group.

Table 4. ^{1}H-NMR assignments for free-ligand cefotaxime and its metal complexes.

^{1}H-NMR Assignments	δ(ppm)			
	Cefotaxime	Ca(II)	Zn(II)	Se(VI)
1H; COOH	12	-	-	-
6H; 2CH_3	1.851	1.864	1.894	1.892
2H; S-CH_2	3.127	3.129	3.131	3.165
2H; O-CH_2	4.741	4.807	4.762	4.745
1H; CO-CH	4.956	4.989	4.975	4.964
1H; N-CH	5.770	5.617	5.575	5.589
2H; NH_2	7.459	7.249	7.255	7.229
1H; NH	9.534	9.535	9.538	9.532
2H; H_2O	-	3.831	3.836	3.808

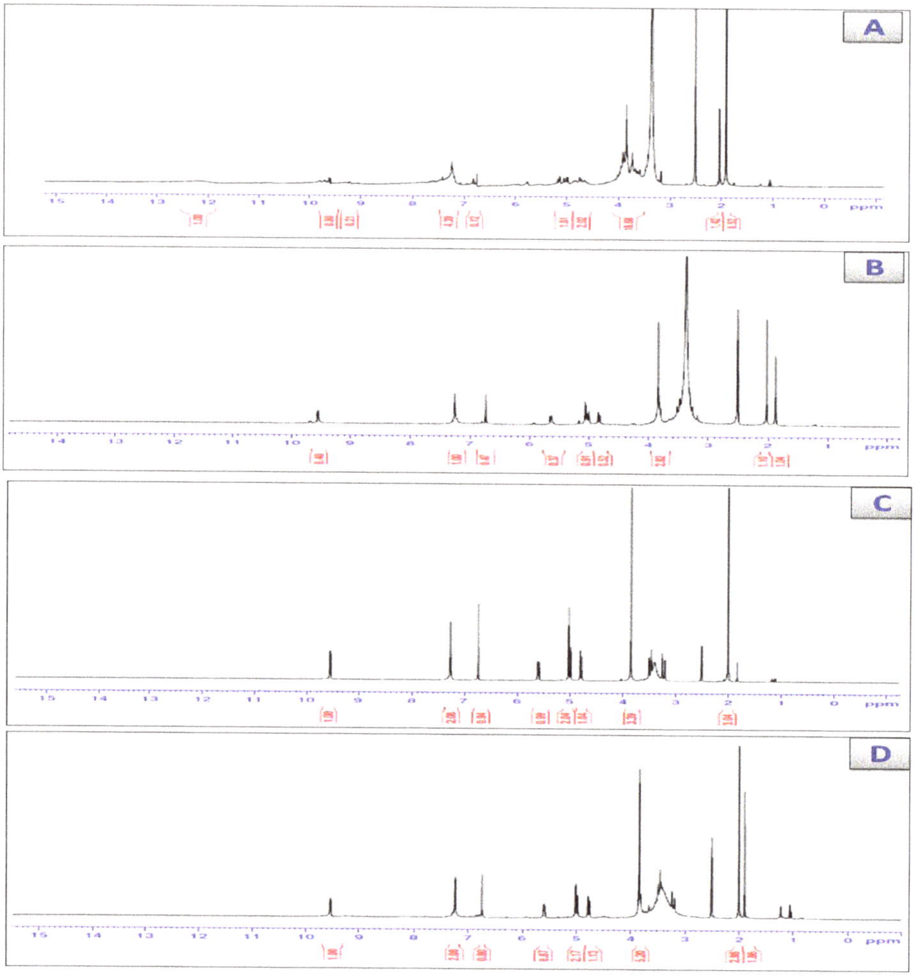

Figure 3. ^{1}H-NMR spectra of (**A**) Cfotax, (**B**) cefotax/Ca, (**C**) cefotax/Zn, and (**D**) cefotax/Se.

3.5. The Magnetic Measurements

According to the magnetic susceptibility values, the corrected magnetic moments were determined by using Pascal's constants. For the cu (II) complex at room temperature, the value of the magnetic moment was 2.03 B.M. which is adequate with systems of d^9 with one unpaired electron. The observation that this was the highest magnetic-moment value found for the cu(II) complex was based on the fact that spin-orbital coupling in the ion can mix with the ground state, confirming that there is no orbital momentum with higher levels of identical multiplicity, leading to small orbital contribution, which was in accordance with square planar or distorted tetrahedral geometries [30–32]. However, the presence of impurities cannot be discarded. The magnetic moment for the chromium(III) cefotaxime was 3.58 B.M., as calculated for system d^3 with three unpaired electrons, in accordance with octahedral geometry [33]. The experimental magnetic susceptibility value for the Se(IV) was 5.40 B.M., which was in agreement with the calculated value for octahedral geometry [33].

3.6. *Thermogravimetric Analyses*

The thermogravimetric and differential thermogravimetric analyses (TGA-DTG) for the Ca^{+2} and Cu^{2+} cefotaxime complexes were carried out up to 1000 °C, as shown in Figure 4. It was observed that the Cefotax/Ca and Cefotax/Cu metal complexes induced losses in weight that were up to 169 °C and 144 °C, owing to the loss of water molecules outside the chelation sphere, which suggested that H_2O molecules were present outside the sphere of chelation. The thermal analysis curves for the Ca(II) and Cu(II) cefotaxime complexes showed losses in weight of up to 980 °C, suggesting that the losses in weight for the Ca(II) and Cu(II) cefotaxime complexes corresponded to the evaporation of the cefotaxime ligand and chlorine atom. The thermal decomposition steps for the divalent calcium and copper complexes were five and three, respectively. As shown on the thermogram of the calcium and copper cefotaxime complexes, CaO and CuO were the most stable final products.

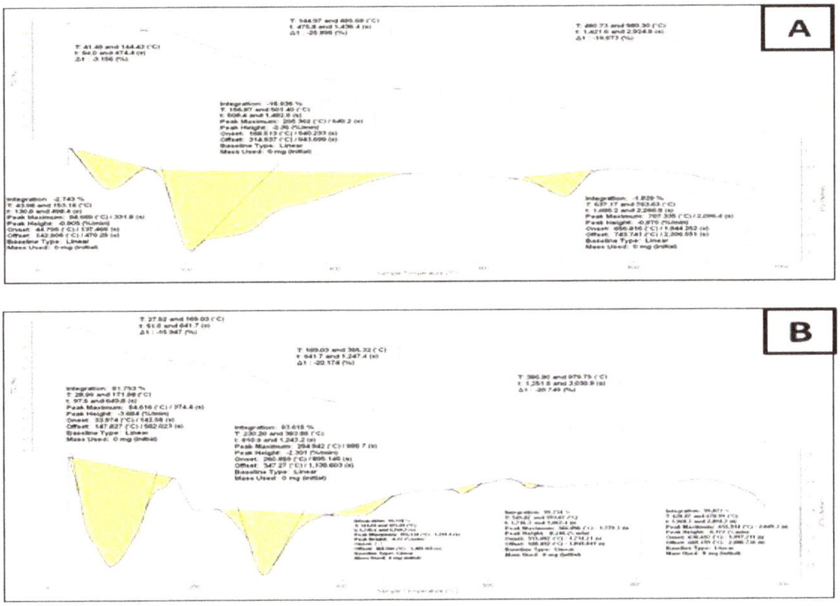

Figure 4. Thermogravimetric analysis of (**A**) cefotax/Ca and (**B**) cefotax/Cu.

3.7. *X-ray Diffraction and Scanning and Transmission Electron Microscopy*

The X-ray diffraction was carried out to confirm the structures of the cefotaxime complexes. For the cefotaxime complexes, the diffractograms produced are shown in Figure 5. The XRD diffractograms confirmed that all the cefotaxime complexes had formulating structures. Many different peaks were well defined at 2θ = 10, 17, 24, 26, 34, 35, 37, and 39 for the Cefotax, 28, 32, 45, 57, and 67 for the Cefotax/Ca (JCPDS = 9–432), 32 and 35 for the Cefotax/Cr (JCPDS = 33–0664), 32, 46, and 71 for the Cefotax/Zn (JCPDS = 36–1451), 21, 24, 29, and 30 for the Cefotax/Cu, and 32, 46, and 57 for the Cefotax/Se. It can be observed that the pattern for free ligand cefotaxime is different from that of its metal complexes, and this can be used as an indication of a well-defined formation and complete crystalline structure. The solid cefotaxime complexes were characterized by X-ray diffraction at room temperature using Cu Kα radiation. The diffraction characterizations of the prepared complexes were detected between 10° and 80°. The crystalline sizes of the synthesized complexes were determined by using the Scherrer equation [34] $D = k\lambda/\beta\cos\theta$, where k is a constant (0.94), λ is the wavelength (0.154 nm), and β is a full-in-width-at-half-maxima peak. The extent of the crystallinity of the cefotaxime complexes was calculated for the divalent cefotax/Ca, cefotax/, Cu, cefotax/Zn, trivalent chromium, and tetravalent selenium with values of 57 nm. 14 nm, 16 nm, 10 nm, and 15 nm respectively.

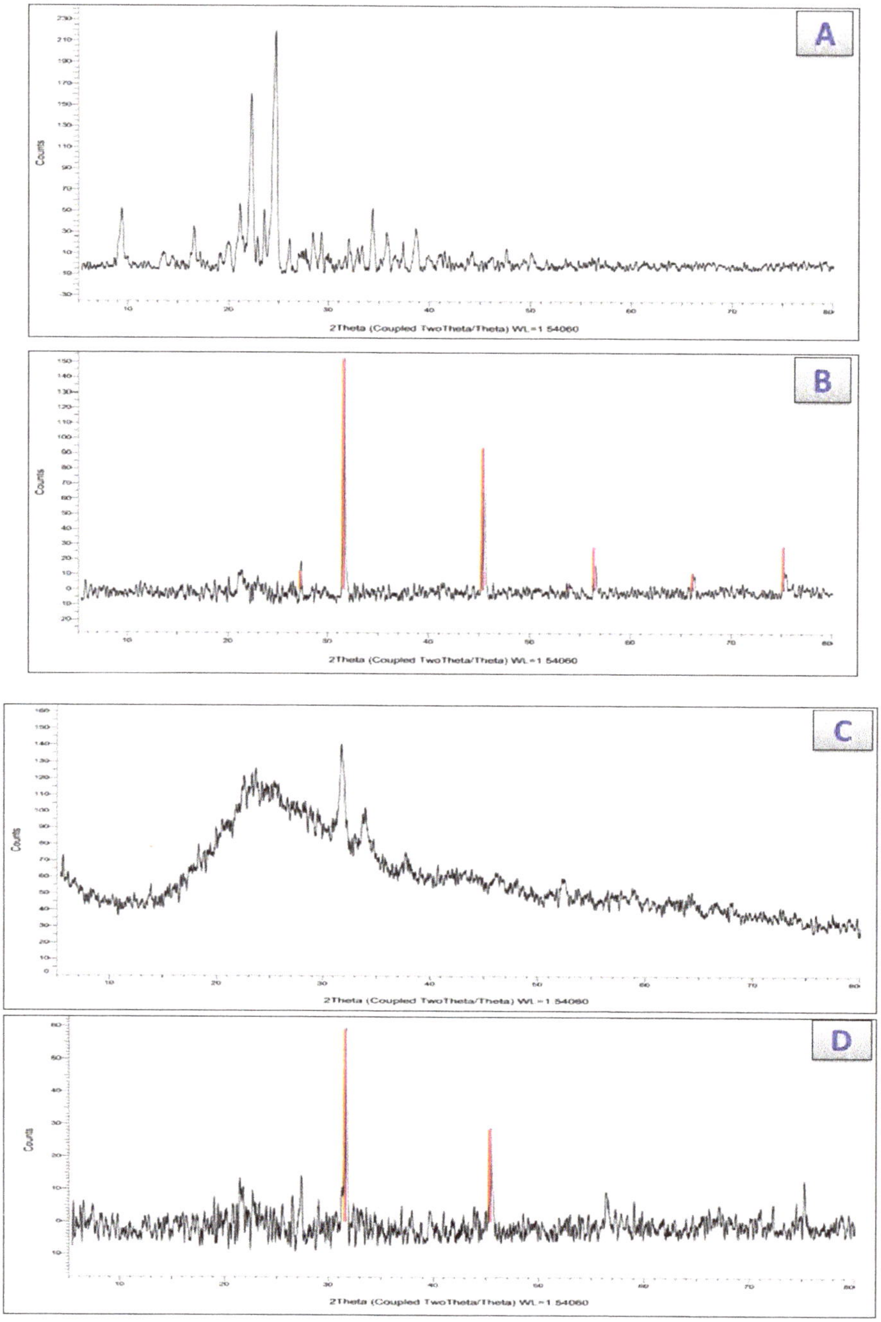

Figure 5. *Cont.*

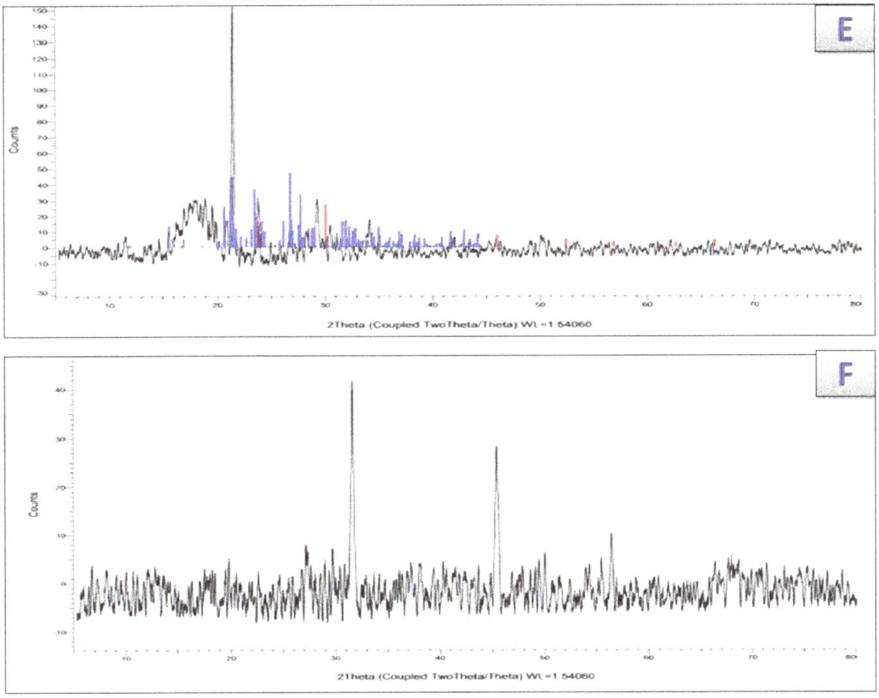

Figure 5. XRDs of (**A**) Cfotax, (**B**) cefotax/Ca, (**C**) cefotax/Cu, (**D**) cefotax/Zn, (**E**) cefotax/Se, and (**F**) cefotax/Cr.

The SEM images offer an explanation for the morphological surfaces of the Ca (II), Cr (II) Cu (II), Zn (II), and Se (VI) cefotaxime complexes that are shown in Figure 6, which give an impression of the SEM images of the NPs formed by the chelation process. The images scanned for the cefotax/Ca, cefotax/Cr, cefotax/Cu, cefotax/Zn, and cefotax/Se by using SEM allowed a full examination of the morphological phases. The SEM images, which were homogenous in size, show that the NPs of the prepared cefotaxime complexes were very close to each other. This can be attributed to the aggregation shape and size.

Figure 6. *Cont.*

Figure 6. Scanning electron microscopy of (**A**) Cfotax, (**B**) cefotax/Ca, (**C**) cefotax/Cr, (**D**) cefotax/Cu, (**E**) cefotax/Zn, and (**F**) cefotax/Se. (1 μm).

The TEM images for the prepared cefotaxime complexes are presented in Figure 7. The orderly matrices in the pictograph for the metal cefotaxime complexes were clarified. This demonstrated that the cefotaxime complexes had phase-material homogeneity. Many spherical black spots appeared in the cefotaxime chelates with particle sizes of 57–85, 10–15, 14–25, 16–33, and 15–25 nm for the cefotaxime complexes with Ca, Cr, Cu, Zn, and Se respectively.

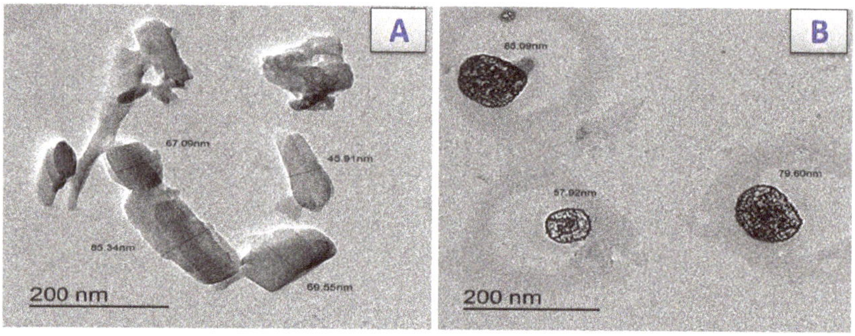

Figure 7. *Cont.*

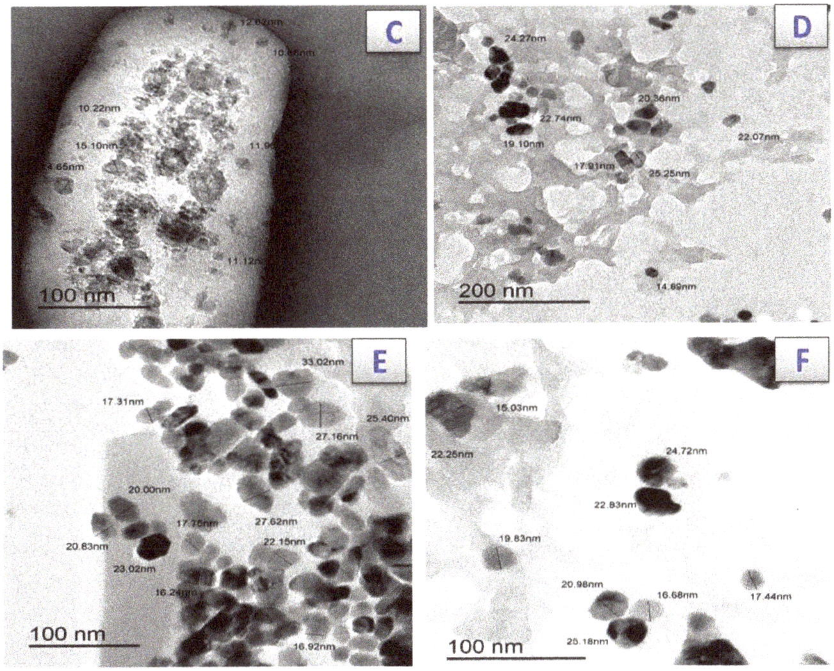

Figure 7. TEM images of (**A**) Cfotax, (**B**) cefotax/Ca, (**C**) cefotax/Cr, (**D**) cefotax/Cu, (**E**) cefotax/Zn, and (**F**) cefotax/Se.

3.8. Structures of Cefotaxime Complexes

The chemistry of the cefotaxime with the calcium, chromium, copper, zinc, and selenium metal ions was studied. Cefotaxime ions contain more than one donation atom; however, due to steric hindrance, cefotaxime can a make maximum of three donation sites (atoms) with a metal center. It has been suggested that the chelation of cefotaxime is carried out via carboxylate, amino, and lactamic carbonyl oxygen atoms. Calcium, copper, and zinc (II) complexes containing two one-chloride anions inside the complex sphere were the general formula for the [M (cefotax)Cl] complexes, where M = ca(II), cu(II), and zn(II), which were tetra-coordinated with cefotaxime as one molecule and one chloride anion. On the other hand, the chromium (III) and selenium (IV) complexes contained two and three chloride anions inside a complex sphere with an octahedral geometry. The proposed structure is shown in Figure 8.

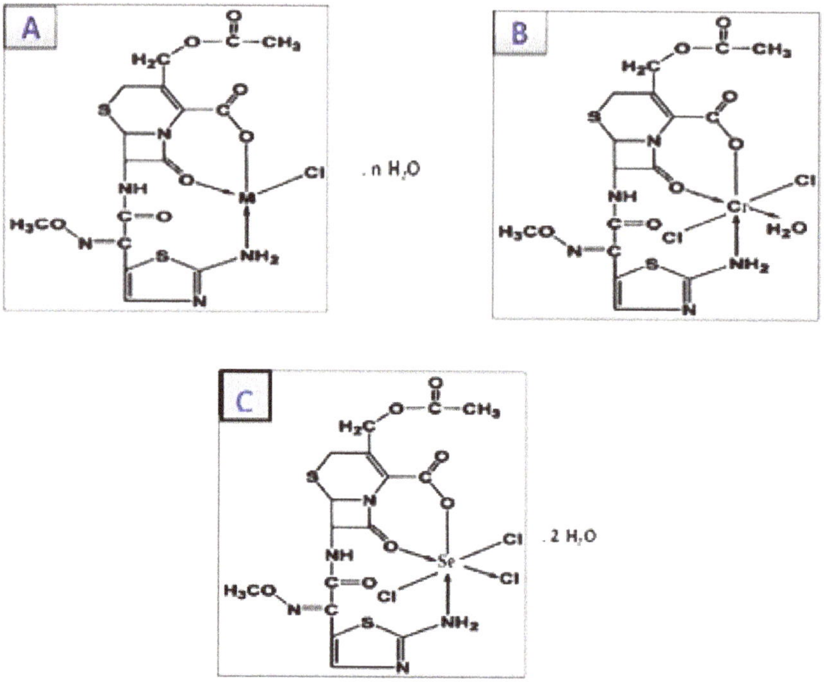

Figure 8. The structures of: (**A**) cefotax/M, where M = Ca (II), Cu(II), and Zn (II), and n = 1, 3, and 2, respectively; (**B**) cefotax/Cr; and (**C**) cefotax/Se.

3.9. Antioxidant Activities of Cefotaxime Metal Complexes

The chelating activity percentages are presented in Table 5. The ORAC assay, the metal chelation percentage, the ABTS assay, the FARAB assay, and the free-radical DPPH scavenging activities were used. The respective capacities of the cefotaxime/Cr and cefotaxime/se complexes to scavenge the free radicals were 320.02- and 303.33-fold, which were greater than that of the cefotaxime alone. Meanwhile, the metal chelating activities of the cefotaxime/Zn, cefotaxime/se, and cefotaxime/Cu were higher than that of cefotaxime alone by 15.42, 13.24, and 10.41% (μM EDTA eq/mg), respectively.

Table 5. Antioxidant activity of cefotaxime and its complexes (cefotaxime/Ca, cefotaxime/Cr, cefotaxime/Cu, cefotaxime/Zn, cefotaxime/Mg, and cefotaxime/Se).

Test Name	ORAC		Metal Chelation		ABTS		FARAB		DPPH	
Sample Name	(μM Trolox eq/mg)		(μM EDTA eq/mg)		(μM Trolox eq/mg)		(μM Trolox eq/mg)		(μM Trolox eq/mg)	
	Mean	SD	Mean	SD	Mean	SD	Mean	SD	Mean	SD
cefotaxime	7925.31 [e]	363.96	2.99 [e]	0.59	288.52 [d]	12.26	34.04 [e]	3.72	19.10 [d]	0.29
cefotaxime/Ca	8504.24 [d]	378.36	2.98 [e]	0.34	296.02 [c]	4.58	24.05 [f]	1.01	21.02 [c]	3.25
cefotaxime/Cr	8852.83 [d]	471.45	6.42 [d]	0.84	320.02 [ab]	3.25	106.02 [d]	9.58	14.25 [e]	2.40
cefotaxime/Cu	9172.406 [c]	532.59	10.41 [c]	2.96	289.14 [d]	4.69	110.87 [c]	10.69	15.05 [e]	1.69
cefotaxime/Zn	15,712.15 [a]	354.84	15.42 [a]	2.52	278.02 [e]	5.02	179.68 [a]	4.58	27.02 [b]	1.58
cefotaxime/Se	13,228.34 [b]	734.90	13.24 [b]	0.24	303.33 [b]	2.64	116.32 [b]	11.16	32.82 [a]	2.05

Results are expressed as mean ± SD. Symbols are different alphabetically (a–f) to indicate a significant comparison compared to the control group and other treated groups ($p < 0.05$) (Similar letters imply partial or complete non-significance).

Additionally, the ferric-reducing abilities of the cefotaxime/Zn, cefotaxime/se, and cefotaxime/Cu were 179.68, 116.32, and 110.87 (μM Trolox eq/mg), respectively. These were greater than that of the cefotaxime alone, which recorded a reducing ability of 34.04 (μM Trolox eq/mg).

On the other hand, the scavenging abilities of both the cefotaxime/se and the cefotaxime/Zn were also the highest when measuring the DPPH stability of the radical: 32.82 and 27.02% (μM Trolox eq/mg), respectively. Thus, the cefotaxime metal complexes exhibited chelation capacities that were higher than that of cefotaxime alone (Figure 9).

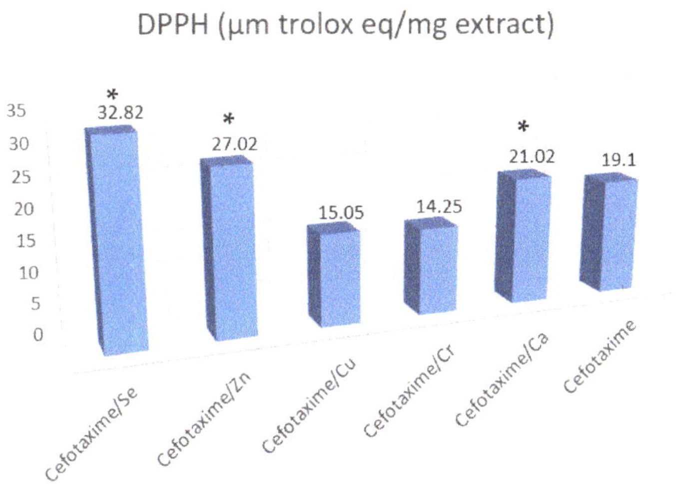

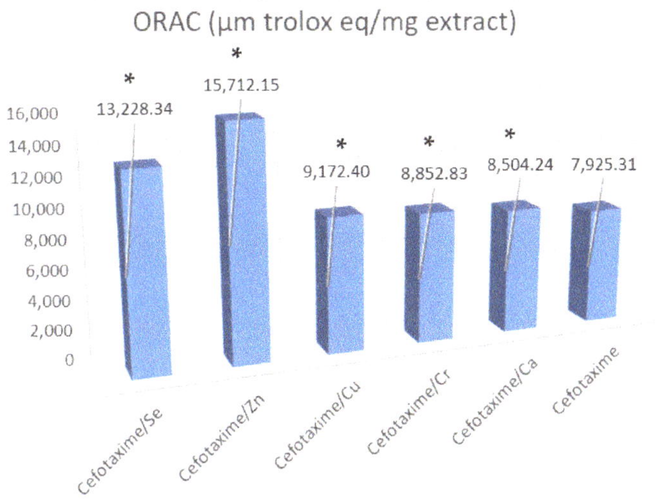

Figure 9. Antioxidant activity of cefotaxime and some of its metal complexes (DPPH and ORAC assays). Results are expressed as mean ± SE. (*) $p \leq 0.05$ vs. cefotax group. Superscript * letters show significant differences ($p < 0.05$) between cefotaxime metal complexes' activities. Trolox eq means Trolox equivalents and SD means standard deviation.

The antioxidant capacities of the cefotaxime metal complexes evaluated by the oxygen radical absorbance (ORAC) are presented in Table 5 and Figure 9. However, in this case, the antioxidant activities of the cefotaxime/Zn, cefotaxime/se, and cefotaxime/Cu were 15,712.15-, 13,228.34-, and 9172.40-fold (µM Trolox eq/mg) higher, respectively, than the antioxidant activity of cefotaxime alone; these results confirmed the greater activity of the novel complexes (cefotaxime/se, cefotaxime/Zn, cefotaxime/Cu, and cefotaxime/Cr) for the absorbance of free radicals than that of the cefotaxime alone.

3.10. Screening of Cytotoxic Activity of Ceftox/Se and Ceftoax/Zn Metal Complexes

The Ceftox/Se and Ceftoax/Zn complexes were tested for cytotoxic activity against HepG2. The results showed that both complexes were active against HepG2; this activity differed at different concentrations, as shown in Figure 10, with an IC50 range of 200–250 µg/mL.

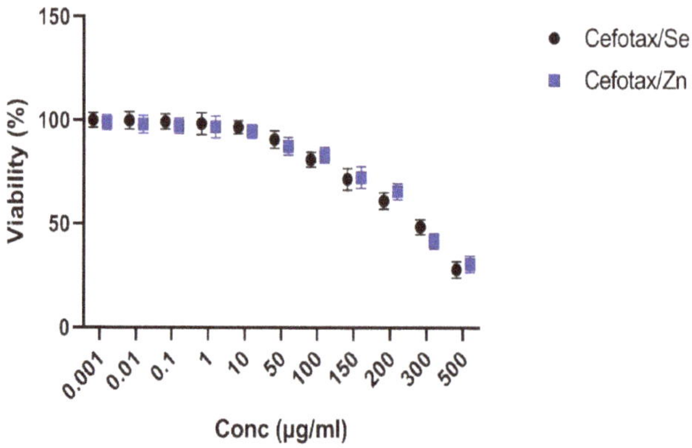

Figure 10. SRB (quick screening) curve of cefotax/Se and cefotax/Zn against HepG2 cells.

3.11. Anticancer Activity against HepG-2

The application of metal drug complexes in the medical field has been confirmed for the treatment of various types of cancer cells. Cefotaxime metal complexes have been evaluated against HepG2 cancer cells. In the present study, the cytotoxic effects of cefotaxime/Se and cefotaxime/Zn were tested on HepG-2 (human hepatocellular cancer cells); the results showed that the biosynthesized cefotaxime/Se and cefotaxime/Zn demonstrated cytotoxicity against the HepG-2 cells. This was also enhanced by increasing the concentrations of the tested complexes. The inhibition of hepatocellular carcinoma growth and the decline in cellular viability in the HepG-2 cells treated with cefotaxime/Se was recorded at concentrations of 100, 200, 300, and 500 µg/mL, as follows: 81.07, 61.11, 48.52, and 28.02 µg/mL, respectively. Meanwhile, the inhibition of the hepatocellular carcinoma growth and decline in cellular viability in the HepG-2 cells treated with cefotaxime/Zn was recorded at concentrations of 100, 200, 300, and 500 µg/mL as follows: 83.25, 65.71, 41.72, and 30.82 µg/mL, respectively. These results are shown in Table 6 and Figure 11.

Table 6. Anticancer activity of cefotaxime and its metal complexes (cefotaxime/Se, and cefotaxime/Zn).

Viability %	HepG-2			
	Cefotaxime/Se (µg/mL)		Cefotaxime/Zn (µg/mL)	
	Mean	SD	Mean	SD
10 µg/mL	96.52	2.05	94.70	1.42
100 µg/mL	81.07	1.06	83.25	0.11
200 µg/mL	61.11	2.60	65.71	2.41
300 µg/mL	48.52	1.80	41.72	2.01
500 µg/mL	28.02	2.68	30.82	1.97

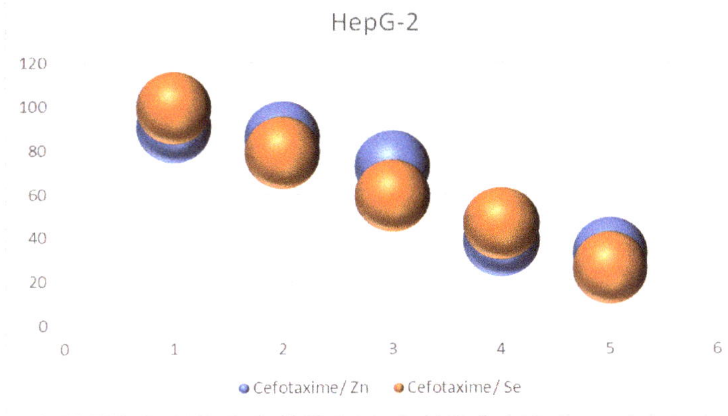

Figure 11. Anticancer activity of cefotaxime/Zn and cefotaxime/Se.

3.12. Antibacterial Activity Evaluation

Biological evaluations of the target complexes were performed on Gram-positive (*Bacillus subtilis*) and Gram-negative (*Escherichia coli*) bacteria. The results of the antimicrobial activities of the cefotaxime and/or metal complexes are presented in Table 7 and Figure 12. The inhibition concentrations of the cefotaxime metal complexes (samples with the same concentration) against both Gram-positive and Gram-negative bacteria (*B. subtilis* and *E. coli*) were found to be high at very low concentrations of 0.009 and 0.03 for the cefotaxime/Se, 0.078 and 0.039 for the cefotaxime/Cu, 1.25 and 0.625 for the cefotaxime/Zn, and 0.625 and 1.25 for the cefotaxime/Cr, respectively. Based on the standard conditions, (Table 7 and Figures 12 and 13), all the cefotaxime complexes were found to be sufficient, with high antimicrobial activity.

Means are expressed as (mean ± SE) and feature different letters. They are significant where $p \leq 0.05$ using Duncan's range test. The highest value has (a) symbol. The declining values are assigned alphabetically.

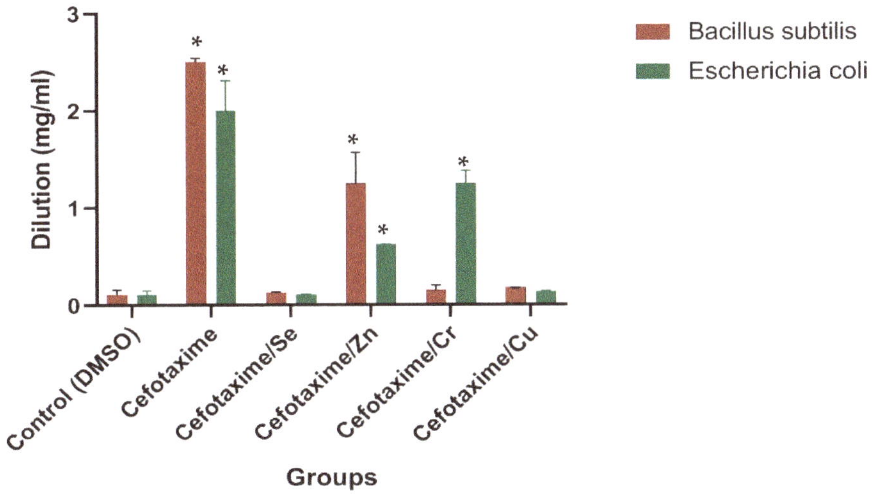

Figure 12. Antibacterial activity of cefotaxime and cefotaxime metal complexes, Results are expressed as mean ± SE. (*) $p \leq 0.05$ vs. cefotax group.

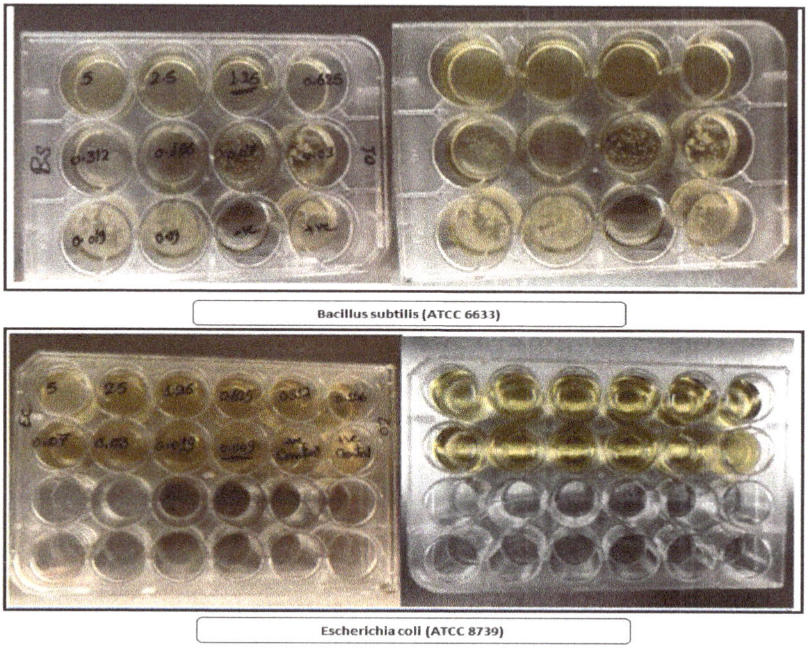

Figure 13. Cefotaxime and cefotaxime metal complexes vs. *Bacillus subtilis* (ATCC 6633) and *Escherichia coli* (ATCC 8739).

Table 7. Dilution concentration (mg/mL sample) and bacterial inhibition concentration of cefotaxime and its metal complexes (three replicates).

Sample	Dilution (mg/mL)	
	Bacillus subtilis (G^+)	*Escherichia coli* (G^-)
Control (DMSO)	0.0 ± 0.00 [e]	0.0 ± 0.00 [f]
Cefotaxime	2.5 ± 0.04 [a]	2.0 ± 0.31 [a]
Cefotaxime/Se	0.03 ± 0.007 [d]	0.009 ± 0.004 [e]
Cefotaxime/Zn	1.25 ± 0.32 [b]	0.625 ± 0.001 [c]
Cefotaxime/Cr	0.156 ± 0.05 [c]	1.25 ± 0.13 [b]
Cefotaxime/Cu	0.078 ± 0.001 [c]	0.039 ± 0.007 [d]

Means are expressed as (mean ± SE) and feature different letters. They are significant where $p \leq 0.05$ using Duncan's range test. The highest value has (a) symbol. The declining values are assigned alphabetically.

4. Discussion

Cefotaxime is an antibiotic drug belonging to the third generation of cephalosporins [35]; it is an active agent against both Gram-negative and Gram-positive bacterial strains via the inhibition of peptidoglycan-layer synthesis by different bacterial cellular walls [36]. The infectious diseases caused by bacterial strains are still a major health problem worldwide due to the rapid increase in bacterial resistance to existing antimicrobial drugs. Oxidative stress is also a serious problem, and it poses a high risk to human health, especially hepatic health problems [37]. Thus, the present study was conducted to prepare novel metal complexes of cefotaxime with different metal ions to clarify the antioxidant potency, antibacterial activity at lower concentrations, and hepatoprotective effects of cefotaxime complexes against HepG-2 cancer cells.

Regarding the multidrug resistance of highly pathogenic bacterial strains, the discovery of novel antimicrobial compounds is of high importance. Many microorganism strains, especially Gram-positive strains, release β-lactamase in large amounts, and they have the ability to destroy antibiotics' β-lactam through the hydrolysis of the β-lactam rings; this is the most prevalent mechanism of antibacterial resistance [38,39].

The oxidative damage to biomolecules is of major interest in diseases such as cancer, and even serious sequences of viruses such as COVID-19. Therefore, it is urgent to examine the role of novel potent metal complexes to prevent severe oxidative injury and the treatment of diseases that involve the risk of oxidative damage.

The findings in this study confirmed the novel antioxidant capacities of cefotaxime complexes with either Se, Zn, Cr, Cu, or Ca by examining their antioxidant capacities after complexation with cefotaxime; this elevated the efficacy of cefotaxime, as it became antioxidant and antibacterial at the same time, which is very important during pandemics such as COVID-19, of which oxidative stress appears to be a catastrophic health effect, since it can destroy different bodily functions of infected patients. Thus, synthesizing a novel cefotaxime as an antibiotic drug with metals will help to alleviate oxidative injury in severe cases.

DPPH assays are used to predict the antioxidant activities through which antioxidants act to inhibit lipid oxidation, scavenge DPPH radicals and, therefore, determine the free-radical scavenging capacity [40]. The findings in this study demonstrated that cefotaxime complexes with Se, Zn, and Cu, respectively, significantly scavenged the DPPH radical, which means that the novel complexes had higher antioxidant capacities than cefotaxime alone.

Another assay that is important for the evaluation of antioxidant capacities is the ABTS assay, which measures the ability of antioxidants to scavenge the ABTS generated in the aqueous phase, compared with a Trolox (analogue vitamin E and water-soluble) standard. The ABTS is generated by reacting a strong oxidizing agent with ABTS salt [40]; our results proved that the cefotaxime complexes with Cr, Se, and Ca exhibited strong antioxidant capacities, confirming our findings.

The ORAC assay uses the ability of AAPH to form peroxyl radicals when heated in the presence of a sufficient quantity of oxygen. These radicals quench the fluorescence of a probe, thereby reducing it. The level of reduction depends on the antioxidant that quenches the produced radicals. The presented results revealed that both cefotaxime/Zn and cefotaxime/Se possessed a high capacity to reduce AAPH; thus, they possessed novel antioxidant capacities.

The liver injuries caused by many drugs may develop into chronic liver failure and hepatocellular carcinoma [41–43]. The majority of adverse liver responses occur due to treatments with antibiotics [44]. Several antibiotics are considered to be a common cause of the induction of hepatic damage. Hepatic impairment occurs due to hepatotoxicity [45]. The current study provided detailed results of the anticancer capacities of both cefotaxime/Zn and cefotaxime/Se in the inhibition of cancer-cell viability and a reduction to about 30 μg/mL of viable cells at a concentration of 500 (μg/mL), which is a promising sign of cefotaxime's prospective anticancer activities.

A previous study reported that in the synthesis of cefotaxime with gold nanoparticles, cefotaxime acted as a reducing agent. The synthesized cefotaxime metal complexes were highly stable. Most importantly, they demonstrated the high antibacterial activity of cefotaxime upon loading onto the gold surfaces; at much lower concentrations, this new formula inhibited the growth of the tested bacterial strains (Gram-negative and Gram-positive) more than cefotaxime alone [46]. These findings are in agreement with those of the current study, which will expand the investigation of this novel formula with high antioxidant and antimicrobial activities.

Due to the recent increase in infectious diseases worldwide and the evolution of many resistant strains of bacteria and viruses, providing novel antibiotics with novel metals that may increase their efficacy, strength, and antioxidant activities and lower their side effects is important. The mechanism that enables bacterial strains to resist β-lactams is essentially the synthesis of β-lactamase enzymes that break the β-lactam ring; this prevents antibiotics from binding to the peptidoglycan layer [39]. We presented the novel metal complexes with cefotaxime and examined their biological activity against both *Bacillus s.* and *Escherichia c.* stains; they had high antibacterial activity against them at very low concentrations, which is a promising result. Previous studies reported the antimicrobial activity of different cephalosporins strains after complexation with metal complexes, confirming our findings [47–49].

5. Conclusions

The present study prepared five cefotaxime metal complexes and chemically characterized the products of the reaction of cefotaxime with Ca^{2+}, Cr^{+3}, Cu^{2+}, Zn^{2+}, and Se^{4+} ions. The cefotaxime complexes' structures were characterized through elemental analysis, molar conductance, FTIR, ^{1}HNMR, electronic spectra, magnetic, scanning and transmission electron microscopy, and X-ray diffraction analyses. The Cr(III) and Se(VI) complexes formed a coordination number equal to six, with distorted octahedral geometry. The acquired results showed that the cefotaxime metal complexes (Cefotax/Zn and Cefotax/Se) greatly inhibited the hepatocellular carcinoma viability percentage in HepG-2 cells and, with other complexes of cefotax/Cr, cefotax/Cu, and cefotax/Ca, enhanced the antioxidant activities tested with DPPH, ABTS, ORAC, FARAB, and metal chelation assays compared with cefotaxime alone. For antibacterial activity, the cefotaxime metal complexes were effective against bacterial strains of *Bacillus subtilis* and *Escherichia coli*. The obtained results are very promising, since they demonstrate the provision of strong protection against hepatocellular carcinoma (HepG-2) and a reduction in the damaging effects and the severe oxidative stress induced by antibiotics, especially during the COVID-19 pandemic.

Author Contributions: Conceptualization, E.H.A.-T., S.M.E.-M., R.Z.H. and B.A.; methodology, E.H.A.-T., S.M.E.-M., R.Z.H. and B.A.; software, E.H.A.-T., S.M.E.-M., R.Z.H. and B.A., validation, E.H.A.-T., S.M.E.-M., R.Z.H. and B.A.; formal analysis, E.H.A.-T., S.M.E.-M., R.Z.H. and B.A.; investigation, E.H.A.-T., S.M.E.-M., R.Z.H. and B.A.; data curation, E.H.A.-T., S.M.E.-M., R.Z.H. and B.A.; writing—original draft preparation, E.H.A.-T., S.M.E.-M., R.Z.H. and B.A.; writing—review and editing, E.H.A.-T., S.M.E.-M., R.Z.H. and B.A.; visualization, E.H.A.-T., S.M.E.-M., R.Z.H. and B.A.; supervision, E.H.A.-T., S.M.E.-M., R.Z.H. and B.A.; project administration, E.H.A.-T., S.M.E.-M., R.Z.H. and B.A.; funding acquisition, E.H.A.-T., S.M.E.-M., R.Z.H. and B.A. All authors have read and agreed to the published version of the manuscript.

Funding: This research received no external funding.

Data Availability Statement: All the Data are available within the text.

Acknowledgments: Authors acknowledge Taif University Researchers Supporting number (TURSP 2020/15), Taif University, Taif, Saudi Arabia.

Conflicts of Interest: The authors declare no conflict of interest.

References

1. Fu, K.P.; Foleno, B.D.; Lafredo, S.C.; LoCoco, J.M.; Isaacson, D.M. In vitro and in vivo antibacterial activities of FK037, a novel parenteral broad-spectrum cephalosporin. *Antimicrob. Agents Chemother.* **1993**, *37*, 301–307. [CrossRef] [PubMed]
2. Gums, J.G.; Boatwright, D.W.; Camblin, M.; Halstead, D.C.; E Jones, M.; Sanderson, R. Differences between ceftriaxone and cefotaxime: Microbiological inconsistencies. *Ann. Pharmacother.* **2008**, *42*, 71–79. [CrossRef] [PubMed]
3. Tripathi, K. A review—Can metal ions be incorporated into drugs? *Asian J. Res. Chem.* **2009**, *2*, 14–18.
4. Anacona, J.R.; Osorio, I. Synthesis and antibacterial activity of copper (II) complexes with sulphathiazole and cephalosporin ligands. *Trans. Met. Chem.* **2008**, *33*, 517–521. [CrossRef]
5. Masoud, M.S.; Ali, A.E.; Elasala, G.S. Synthesis, spectral, computational and thermal analysis studies of metallocefotaxime antibiotics. *Spectrochim. Acta Part A Mol. Biomol. Spectrosc.* **2015**, *149*, 363–377. [CrossRef]
6. Reiss, A.; Chifiriuc, M.C.; Amzoiu, E.; Spînu, C.I. Transition metal (II) complexes with cefotaxime-derived Schiff base: Synthesis, characterization, and antimicrobial studies. *Bioinorg. Chem. Appl.* **2014**, *17*, 926287. [CrossRef]
7. Williams, D.R. *The Metals of Life*; Van Nostrand Reinhold: London, UK, 1971.

8. Sorenson, J.R.J. Copper chelates as possible active forms of the antiarthritic agents. *J. Med. Chem.* **1976**, *19*, 135. [CrossRef]
9. Brown, D.H.; Smith, W.E.; Teape, J.W. Antiinflammatory effects of some copper complexes. *J. Med. Chem.* **1980**, *23*, 729. [CrossRef]
10. Sorenson, J.R.J.; Nraign, J.O. (Eds.) *Copper in the Environment*; Wiley-Interscience: New York, NY, USA, 1981; Part 2, Chapter 5.
11. Anacona, J.R.; Toledo, C. Synthesis and antibacterial activity of metal complexes of ciprofloxacin. *Transit. Met. Chem.* **2001**, *26*, 228–231. [CrossRef]
12. Anacona, J.R.; Alvarez, P. Synthesis and antibacterial activity of metal complexes of cefazolin. *Transit. Met. Chem.* **2002**, *27*, 856–860. [CrossRef]
13. Anacona, J.R.; Serrano, J. Synthesis and antibacterial activity of metal complexes of cephalothin. *J. Coord. Chem.* **2003**, *56*, 313. [CrossRef]
14. Anacona, J.R.; Rodriguez, I. Synthesis and antibacterial activity of cephalexin metal complexes. *J. Coord. Chem.* **2004**, *57*, 1263. [CrossRef]
15. Liang, Z.; Cheng, L.; Zhong, G.Y.; Liu, R.H. Antioxidant and Antiproliferative activities of Twenty-Four Vitis vinifera Grapes. *PLoS ONE* **2014**, *9*, e105146. [CrossRef] [PubMed]
16. Santos, J.S.; Brizola, V.R.A.; Granato, D. High-throughput assay comparison and standardization for metal chelating capacity screening: A proposal and application. *Food Chem.* **2017**, *214*, 515–522. [CrossRef] [PubMed]
17. Boly, R.; Lamkami, T.; Lompo, M.; Dubois, J.; Guissou, I. DPPH free radical scavenging activity of two extracts from Agelanthus sosoneifolius (Loranthaceae) Leaves. *Int. J. Toxicol. Pharmacol. Res.* **2016**, *8*, 29–34.
18. Arnao, M.B.; Cano, A.; Acosta, M. The hydrophilic and lipophilic contribution of total antioxidant activity. *Food Chem.* **2001**, *73*, 239–244. [CrossRef]
19. Benzie, I.F.; Strain, J.J. The ferric reducing ability of plasma (FARAP) as a measure of "antioxidant power": The FARAP assay. *Anal. Biochem.* **1996**, *239*, 70–76. [CrossRef]
20. Algehani, R.A.; Abou Khouzam, R.; Hegazy, G.A.; Alamoudi, A.A.; El-Halawany, A.M.; El Dine, R.S.; Ajabnoor, G.A.; Al-Abbasi, F.A.; Baghdadi, M.A.; Elsayed, I.; et al. Colossolactone-G synergizes the anticancer properties of 5-fluorouracil and gemcitabine against colorectal cancer cells. *Biomed. Pharmacother.* **2021**, *140*, 111730. [CrossRef]
21. Abdel-Hameed, E.S.S.; Bazaid, S.A.; Shohayeb, M.M.; El-Sayed, M.M.; El-Wakil, E.A. Phytochemical studies and evaluation of antioxidant, anticancer and antimicrobial properties of *Conocarpus erectus* L. growing in Taif, Saudi Arabia. *Eur. J. Med. Plants* **2012**, *2*, 93–112. [CrossRef]
22. Houdkova, M.; Chaure, A.; Doskocil, I.; Havlik, J.; Kokoska, L. New Broth Macrodilution Volatilization Method for Antibacterial Susceptibility Testing of Volatile Agents and Evaluation of Their Toxicity Using Modified MTT Assay In Vitro. *Molecules* **2021**, *26*, 4179. [CrossRef]
23. Liu, D.; Kwasniewska, K. An improved agar plate method for rapid assessment of chemical inhibition to microbial populations. *Bull. Environ. Contam. Toxicol.* **1981**, *27*, 289–294. [CrossRef] [PubMed]
24. Geary, W.J. The Use of Conductivity Measurements in Organic Solvents for the Characterisation of Coordination Compounds. *Coord. Chem. Rev.* **1971**, *7*, 81. [CrossRef]
25. Maeda, Y.; Okawara, R.J. Vibrational Spectra of Organometallics: Theoretical and Experimental Data. *Organometal. Chem.* **1967**, *10*, 247. [CrossRef]
26. Nakamoto, K. *Infrared and Raman Spectra of Inorganic and Coordination Compounds*, 4th ed.; John Wiley: New York, NY, USA, 1986.
27. Franchini, G.C.; Giusti, A.; Preti, C.; Tassi, L.; Zannini, P. Coordinating ability of methylpiperidine dithiocarbamates towards platinum group metals. *Polyhedron* **1985**, *4*, 1553–1558. [CrossRef]
28. Hadjikostas, C.C.; Katsoulos, G.A.; Shakhatreh, S.K. Synthesis and spectral studies of some new palladium(II) and 8 International Journal of Inorganic Chemistry platinum(II) dithiocarbimato complexes. Reactions of bases with the corresponding N-alkyldithiocarbamates. *Inorg. Chim. Acta* **1987**, *133*, 129–132. [CrossRef]
29. Castillo, M.; Criado, J.J.; Macias, B.; Vaquero, M.V. Chemistry of dithiocarbamate derivatives of amino acids. I. Study of some dithiocarbamate derivatives of linear α-amino acids and their nickel(II) complexes. *Inorg. Chim. Acta* **1986**, *124*, 127–132. [CrossRef]
30. Oztürk, O.F.; Şekerci, M.; Ozdemir, E. Synthesis of 5, 6-O-Cyclohexylidene-l-amino-3-azahexane and Its Co (II), Ni (II), and Cu (II) Complexes. *Russ. J. Coord. Chem.* **2005**, *31*, 687–690. [CrossRef]
31. El-Megharbel, S.M.; Qahl, S.H.; Alaryani, F.S.; Hamza, R.Z. Synthesis, Spectroscopic Studies for Five New Mg (II), Fe (III), Cu (II), Zn (II) and Se (IV) Ceftriaxone Antibiotic Drug Complexes and Their Possible Hepatoprotective and Antioxidant Capacities. *Antibiotics* **2022**, *11*, 547. [CrossRef]
32. Lever, A.B.P. *Inorganic Electronic Spectroscopy*; Elsevier: Amsterdam, The Netherlands, 1984.
33. Figgis, B.N. *Introduction to Ligand Fields*; John Wiley: New York, NY, USA, 1976.
34. Cullity, B.D. *Elements of X-ray Diffraction*; Addison-Wesley: Reading, MA, USA, 1972; p. 102.
35. Anacona, J.R.; Da Silva, G. Synthesis and antibacterial activity of cefotaxime metal complexes. *J. Chil. Chem. Soc.* **2005**, *50*, 447–450. [CrossRef]
36. Chin, N.-X.; Gu, J.-W.; Fang, W.; Neu, H.C. In vitro activity and β-lactamase stability of GR69153, a new long-acting cephalosporin. *Antimicrob. Agents Chemother.* **1991**, *35*, 259–266. [CrossRef]

37. Hamza, R.Z.; Al-Malki, N.A.; Alharthi, S.; Alharthy, S.A.; Albogami, B.; El-Megharbel, S.M. Chemical Characterization of Taif Rose (*Rosa damascena*) Methanolic Extract and Its Physiological Effect on Liver Functions, Blood Indices, Antioxidant Capacity, and Heart Vitality against Cadmium Chloride Toxicity. *Antioxidants* **2022**, *11*, 1229. [CrossRef]
38. Hamza, R.Z.; Alaryani, F.S.; Alotaibi, R.E.; Al-Harthi, M.A.; Alotaibi, G.S.; Al-Subaie, N.A.; Al-Talhi, A.A.; Al-Bogami, B.; Al-Baqami, N.M.; El-Megharbel, S.M.; et al. Efficacy of Prednisolone/Zn Metal Complex and Artemisinin Either Alone or in Combination on Lung Functions after Excessive Exposure to Electronic Cigarettes Aerosol with Assessment of Antibacterial Activity. *Crystals* **2022**, *12*, 972. [CrossRef]
39. Anacona, J.R.; Estacio, J. Synthesis and antibacterial activity of cefixime metal complexes. *Transit. Met. Chem.* **2006**, *31*, 227–231. [CrossRef]
40. El-Megharbel, S.M.; Hamza, R.Z. Synthesis, spectroscopic characterizations, conductometric titration and investigation of potent antioxidant activities of gallic acid complexes with Ca (II), Cu (II), Zn(III), Cr(III) and Se (IV) metal ions. *J. Mol. Liq.* **2022**, *358*, 119196. [CrossRef]
41. Hamza, R.Z.; Al-Yasi, H.M.; Ali, E.F.; Fawzy, M.A.; Abdelkader, T.G.; Galal, T.M. Chemical Characterization of Taif Rose (Rosa damascena Mill var. trigentipetala) Waste Methanolic Extract and Its Hepatoprotective and Antioxidant Effects against Cadmium Chloride (CdCl2)-Induced Hepatotoxicity and Potential Anticancer Activities against Liver Cancer Cells (HepG2). *Crystals* **2022**, *12*, 460.
42. Hamza, R.Z.; Al-Eisa, R.A.; El-Shenawy, N.S. Possible Ameliorative Effects of the Royal Jelly on Hepatotoxicity and Oxidative Stress Induced by Molybdenum Nanoparticles and/or Cadmium Chloride in Male Rats. *Biology* **2022**, *11*, 450. [CrossRef]
43. El-Megharbel, S.M.; Al-Thubaiti, E.H.; Qahl, S.H.; Al-Eisa, R.A.; Hamza, R.Z. Synthesis and Spectroscopic Characterization of Dapagliflozin/Zn (II), Cr (III) and Se (IV) Novel Complexes That Ameliorate Hepatic Damage, Hyperglycemia and Oxidative Injury Induced by Streptozotocin-Induced Diabetic Male Rats and Their Antibacterial Activity. *Crystals* **2022**, *12*, 304.
44. Alhumaidha, K.A.; El-Awdan, S.A.; El-Iraky, W.I.; Ezz-El-Din, S. Protective effects of ursodeoxycholic acid on ceftriaxone-induced hepatic injury in rats. *Bull. Fac. Pharm. Cairo Univ.* **2014**, *52*, 45–50. [CrossRef]
45. Andrade, R.; Lopez-Vega, M.; Robles, M.; Cueto, I.; Lucena, M.I. Idiosyncratic drug hepatotoxicity: A 2008 update. *Expert Rev. Clin. Pharmacol.* **2008**, *1*, 261–276. [CrossRef]
46. Al Hagbani, T.; Rizvi, S.M.D.; Hussain, T.; Mehmood, K.; Rafi, Z.; Moin, A.; Abu Lila, A.S.; Alshammari, F.; Khafagy, E.-S.; Rahamathulla, M.; et al. Cefotaxime Mediated Synthesis of Gold Nanoparticles: Characterization and Antibacterial Activity. *Polymers* **2022**, *14*, 771. [CrossRef]
47. Chohan, Z.H.; Supuran, C.T. In-vitro antibacterial and cytotoxic activity of cobalt (ii), copper (ii), nickel (ii) and zinc (ii) complexes of the antibiotic drug cephalothin (keflin). *J. Enzym. Inhib. Med. Chem.* **2005**, *20*, 463–468. [CrossRef] [PubMed]
48. Chohan, Z.H.; Pervez, H.; Khan, K.M.; Rauf, A.; Maharvi, G.M.; Supuran, C.T. Antifungal cobalt(II), copper(II), nickel(II) and zinc(II) complexes of furanyl-thiophenyl-, pyrrolyl-, salicylyl- and pyridyl-derived cephalexins. *J. Enzym. Inhib. Med. Chem.* **2004**, *19*, 85–90. [CrossRef] [PubMed]
49. Iqbal, M.S.; Bukhari, I.H.; Arif, M. Preparation, characterization and biological evaluation of copper(II) and zinc(II) complexes with Schiff bases derived from amoxicillin and cephalexin. *Appl. Organomet. Chem.* **2005**, *19*, 864–869. [CrossRef]

antibiotics

MDPI

Article

Synthesis, Spectroscopic Studies for Five New Mg (II), Fe (III), Cu (II), Zn (II) and Se (IV) Ceftriaxone Antibiotic Drug Complexes and Their Possible Hepatoprotective and Antioxidant Capacities

Samy M. El-Megharbel [1,2,*], Safa H. Qahl [3], Fatima S. Alaryani [3] and Reham Z. Hamza [4,5,*]

1 Chemistry Department, Faculty of Science, Zagazig University, Zagazig 44519, Egypt
2 Chemistry Department, College of Sciences, Taif University, Taif 21944, Saudi Arabia
3 Department of Biology, College of Science, University of Jeddah, P.O. Box 80327, Jeddah 21589, Saudi Arabia; shqahal@uj.edu.sa (S.H.Q.); fsalaryani@uj.edu.sa (F.S.A.)
4 Biology Department, College of Sciences, Taif University, Taif 21944, Saudi Arabia
5 Zoology Department, Faculty of Science, Zagazig University, Zagazig 44519, Egypt
* Correspondence: selmegharbl@zu.edu.eg or s.megherbel@tu.edu.sa (S.M.E.-M.); reham.z@tu.edu.sa or dr_reham_z@yahoo.com or reham_z@zu.edu.eg (R.Z.H.)

Abstract: Magnesium, copper, zinc, iron and selenium complexes of ceftriaxone were prepared in a 1:1 ligand to metal ratio to investigate the ligational character of the antibiotic ceftriaxone drug (CFX). The complexes were found to have coordinated and hydrated water molecules, except for the Se (IV) complex, which had only hydrated water molecules. The modes of chelation were explained depending on IR, ^{1}HNMR and UV–Vis spectroscopies. The electronic absorption spectra and the magnetic moment values indicated that Mg (II), Cu (II), Zn (II), Fe (III) and Se (VI) complexes form a six-coordinate shape with a distorted octahedral geometry. Ceftriaxone has four donation sites through nitrogen from NH_2 amino, oxygen from triazine, β-lactam carbonyl and carboxylate with the molecular formulas $[Mg(CFX)(H_2O)_2]\cdot 4H_2O$, $[Cu(CFX)(H_2O)_2]\cdot 3H_2O$, $[Fe(CFX)(H_2O)(Cl)]\cdot 5H_2O$, $[Zn(CFX)(H_2O)_2]\cdot 6H_2O$ and $[Se(CFX)(Cl)_2]\cdot 4H_2O$ and acts as a tetradentate ligand towards the five metal ions. The morphological surface and particle size of ceftriaxone metal complexes were determined using SEM, TEM and X-ray diffraction. The thermal behaviors of the complexes were studied by the TGA(DTG) technique. This study investigated the effect of CFX and CFX metal complexes on oxidative stress and severe tissue injury in the hepatic tissues of male rats. Fifty-six male rats were tested: the first group received normal saline (1 mg/kg), the second group received CFX orally at a dose of 180 mg/kg, and the other treated groups received other CFX metal complexes at the same dose as the CFX-treated group. For antibacterial activity, CFX/Zn complex was highly effective against *Streptococcus pneumoniae*, while CFX/Se was highly effective against *Staphylococcus aureus* and *Escherichia coli*. In conclusion, successive exposure to CFX elevated hepatic reactive oxygen species (ROS) levels and lipid peroxidation final marker (MDA) and decreased antioxidant enzyme levels. CFX metal complex administration prevented liver injury, mainly suppressing excessive ROS generation and enhancing antioxidant defense enzymes and in male rats.

Keywords: ceftriaxone; hepatotoxicity; metal complexes; oxidative stress

Citation: El-Megharbel, S.M.; Qahl, S.H.; Alaryani, F.S.; Hamza, R.Z. Synthesis, Spectroscopic Studies for Five New Mg (II), Fe (III), Cu (II), Zn (II) and Se (IV) Ceftriaxone Antibiotic Drug Complexes and Their Possible Hepatoprotective and Antioxidant Capacities. *Antibiotics* **2022**, *11*, 547. https://doi.org/10.3390/antibiotics11050547

Academic Editors: Yaojun Tong, Linquan Bai and Zixin Deng

Received: 28 March 2022
Accepted: 14 April 2022
Published: 20 April 2022

1. Introduction

The antibiotic ceftriaxone drug (CFX) (Figure 1) is the third generation of antibiotic cephalosporin drugs. It is a parenteral cephalosporin that shows a high antibacterial activity [1]. This effect decreases urinary tract and respiratory infections, skin infections and skin structure, infections of bones and joints, pelvic inflammatory disease, non-enlarged gonorrhea, intra-abdominal infections and acute otitis media due to its surgical prophylaxis [2]. The drug shows high antibacterial activity and rare side effects, a long half-life of serum and is currently recommended for newborns that have Neisseria gonorrhea during childbirth [3]. CFX can be used as a stable mediator for acyl enzyme, preventing peptidoglycan cross-linking and thus disrupting the cell wall's structural integrity [4].

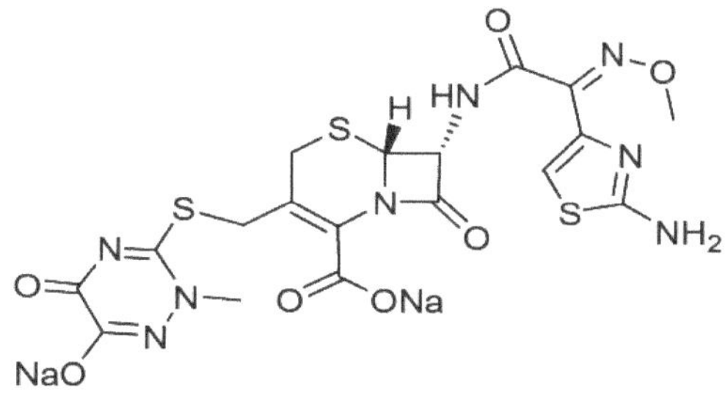

Figure 1. Chemical structure of sodium salt from the antibiotic ceftriaxone drug.

CFX is also stable in relation to beta lactamases, which are formed using two bacterial types—Gram-positive and Gram-negative—and so can be used in the treatment of neonates [5]. Cefotaxime complexes with the general formula of MLCl (where L = cefotaxime drug; M = manganese $(II)^+$, cobalt (II), nickel (II), cobber (II) and cadmium (II)) were prepared, and the ratio of metal to cefotaxime was 1:2, where cefotaxime was chelated via atoms of oxygen and nitrogen from groups of carboxylates, beta-lactam and aminothiazole. The antimicrobial activity of Cu (II) complexes is greater than free cefotaxime ligand [6].

CFX is an antibiotic that is commonly used for the treatment of bacterial infections such as abdominal and joint infection, skin and pelvic inflammatory diseases and bone and middle ear infection [7]. CFX vials are among the most prevalent types of antibiotics [8]. However, CFX produces a great deal of side effects, such as elevated liver enzymes and urea levels and diarrhea, and sometimes it induces thrombocytosis. Given the side effects of using this antibiotic drug in today's health care system, it is essential to develop new drug complexes to elevate its wide activity and reduce any possible side effects [7].

Recently, some studies revealed that a novel nano-formula of the CFX drug had higher antibacterial activity against *E. coli* Gram-negative bacteria compared to the CFX drug alone. The greater antibacterial effect of the CFX nano-formula at a lower dose is another important finding with regard to the reduction of the antibiotic dose and to the cost-effective treatment of resistant microbes [9].

CFX complexes of Mn (II), Co (II), Cu (II), and Cd (II) were prepared in a molar ratio of 1:1 (M: CFX) and acted as pentadentate chelator with metal ions [10]. The antimicrobial activity of cadmium (II) complexes is more than free ceftriaxone ligand, while other complexes have almost the same effect as ceftriaxone. Ceftriaxone complexes of Fe (III), Co (II), Ni (II) and Cu (II) were prepared with octahedral geometry and molar ratios of 1:3 (CFX:M) [11]. Cefixime complexes with Mn (II), Co (II), Ni (II) and Cd (II) were prepared with a 1:1 molar ratio [12,13]. In addition, Fe (III) ceftriaxone complex was prepared with an octahedral geometry and was found to have high activity against bacterial species such as *Pseudomonas aeruginosa* [14].

Recently, there has been a great correlation between SARS-CoV-2 severity and hepatotoxicity especially induced by antibiotics. The severity of COVID-19 may be related with the risk of liver injury development [15]. There is increasing evidence that indicates that hepatotoxicity has been associated with the use of some medications in the treatment of patients infected with SARS-CoV-2 during the COVID-19 pandemic [15]. Recent epidemiological studies indicate different degrees of elevated liver hepatic enzymes with an incidence of 24.4%, particularly in liver transaminases, AST and ALT in COVID-19 patients. Liver injury associated with COVID-19 is defined as any damage that occurred in about 20–46.9% of COVID-19 patients to the liver due to either the treatment or pathogenesis of COVID-19 [16].

It has now been concluded that the severity of COVID-19 is correlated with the risk of liver injury. Additionally, it has been suggested that SARS-CoV-2 is greatly associated with liver injury and infection, which is still a matter of debate [17]. Meanwhile, in most cases, some treatments with antibiotics can cause liver damage during infection and can potentially cause some adverse effects, from severe bleeding to liver failure and even death. Hence, it is essential to find out novel antibiotic drug complexes with high antioxidant efficacy and low hepatic dysfunction to prevent such adverse effects on the hepatic tissues in COVID-19 patients, which is of great importance to alleviate pandemic diseases and enhance antioxidant defense systems [18].

A CFX complex of lead (II) was prepared and characterized, and the antibacterial activity (Gram-positive and Gram-negative bacteria) was evaluated [19]. Five CFX complexes were prepared with Ca (II), Zn (II), Fe (III), Au (III) and Pd (II) [20].

CFX metal complexes of Ca (II), Zn (II), Fe (III), Au (III) and Pd (II) metal ions were prepared, and all chemical characterizations were performed. Calcium (II), zinc (II), and iron (III) complexes have a distorted octahedral geometry, while Au (III) and Pd (II) are in the four-coordinate mode. The CFX ligand acts as a tetradentate towards the five metal ions through N (NH_2) and O (triazine, β-lactam carbonyl, and COO groups). The assessment of the cytotoxicity of the Au (III) complex against HCT-116 and HepG-2, known as colon and hepatocellular carcinoma cells, showed that the IC_{50} of CFX/Au against HepG-2 cell line is 8.53 μg higher than that of HCT-116 cell line, at 20.5 μg [20].

Reactions of CFX with transition metal (II) ions with the general formula of [M(CFX)] (M = Mn, Co, Cu and Cd) and [Fe (CFX)Cl] were characterized using physicochemical and spectroscopic methods, where ceftriaxone acted as a dianionic pentadentate chelating agent through N_2O_3. The antibacterial activity was screened against several bacteria [21].

A CFX/Ca (II) complex was prepared and characterized [22] using elemental, TGA, IR spectroscopy and density functional theory calculations. The antibacterial and luminescence of ceftriaxone and the calcium complex were studied. The Ca (II) complex has a crystalline form. Cell parameters of the compound were determined. Ceftriaxone was chelated with calcium ion through oxygen (triazine cycle, lactam carbonyl and carboxylate groups) and nitrogen from the amino group of the thiazole ring. The antibacterial activity of the Ca (II) complex was screened against *Staphylococcus aureus*, *Escherichia coli* and *Pseudomonas aeruginosa*, and the results were compared with the activity of ceftriaxone disodium salt.

2. Experimental

2.1. Chemicals

All chemicals used were pure, and no further purifications were performed. Sodium salt of ceftriaxone ligand (Figure 1), $MgCl_2$, $CuCl_2$, $FeCl_3{\cdot}6H_2O$, $ZnCl_2$ and $SeCl_4$ were from Sigma-Aldrich Chemical Company, Saint Louis, MO, USA.

2.2. Synthesis

CFX complexes were prepared by adding 1.0 mmol of $MgCl_2$, $CuCl_2$, $FeCl_3{\cdot}6H_2O$, $ZnCl_2$ and $SeCl_4$ in CH_3OH (40 mL solvent) with sodium salt to ceftriaxone (1.0 mmol) in CH_3OH (40 mL). Then, refluxing was performed for about 4 h until colored precipitates were produced. After that, cooling, filtration for the solid complexity and washing using with hot methanol were performed; finally, the complexes were dried in a desiccator using dry $CaCl_2$. All synthesized complexes were fully characterized as shown in (Table 1).

Table 1. Instrumentations and experimental analyses.

Instrument	Measurement
Perkin Elmer CHN 2400 (USA)	Contents of C, H and N
Jenway 4010 conductivity meter	Electrolytic or non-electrolytic character
Bruker FTIR Spectrophotometer (4000–400 cm^{-1})	IR measurements
UV2 Unicam UV/Vis Spectrophotometer	Electronic spectra
varian mercury VX-300 NMR spectrometer	The ^{1}H-NMR
Sherwood scientific magnetic balance using Gouy method	Magnetic measurements
Quanta FEG 250 equipment	Scanning electron microscopy (SEM) images
X'Pert PRO PAN analytical X-ray powder diffraction, target copper with secondary monochromate	X-ray diffraction patterns
JEOL 100s microscope	Transmission electron microscopy images (TEM)

2.3. Experimental Animals

Fifty-six two-month-old male rats were used in this study. They were housed under standard conditions of temperature and supplied food ad libitum, and the study was ethically approved following all the international ethics guidelines for animal care. The treated groups were then divided into seven treated groups (eight rats in each group): Group 1 received 1 mL/kg saline solution (control group); Group 2 received CFX (180 mg/Kg) [23] orally for 30 consecutive days; Group 3, 4, 5, 6 and 7 received 180 mg/kg of Mg (II), Fe (III), Cu (II), Zn (II) and Se (IV) dissolved in saline solution for the same period of time.

Blood samples were collected, and serum samples were obtained after centrifugation at 10,000 rpm for approximately 25 min for biochemical tests. The male rats were gently dissected after light anesthesia by xylene/ketamine, and hepatic tissues were collected. Tissue samples were fixed in approximately 6% of neutral buffered formalin for the examination of histopathological sections.

2.4. Hepatic Functions and Antioxidant Assay

ALT and AST were evaluated in serum using a kit (Spinreact, Sant Esteve de Bas, Spain). Malondialdehyde, a final lipid peroxidation marker (MDA), was assayed in the hepatic tissues [24]. Superoxide dismutase (SOD) [25] and catalase (CAT) antioxidant enzymes were assessed in the homogenates of the liver tissues [26].

2.5. Histopathological Study

Liver tissue pieces were fixed in 6% neutral buffered formalin for 48 h, further processed for examination by hematoxylin and eosin (H&E) staining [27] and examined under a microscope (Leica Microsystems, New York, NY, USA).

2.6. Antibacterial Activities of CFX and Its Metal Complexes

The antimicrobial activity of the tested samples was determined by a modification of the Kirby–Bauer disc diffusion method. Antibacterial activity was tested in triplicate, and then the mean was calculated. In brief, 100 µL of the best bacteria was grown in 10 mL of fresh media until reaching an amount of approximately 108 cells/mL [28]. Then, 100 µL of the microbial suspension was spread into agar plates corresponding to the broth in which they were maintained. Isolated colonies of each organism that may play a pathogenic role were selected from the primary agar plates and tested for susceptibility by the disc diffusion method [29]. The Gram-positive bacteria *Bacillus subtilis* (Ehrenberg 23857™), *Streptococcus pneumonia* (Klein) Chester (6303™) and *Staphylococcus aureus* (23235™) and the Gram-negative bacteria *Escherichia coli* (BAA-2471™) and *Pseudomonas aeruginosa* (BAA-1744™) were incubated at 35–37 °C for 24–48 h. Afterwards, the inhibition zones' diameters were measured in millimeters [30]. Standard discs of tetracycline drug served as positive controls for the antimicrobial activity, and a filter disc impregnated with 10 μL solvent (dist. H_2O, DMSO) was used as a negative control.

The agar used was the Mueller–Hinton agar, which was tested continuously in terms of its pH. Furthermore, the depth of the agar in the plates was considered in the disc diffusion method [30].

2.7. Statistical Analysis

The results were presented as mean ± standard error. Comparisons within groups were conducted with a one-way ANOVA followed by post hoc analysis using SPSS version 17 (IBM® SPSS®, Armonk, NY, USA).

3. Results and Discussions

3.1. Microanalytical and Conductance Measurements

Equal molar ratios for the metal ions ($MgCl_2$, $CuCl_2$, $FeCl_3{\cdot}6H_2O$, $ZnCl_2$ and $SeCl_4$) and ligand ceftriaxone sodium salt (Na_2CFX) produced colored metal complexes. C, H and N analysis data, magnetic susceptibility values and molar conductance ($\Lambda m = 15–25\ \Omega^{-1}{\cdot}cm^2{\cdot}mol^{-1}$) for ceftriaxone metal complexes are in Table 2. White, black, brown, white and yellowish white colors of the Mg (II), Cu (II), Fe (III), Zn (II) and Se (VI) complexes were shown, respectively. The data of the conductivity measurements prove the non-electrolytic character of Mg (II), Cu (II), Fe (III), Zn (II) and selenium (IV) complexes. Hence, CFX metal complex structures can be written as $[Mg(CFX)(H_2O)_2]{\cdot}4H_2O$, $[Cu(CFX)(H_2O)_2]{\cdot}3H_2O$, $[Fe(CFX)(H_2O)(Cl)]{\cdot}5H_2O$, $[Zn(CFX)(H_2O)_2]{\cdot}6H_2O$ and $[Se(CFX)Cl_2]{\cdot}4H_2O$. The ceftriaxone complexes are insoluble in most organic and inorganic solvents, such as H_2O, CH_3OH, C_2H_5OH, $CHCl_3$, CH_2Cl_2 and CCl_4, but they are soluble in DMSO and DMF. The contents of metal were measured gravimetrically [31]. The produced complexes were elucidated using different tools of analysis such as C, H and N, molar conductance, IR, ^{1}HNMR, electronic spectra, magnetic, SEM, TEM and XRD analyses.

Table 2. Elemental analysis and conductivity measurements for ceftriaxone complexes.

Complexes	M.Wt	Color	Elemental Analysis			Λm ($\Omega^{-1}cm^2mol^{-1}$)	Magnetic Moment (BM)
			C	H	N		
$[Mg(CFX)(H_2O)_2]{\cdot}4H_2O$ $C_{18}H_{28}N_8O_{13}S_3$ Mg	684.30	White	(31.56) 31.78	(4.09) 4.12	(16.36) 16.16	15	-
$[Cu(CFX)(H_2O)_2]{\cdot}3H_2O$ $C_{18}H_{26}N_8O_{11}S_3Cu$	705.546	black	(30.61) 30.58	(3.68) 3.94	(15.82) 15.59	17	2.31
$[Fe(CFX)(H_2O)(Cl)]{\cdot}5H_2O$ $C_{18}H_{28}ClN_8O_{13}S_3Fe$	751.98	Greenish black	(28.72) 28.46	(3.72) 3.49	(14.89) 14.65	21	5.92
$[Zn(CFX)(H_2O)_2]{\cdot}6H_2O$ $C_{18}H_{30}N_8O_{14}S_3Zn$	762.09	White	(28.34) 28.61	(3.93) 3.94	(14.96) 14.32	16	-
$[Se(CFX)Cl_2]{\cdot}4H_2O$ $C_{18}H_{26}Cl_2N_8O_{12}S_3Se$	694.96	Yellowish white	(31.08) 31.37	(3.74) 384	(16.11) 16.57	25	5.98

3.2. FTIR Spectral Studies

Infrared spectroscopy is an essential tool for identifying the main functional groups of organic molecules. The CFX free ligand has more than one donor atom, such as the O atom from the thiazole cycle, N atom of the NH_2 group and atoms of O from carboxylate, lactam and amide carbonyl groups. The mode of chelation for free ceftriaxone drug ligand towards metal ions Mg (II), Cu (II), Fe (III), Zn (II) and Se (VI) was studied. The IR for ceftriaxone and its metal complex are similar and are recorded in Table 3 and Figure 2. Generally, the absorption frequency for carbonyl ring groups for free ceftriaxone ligands will be shifted to lower wave numbers after complexation.

Table 3. Infrared frequencies (cm^{-1}) for ceftriaxone and its complexes.

Assignments	Compounds					
	Na_2CFX	Mg (II)	Cu (II)	Fe (III)	Zn (II)	Se (VI)
ν(N–H)	3410	3385	3395	3390	3380	33,385
ν(O–H); H_2O	-	3264	3280	3270	3265	3290
ν(C=O); lactam ring	1782	1744	1769 1670	1766 1657	1744 1649	1766 1660
ν_{as}(C–N) + $\nu(C{=}O)_{OCO}$ ν(COO)	1604	1537 1503	1550	1543	1537 1503	1545
$\delta(CH_2) + \delta(CH_3)$	1416	1408	1403	1404	1400	1409
δ(CH)lactam + ν_{as}(COO)	1374	1367	1359	1352	1366	1309
ν_s(C–N)triazine	1281	1286	1288	1287	1266	1242
δ(CH)lactam + $\delta_w(CH_3)$	1260	1250	1245	1232	1212	1242
$\delta_r(CH_3)$	1178	1171	1138	1146	1106	1135
δ(CH)aminothiazol	1040	1033	1045	1040	1034	1039
ν(N–O)˙	921 864	889 805	898 864	895 805	890 805	918 760
ν(M–O)	-	645 606	657 620 552	618 541	678 645 610	636 619
ν(M–N)	-	510 492	510 487	485 461	507 482	513 475

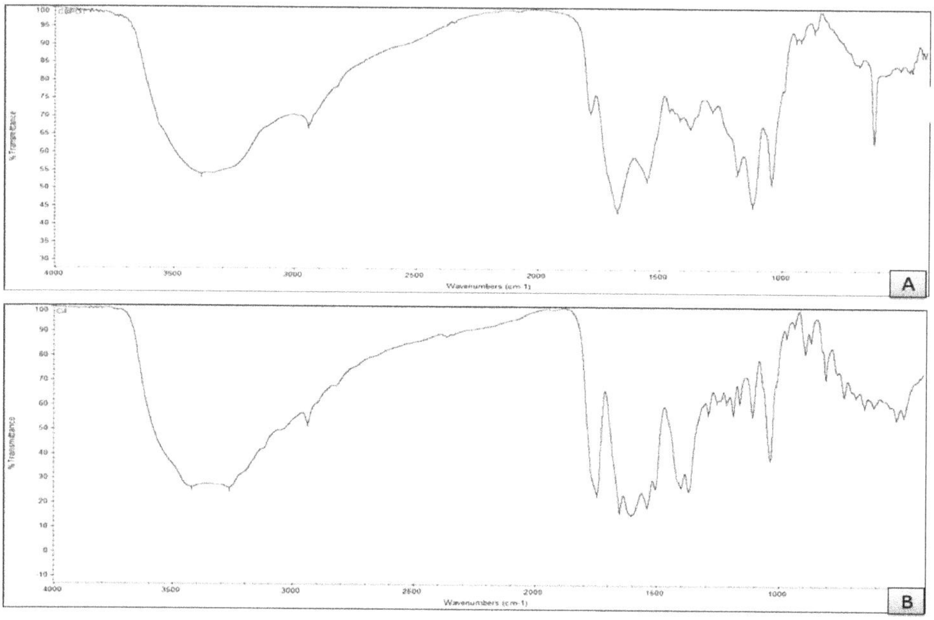

Figure 2. *Cont.*

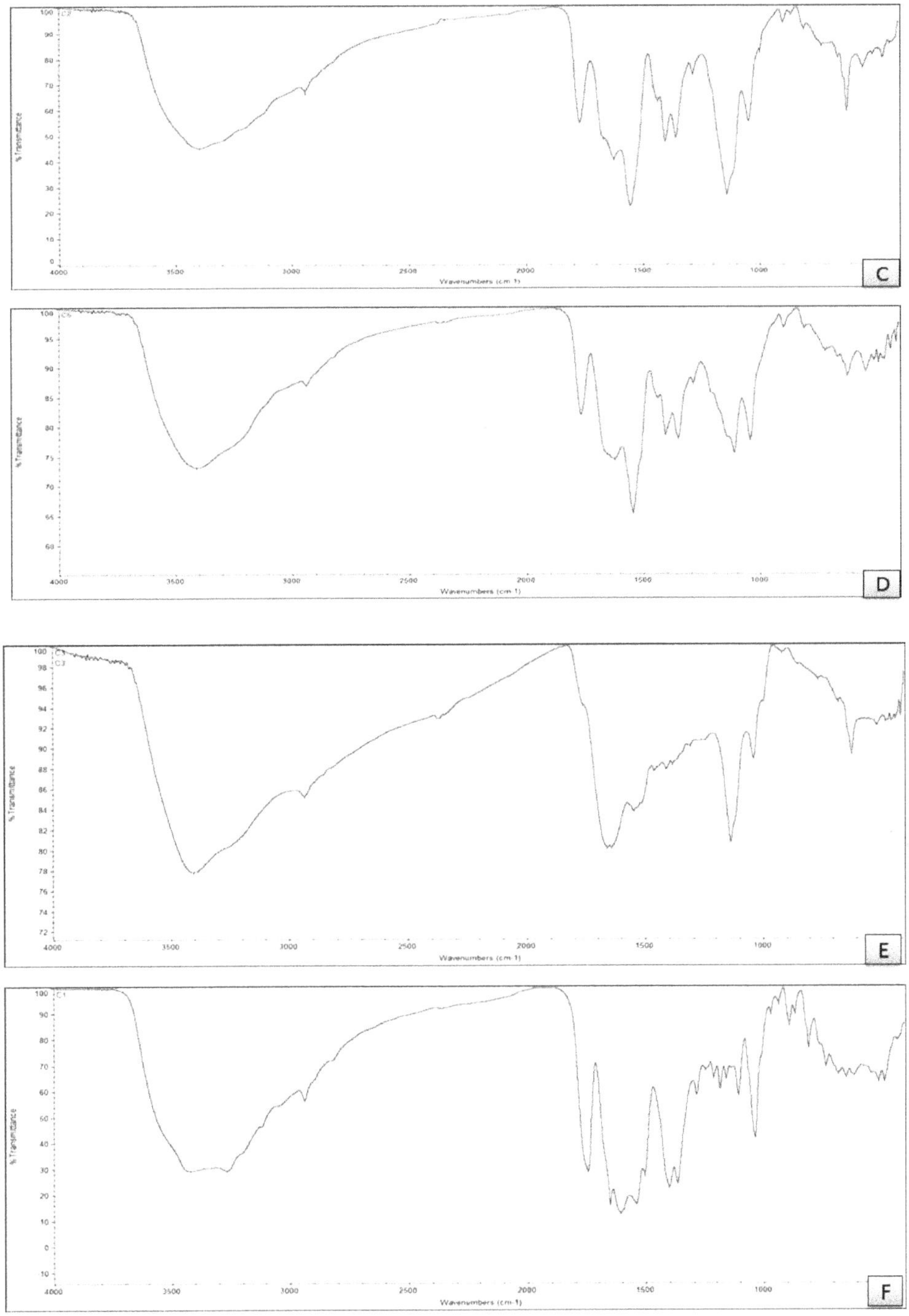

Figure 2. FT-IR of (**A**) CFX, (**B**) CFX/Mg, (**C**) CFX/Cu, (**D**) CFX/Fe, (**E**) CFX/Se and (**F**) CFX/Zn.

After the reaction of CFX with metal ions, there are shifts in the stretching vibrations of ν(C=O) βlactam and ν(C=O) triazine to 1766–1621 cm^{-1} and 1552–1536 cm^{-1}, respectively [32]. These shifts can be attributed to the contribution of oxygen atoms to the chelation with metal ions. The frequencies of symmetric stretching for the carboxylate group vs. (COO^-) shift to 1399–1309 cm^{-1} [10,28]. Based on frequencies of the FTIR spectra of Na_2CFX and its metal complexity, a shift in the band appeared at 3410 cm^{-1} assigned to a stretching vibration ν(N–H) of the amino group to wavenumbers 3380–3395 cm^{-1}, confirming the participation of N atoms of the NH_2 group in the coordination with metal ions. For the monodentate coordination of the COO- group, according to the explanations of Deacon and Phillips [31,32], a difference larger than >200 cm^{-1} disproves this, whereas one smaller than 200 cm^{-1} indicates that coordination is monodentate. These shifts confirm the involvement of the oxygen from the (COO)carboxylate group, the oxygen from the (C=O) carbonyl group of β-lactam, nitrogen from the amine group and the oxo group of the triazine ring in the coordination. All these data are in agreement with previous studies showing a tetradentate behavior of ceftriaxone ligand [10,31]. The spectral band, which is broad in all ctx complexes that appear at 3264–3290 cm^{-1}, is due to the ν(OH) of hydrated and coordinated water molecules [33]. There are new bands that appear at the range of 513–461 cm^{-1} corresponding to stretching vibration bands ν(M–N) for the metal complexes (with no free ligand), confirming that the -NH_2 group of the thiazole moiety is chelated with metal ions. The chelation of the group of -NH_2 with metal ions is not the only explanation for these absorption bands. For CONH and C=N-OCH_3 groups, nitrogen atoms could react with metal ions in solid complexes; however, coordination through nitrogen atoms and COO and lactam CO groups is prevented due to steric constraints. In addition, the stretching vibration for CONH moiety and C=N of C=N-OCH_3 appeared in free ligands of ceftriaxone at 1178 and 1551 cm^{-1}, respectively, and did not shift for all ceftriaxone metal chelates, confirming that these groups did not participate in coordination. The new bands appear in the range of 541–678 cm^{-1} for ceftriaxone complexes and are absent for free ceftriaxone; these are assigned to stretching vibrations of ν(M–O). At the range of 1700–1600 cm^{-1} in ceftriaxone metal complexesm there are broad bands that have high intensity and low resolution regarding the overlap between several vibrational modes, such as ν(C=O)-amide, ν(C=O)-triazine, $\nu_{as}(COO^-)$, ν(C=C) and ν(C=N). This is in an agreement with previous data for polydentate ceftriaxone ligands [34,35].

3.3. Electronic Spectra

The u.v.-vis. spectra for sodium salt of the CFX free ligand and its metal complexes measured within 200–800 nm using DMSO as a solvent are shown in Figure 3.

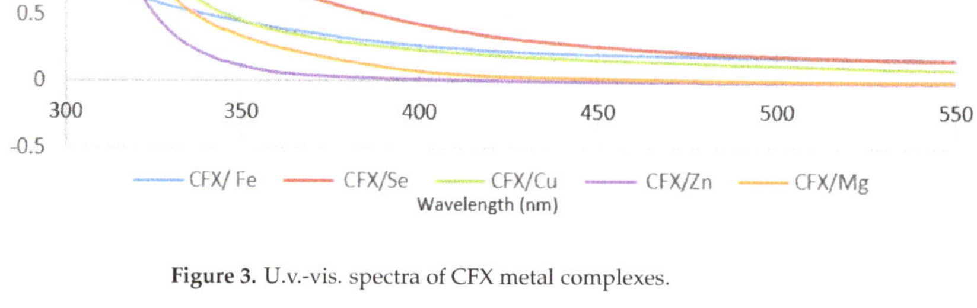

Figure 3. U.v.-vis. spectra of CFX metal complexes.

The u.v.-vis. spectra for ceftriaxone and its complexes give an absorption maximum at 250–270 nm assigned to the $\pi \rightarrow \pi^*$ transition due to orbital molecular energy levels of the nitrogen–carbon–sulfur moiety [36] at 285–300 nm, due to transitions of the $\pi \rightarrow \pi^*$ band of intraligands in moieties of triazole and 1,3-thiazole. The appearance of bands in the region of 350–390 nm is related to the $n \rightarrow \pi^*$ type of transition in intraligands, and this is in agreement with literature data for sulfur atom transitions [37]. The bands related to sulfur atoms are not shifted, confirming that S atoms do not participate in chelation with metal ions. The Fe(III) complex gives very weak absorption bands, and this may be attributed to spin-orbit forbidden transitions. The selenium(VI) complex gives a weak band at around 500 nm, while the Cu(II) complex exhibits a transition of d–d, which appears as a weak band around at 400 nm, suggesting that copper(II) and Se(IV) complexes form six coordinate chelates [37]. The difference in wavelength values in CFX complexes is more than that for free ceftriaxone ligand, confirming the participation of Mg (II), Cu (II), iron (III), Zn(II) and Se (VI) with CFX complexes [38].

3.4. Magnetic Measurements

The value of magnetic moment μeff for the Fe (III) complex is 5.92 B.M., which is consistent with d^5 high spin systems with five electrons unpaired. This value of effective magnetic moments is located within the high spin octahedral geometry. The magnetic moment value for copper(II) ceftriaxone complex at room temperature is 2.31 B.M., confirming that Cu metal ions are present in an excess amount inside the chelation sphere. The lowered values for magnetic moments are related to antiferromagnetic interactions between the ions, while the higher values for magnetic moments show that ferromagnetic interactions rarely occurred. The value of magnetic moment μeff for the Se (VI) complex is 5.98 B.M.—this value of effective magnetic moments is located within the high spin octahedral geometry.

3.5. ^{1}H-NMR Study

For free ceftriaxone Na_2CFX, the ^{1}HNMR spectrum data obtained can be summarized as follows.

At 3.368 [CH_2 of thiazine, 2H] at δ 3.489, [N-CH_3 of triazine, 3H] at 3.889, [=N-O-CH_3, 3H] at 3.960 [S-methylen, 2H], at 5.069 [β-lactam, 1H] and 6.910 [thiazol ring, 1H]. The spectra of proton nuclear magnetic resonance for $[Mg(CFX)(H_2O)_2]\cdot 4H_2O$, $[Zn(CFX)(H_2O)_2]\cdot 6H_2O$ and $[Se(CFX)Cl_2]\cdot 4H_2O$ complexes in Table 4 show occurrences of upward shifts of the field in the reading signals for Na_2CFX, confirming the coordination between Mg (II), Zn (II) and Se (VI) metal ions and ceftriaxone ligand. Based on C, H and N, molar conductance, IR, ^{1}HNMR, electronic spectra and magnetic analyses, the suggested structures for ceftriaxone complexes are shown in Figure 4.

Table 4. HNMR spectral assignments of ceftriaxone and its complexes.

Signals	Na_2CFX Ligand	Mg (II)	Zn (II)	Se (VI)
[2H, CH_2 of thiazine]	3.368	3.352	3.342	3.318
[3H, N-CH_3 of triazine ring]	3.489	3.375	3.312	3.254
[3H, =N-O-CH_3]	3.889	3.879	3.785	3.547
[2H, S-CH_2]	3. 960	3.864	3.687	3.758
[1H, β-lactam]	5.069	4.652	4.758	4.989
[1H, of thiazol ring]	6.910	6.897	6.874	6.987

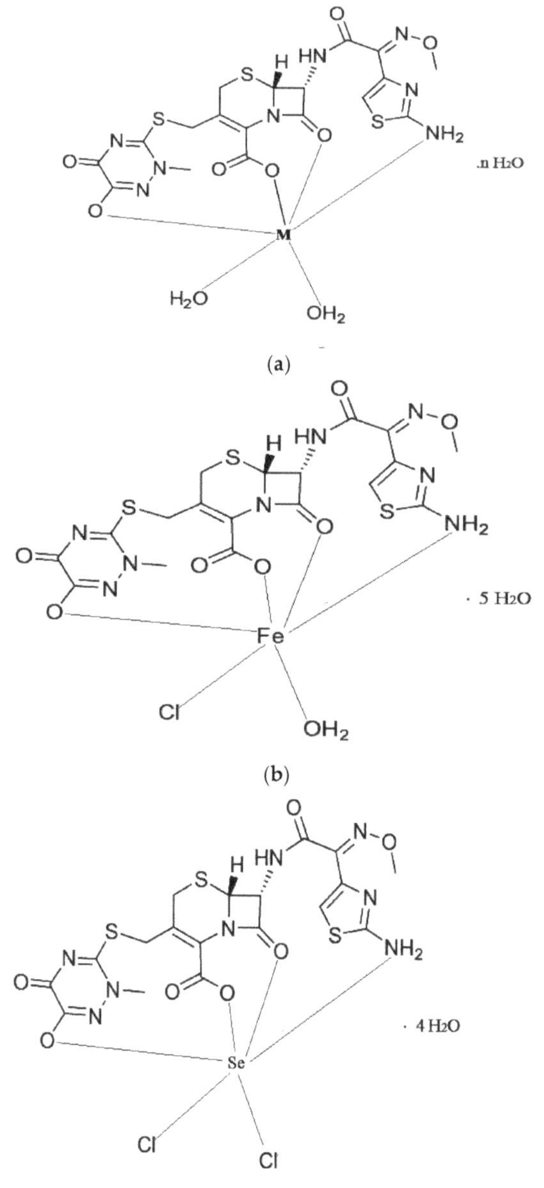

(**a**)

(**b**)

(**c**)

Figure 4. The structure of the prepared CFX complexes. (**a**) M = Mg, Cu and Zn/CFX, n = 4,3 and 6 respectively. (**b**) Fe/CFX complex. (**c**) Se/CFX complex.

3.6. XRD Analysis

XRD analysis is an essential technique for identifying the nature of crystallinity in metal complexes. The patterns of X-ray diffraction powder for $[Zn(CFX)(H_2O)_2]{\cdot}6H_2O$, $[Cu(CFX)(H_2O)_2]{\cdot}3H_2O$, $[Fe(CFX)(H_2O)_2]{\cdot}5H_2O$ and $[Se(CFX)Cl_2]{\cdot}4H_2O$ complexes were characterized at room temperature using X-ray diffraction by Cu Kα radiation and are shown in Table 5. The X-ray diffraction patterns for Zn (II), Cu (II), Fe (III) and Se (VI) CFX complexes were measured in the range of 2θ = 10–70° and are shown in Figure 5, and they have an amorphous behavior with a nano-form structure, as shown in Figure 5. Depending on the Scherrer relationship, the sizes of particles were detected [39] with the help of the full width at half maximum (FWHM) and have a 15–57 nm range.

Table 5. XRD analysis data.

Compound	Pos. [2Th.]	Height [cts]	FWHM [2Th.]	d-Spacing [Å]	Rel. Int. [%]
Zn (II)	22.7560	169.85	0.1279	3.90457	100.00
Cu (II)	31.8381	32.61	0.1092	2.80845	100.00
Fe (III)	19.767	40.99	0.1535	4.48754	100.00
Se (IV)	19.7678	40.99	0.1535	2.80542	100.00

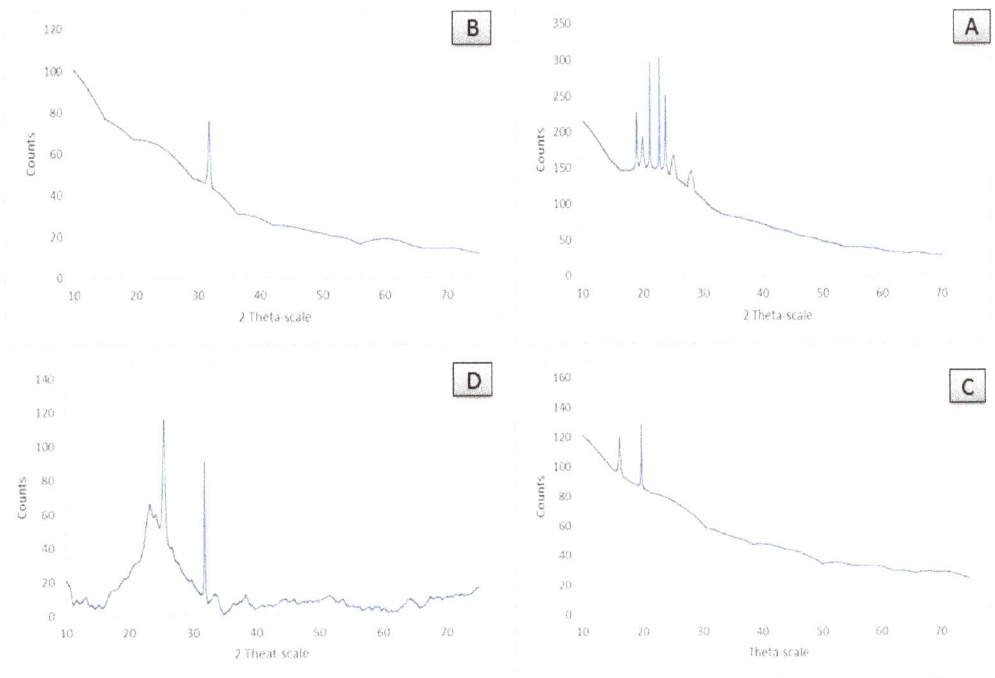

Figure 5. XRD for (**A**): [Zn(CFX)(H_2O)$_2$]·6H_2O; (**B**): [Cu(CFX)(H_2O)$_2$]·3H_2O; (**C**): [Fe(CFX)(H_2O)$_2$]·5H_2O; (**D**): [Se(CFX)Cl_2]·4H_2O.

3.7. SEM and TEM Investigations

TEM and SEM tools are essential techniques for identifying the surface morphology of synthesized metal complexes. Images of SEM for ceftriaxone metal complexes are shown in Figure 6. SEM images show homogeneity and uniform aggregation of Mg (II), Cu (II), Fe (III), Zn (II) and Se (VI) CFX complexes.

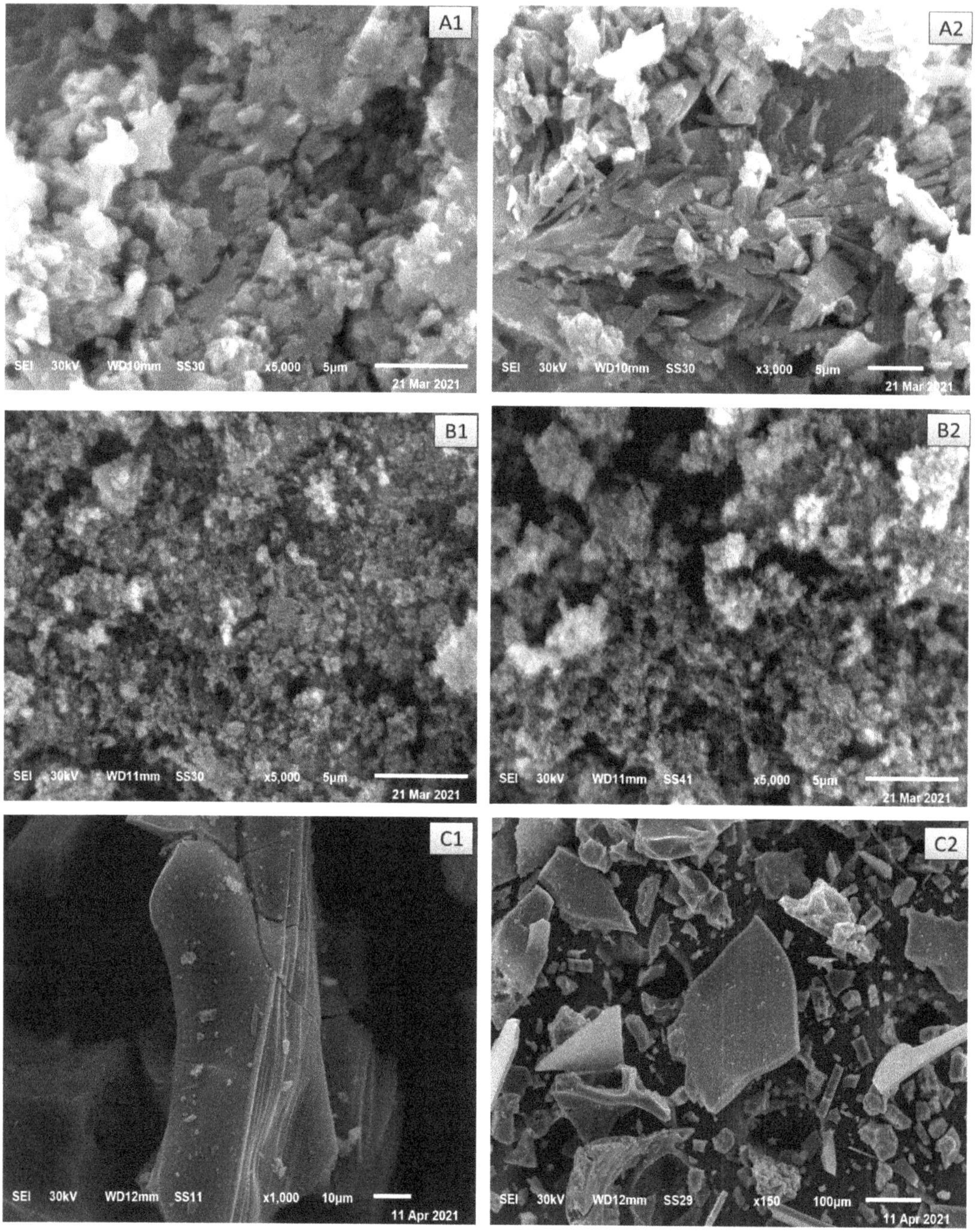

Figure 6. *Cont.*

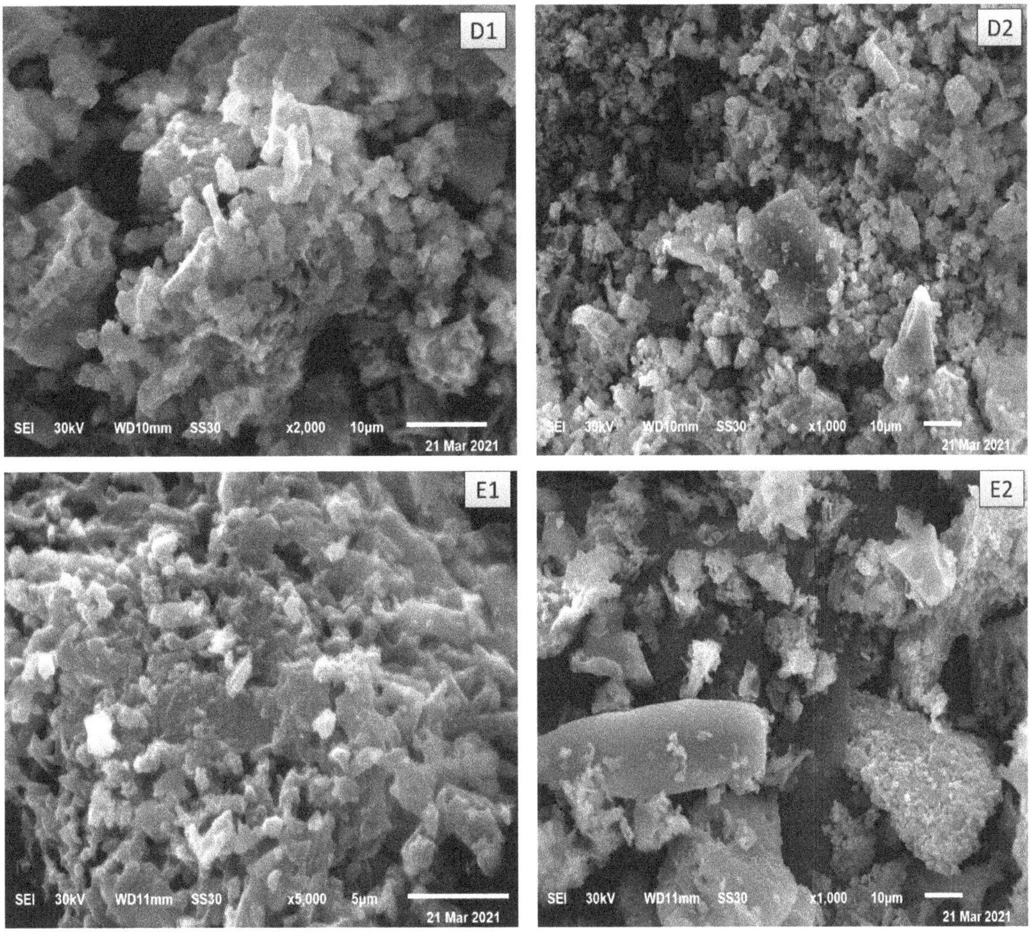

Figure 6. SEM (**A1,A2**) Mg, (**B1,B2**) Cu, (**C1,C2**) Fe, (**D1,D2**) Se and (**E1,E2**) Zn CFX complexes.

Images of TEM for $[Mg(CFX)(H_2O)_2]\cdot 4H_2O$, $[Cu(CFX)(H_2O)_2]\cdot 3H_2O$, $[Fe(CFX)(H_2O)_2]\cdot 5H_2O$, $[Zn(CFX)(H_2O)_2]\cdot 4H_2O$ and $[Se(CFX)Cl_2]\cdot 4H_2O$ complexes are shown in Figure 7 and refer to the formation of spherical black spots with nanoparticles in the ranges of 33–57 nm, 16.18–35.91 nm, 15.03–26.87 nm, 22.83–26.87 nm and 16.09–55.63 nm, respectively. The nano-sized CFX metal complexes were observed with TEM to have a particle size of 15.03–57 nm, which is in agreement with XRD data.

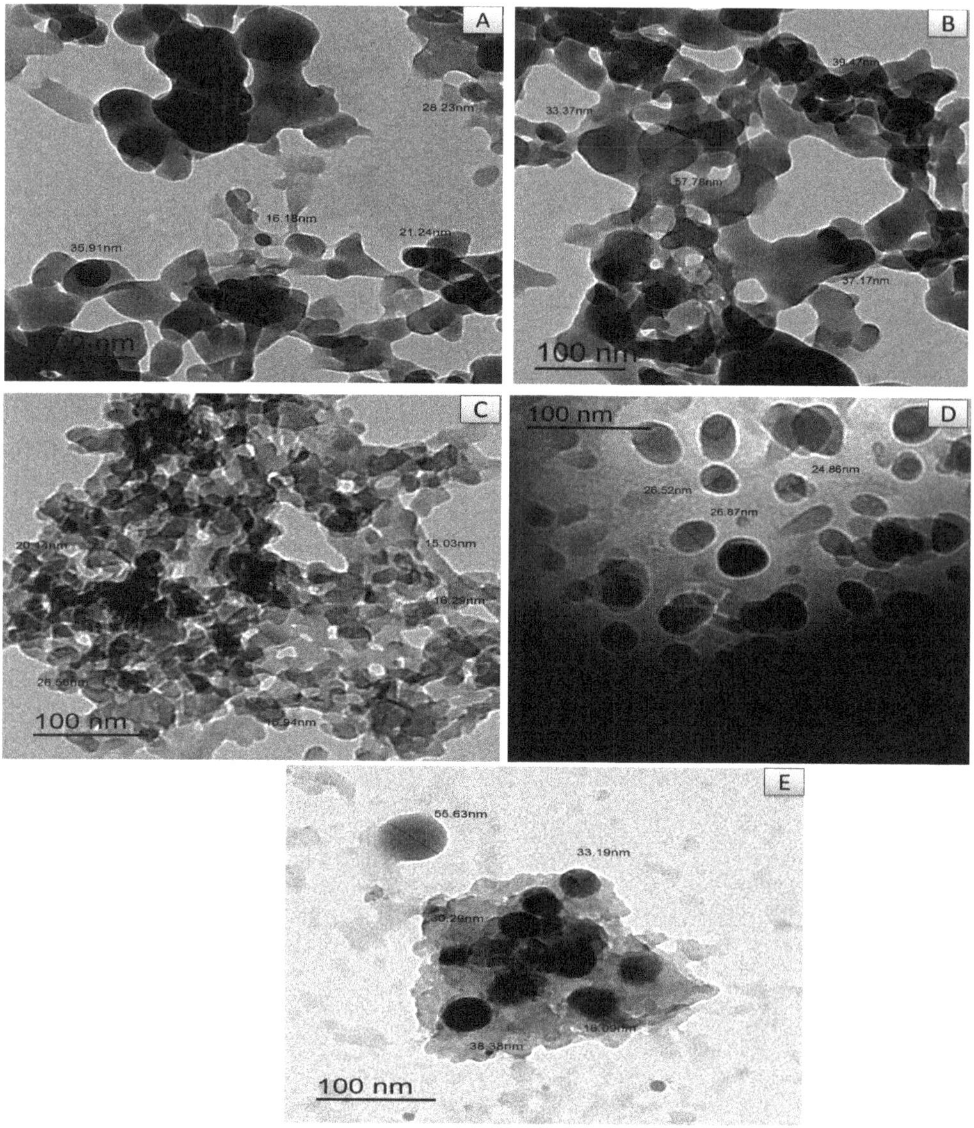

Figure 7. TEM (**A**) Cu, (**B**) Mg, (**C**) Fe, (**D**) Zn and (**E**) Se CFX complexes.

3.8. Thermal Analysis

Thermal analysis is an essential tool used for the characterization of metal complexes where, when heating a compound, its weight loss increases. In addition, DTG differential thermogravimetric analysis (DTGA) is used to study the thermal stability of compounds and the weight loss at different temperatures. TGA confirmed the successful chelation of CFX with different metal ions. Thermogravimetric analysis (TGA) and differential (DTGA) analysis for Mg (II), Cu (II), Fe (III), Zn (II) and Se (IV) CFX complexes were carried out in a temperature range of 30–800 °C under an N_2 atmosphere. The first decomposition endothermic step occurred in a temperature range 30–130 °C and corresponded to a loss of molecules of crystalline water. The second cracking step was carried out between 130–380 °C owing to a loss of ceftriaxone ligand. The third stage of decomposition occurred in temperature ranges of 380–530 °C and 530–620 °C, representing the evolution of NH_3, CO_2 and HNCO. The final residual products for all ceftriaxone metal complexes are metal oxides.

3.9. Ceftriaxone Metal Complexes Alleviate Hepatic Injury in Male Rats

Excessive CFX exposure for 30 consecutive days afforded a significant increment in serum activities of ALT and AST in male rats. However, in contrast, the supplementation of CFX complexes induced significant improvements in liver enzymes, and the best ameliorations were recorded in CFX/Mg, CFX/Zn and CFX/Se, respectively, as compared to the CFX treated group alone, as shown in Table 5.

3.10. Ceftriaxone Metal Complexes Alleviate Oxidative Injury in the Hepatic Tissues and Structural Alterations of Male Rats Exposed to Ceftriaxone

Reactive oxygen species levels manifested a marked elevation in the hepatic tissues of CFX-exposed male rats. Supplementation of CFX metal complexes reduced the excessive hepatic generation of free radicals, as recorded in the CFX-treated group. Hepatic MDA levels were significantly elevated in the CFX-treated group only. These changes were mainly inverted in the group to which CFX metal complexes were administered, especially CFX/Zn and CFX/Mg, respectively.

On the contrary, male rats exposed to CFX exhibited a high decline in hepatic antioxidant enzyme (GSH, SOD and CAT) activities. GSH, SOD activity and CAT declined markedly as compared with the control group. Oral supplementation of CFX metal complexes elevated the hepatic antioxidant enzymes as compared to CFX alone, especially for CFX/Mg, CFX/Zn and CFX/se, CFX/Cu and CFX/Fe, in the same order as shown in Table 6.

Table 6. Antioxidant enzyme activities in groups treated with either ceftriaxone or ceftriaxone metal complexes.

Group Items	Control	CFX	CFX/Mg	CFX/Zn	CFX/Se	CFX/Cu	CFX/Fe
ALT (U/L)	9.68 ± 1.02 ^{g}	89.38 ± 4.69 ^{a}	12.98 ± 1.98 ^{f}	15.58 ± 2.36 ^{e}	19.36 ± 2.56 ^{d}	22.39 ± 1.69 ^{c}	23.69 ± 2.02 ^{bc}
AST (U/L)	29.36 ± 2.36 ^{g}	203.36 ± 4.69 ^{a}	50.69 ± 3.69 ^{f}	55.98 ± 4.25 ^{e}	72.25 ± 3.69 ^{d}	79.58 ± 3.69 ^{c}	82.35 ± 2.69 ^{bc}
MDA (U/g)	9.68 ± 1.02 ^{g}	123.65 ± 6.25 ^{a}	18.52 ± 2.99 ^{e}	16.52 ± 2.69 ^{f}	24.28 ± 2.69 ^{d}	29.68 ± 3.69 ^{c}	31.02 ± 2.69 ^{bc}
GSH (nmol/100 mg)	18.69 ± 1.69 ^{a}	6.98 ± 0.98 ^{g}	15.68 ± 2.05 ^{bc}	14.69 ± 1.69 ^{c}	12.97 ± 2.58 ^{d}	9.01 ± 1.69 ^{f}	10.39 ± 1.69 ^{ef}
SOD (U/g)	13.69 ± 2.25 ^{ab}	5.25 ± 0.35 ^{e}	12.66 ± 1.69 ^{b}	11.36 ± 1.58 ^{c}	10.98 ± 1.69 ^{d}	10.32 ± 1.69 ^{d}	10.02 ± 0.69 ^{d}
CAT (U/g)	15.69 ± 1.69 ^{ab}	4.36 ± 0.98 ^{f}	13.25 ± 1.65 ^{b}	13.05 ± 1.69 ^{b}	12.69 ± 2.68 ^{c}	10.39 ± 2.69 ^{e}	11.65 ± 1.69 ^{d}

Values are means ± standard error. Mean values with different letters in the same row show significance at $p \leq 0.05$, where the highest mean value has the symbol ^{a} and $^{b-g}$: Decreases in value were assigned alphabetically, with similar letters implying partial or complete non-significance and different letters implying significance. CFX = group supplemented with ceftriaxone; CFX/Mg = group supplemented with ceftriaxone/Mg complex; CFX/Zn = group supplemented with ceftriaxone/Zn complex; CFX/Se = group supplemented with ceftriaxone/Se complex; CFX/Cu = group supplemented with ceftriaxone/Cu complex.

Histological sections of hepatic tissues of different treated groups as shown in (Figure 8) which showed marked hepatic alterations and some atrophy in CFX treated group only, while there was marked improvement recorded in hepatic tissues of other treated groups with either CFX/Mg, CFX/Zn, CFX/Se, CFX/Cu and CFX/Fe, Meanwhile the marked improvement was recorded in CFX/Mg, CFX/Zn and CFX/Cu treated groups respectively.

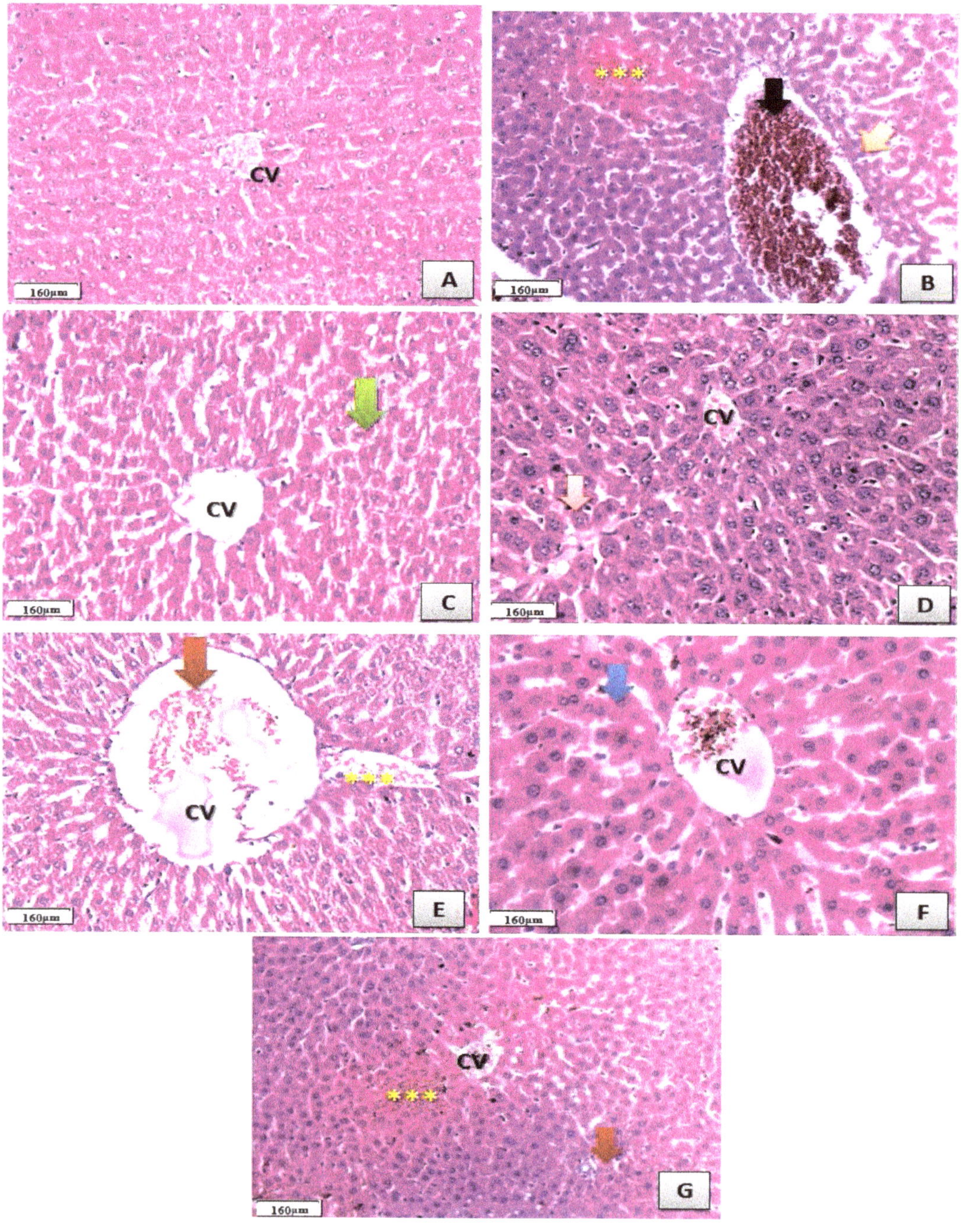

Figure 8. Photomicrographs showing histological sections of the hepatic tissues. (**A**) Control group showing normal hepatic structure and normal central vein (CV) (H&EX400). (**B**) CFX-treated group showing severe toxicity in the form of the hypertrophy of hepatocytes (orange arrow) with the appearance

of binucleated hepatocytes and increased eosinophilia, granular cytoplasm and vesicular nuclei (***); the central vein is dilated and filled with hemorrhage and necrotic tissue (black arrow), with focal necrosis in some hepatocytes with increased eosinophilia and nuclear disappearance and the accumulation of a few mononuclear inflammatory cells in blood sinusoids (H&EX400). (**C**) CFX/Mg-treated group showing an almost normal hepatic structure (green arrow) with a normal sized central vein (CV) (H&EX400). (**D**) CFX/Zn-treated group showing very mild toxicity in the form of the hypertrophy of hepatocytes with granular eosinophilic cytoplasm and vesicular nuclei and the appearance of some binucleated cells (orange arrow), with a mild congested central vein which contains (CV) mild brown particles of bilirubin and single hepatocyte necrosis (H&EX400). (**E**) CFX/Se-treated group showing an almost normal hepatic structure with normal hepatocytes with a moderately enlarged central vein (CV) filled with some red blood cells (red arrow) with some mild detachment (***) (H&EX400). (**F**) CFX/Cu-treated group showing an almost normal hepatic structure with ballooning cytoplasm (blue arrow) and a central vein (CV) filled with some red blood cells (H&EX400). (**G**) CFX/Fe-treated group showing mild toxicity in the form of the hypertrophy of hepatocytes with the. appearance of binucleated hepatocytes and increased eosinophilia, granular cytoplasm and vesicular nuclei (red arrow); the central vein is normal in diameter with mild congestion (CV) and ballooning degeneration in some hepatocytes (***), with increased eosinophilia around the central vein indicating the beginning of necrosis (H&EX400).

3.11. Antibacterial Activity Evaluation

Biological evaluations were performed for CFX complexes against Gram-positive (*Bacillus subtilis, Streptococcus pneumoniae* and *Staphylococcus aureus*) and Gram-negative (*Escherichia coli* and *Pseudomonas aeruginosa*) bacteria. Results from the agar disc diffusion tests for the antimicrobial activities are presented in Table 6 and demonstrated in Figure 9. The diameters of the zone of inhibition (in mm) of the standard drug Amikacin (Aminoglycoside) ($C_{22}H_{43}N_5O_{13}$) against Gram-positive bacteria *B. subtilis* and *S.aureus* and Gram-negative bacteria *E. coli* and *P. aeruginosa* were found to be 36, 30, 31 and 35 mm, respectively. Under identical conditions, Table 7 and Figure 9 show that all complexes were found to be efficient, with a high antimicrobial activity that exceeded CFX itself.

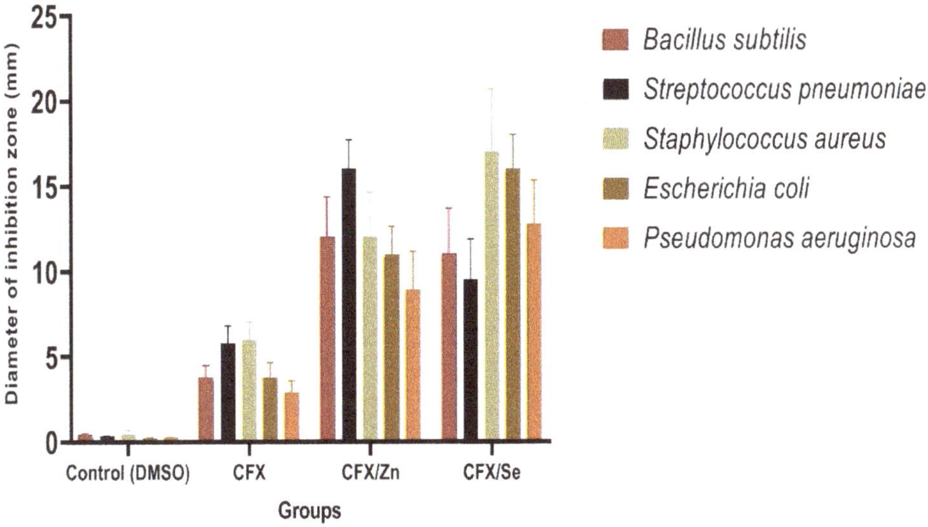

Figure 9. Antibacterial activity of CFX and Zn (II) and Se (IV) metal complexes.

Table 7. Inhibition zone diameter (mm/mg sample) of CFX and CFX metal complexes.

Sample	Inhibition Zone Diameter (mm/mg Sample)				
	Bacillus subtilis (G^+)	*Streptococcus pneumoniae* (G^+)	*Staphylococcus aureus* (G^+)	*Escherichia coli* (G^-)	*Pseudomonas aeruginosa* (G^-)
Control (DMSO)	0.0 ± 0.0 [c]	0.0 ± 0.0 [d]	0.0 ± 0.0 [e]	0.0 ± 0.0 [d]	0.0 ± 0.0 [d]
Ceftriaxone (CFX)	3.80 ± 0.11 [b]	5.8 ± 0.73 [c]	6 ± 0.89 [d]	3.76 ± 0.31 [c]	2.89 ± 0.45 [c]
Zn (II)–CFX	12 ± 0.62 [a]	16 ± 0.21 [a]	12 ± 0.58 [b]	10.98 ± 0.96 [a]	8.90 ± 0.85 [a]
Se (III)–CFX	11 ± 0.64 [a]	9.50 ± 0.91 [b]	17 ± 0.37 [c]	15.98 ± 0.85 [a]	12.76 ± 0.59 [a]

Means within the same column (mean ± SE) carrying different letters are significant at $p \leq 0.05$ using Duncan's multiple range test, where the highest mean value has the symbol [a] and [b–e] those decreasing in value are assigned alphabetically.

4. Discussion

A serious problem now facing humanity is oxidative damage and its dangerous consequences for human health and hepatic vitality. The world now needs powerful antibiotics with potent antioxidant activities and fewer side effects for the liver tissues to reinforce the immune system and fight against resistant microbes and microorganism. Thus, the current study was designed to synthesize novel metal complexes of ceftriaxone (CFX) with Mg (II), Fe (III), Cu (II), Zn (II) and Se (IV) to investigate the potency of the hepatoprotective effects and antioxidant capacities of CFX complexes and determine if they succeed in the elevation of antioxidant capacities to investigate novel compounds with high antioxidant capacities.

The oxidative injury of biomolecules is the major concern in the pathogenesis of a large number of diseases such as cancer, degenerative diseases, metabolic diseases and even dangerous instances of SARS-CoV-2. Thus, it is very important to investigate the role of potent novel metal complexes to prevent oxidative stress and the treatment of side effects and diseases induced by such oxidative stress.

Liver injury caused by drugs ranges from mild biochemical abnormalities to chronic liver failure. The majority of adverse liver reactions occur in most antibiotic treatments [23].

Some antibiotics are considered to be a common cause of liver injury. The hepatotoxicity that occurs is usually associated with hepatic impairment [40]. CFX is a broad-spectrum antibiotic with potent activity against Gram-positive and Gram-negative bacteria [41].

The hepatotoxicity caused by CFX appears after an average of 9–11 days [42,43]. Previous studies have reported high aspartate aminotransferase (ALT) and alanine aminotransferase (AST) activities with the administration of CFX [44,45], and this concept coincides with the current findings that confirmed that CFX causes hepatic damage as a result of elevations in some biochemical parameters such as AST, ALT and low-density lipoprotein (LDL) as well as a decline in high-density lipoprotein (HDL) concentrations [46–48].

Based on the previous background, researchers have demonstrated that CFX is widely used as a third-generation cephalosporin antibiotic that has a broad spectrum of bactericidal activity. However, previous evidence has indicated that CFX carries a risk of the elevation of liver enzymes, with liver injury as an adverse effect [48].

In agreement with previous findings [23] that reported that the oxidative stress mediated by oxygen free radicals (ROS) has been implicated as a common link between chronic liver damage and hepatic fibrosis, the current study demonstrated that the administration of CFX resulted in a significant decline in hepatic GSH. Conversely, a significant increment in the level of hepatic MDA (a marker of lipid peroxidation) was shown. The increase in MDA level was more pronounced in rats in CFX-treated groups.

The novel finding of our current study is the improvement effects of CFX metal complexes, especially in groups treated with CFX with Mg, Zn, Se and Cu complexes, respectively, as they improved hepatic function serum levels and ameliorated the hepatic structure greatly.

Our explanation depends on the functional activity of the used metals, as previous researchers demonstrated the importance of Se metal in enhancing the glutathione levels of hepatic tissues, leading to enhancements of the whole antioxidant defense system [49–52] and a decline in the excessive generation of reactive oxygen species.

Metal ions are required to keep the human body healthy because several critical biological functions in humans depend upon their presence, and their absence or scarcity may lead to diseases. Magnesium (Mg) belongs to the main group of elements and is required mainly for fat and carbohydrate metabolism. Meanwhile, zinc (Zn) belongs to the transition metal group of elements. Zn is a vital element due to its strong binding to proteins. Zn also is exceptionally stable with respect to oxidation and reduction, and thus it does not participate in redox reactions. Additionally, Zn^{2+} shows a strong preference in enzymes for a tetrahedral coordination over an octahedral one [53], and all these previous aspects confirm the findings of the current study and the great improvement activities of CFX metal complexes, especially with Zn^{2+}, in the alleviation of the hepatic side effects induced by CFX alone.

A deficiency of Fe leads to anemia, as it is known that Fe is a part of hemoglobin; besides this, a deficiency of Cu leads to heart diseases and anemia. The importance of all of these metals and especially in conjugation with CFX has been proved chemically in alleviating oxidative stress and providing high hepatic protection. The Gram-negative strains are exposed to various stress conditions during pathogenesis, of which the stress of acids serves as a major defense mechanism in the host [54]. Such environments are encountered, and this supports the main hypothesis of the study and proved the hepatoprotective and antioxidant capacities of two synthesized CFX complexes with either Zn (II) or Se (IV); in addition, the two complexes showed antibacterial activity besides their abilities to reduce free radicals' production.

5. Conclusions

The current study aimed to synthesize five ceftriaxone complexes by the reaction of ceftriaxone sodium salt with (Mg^{2+}, Zn^{2+}, Fe^{3+}, Cu^{2+} and Se^{4+}) ions. The structures of the CFX complexes have been explained using microanalytical, molar conductance, IR, ^{1}HNMR, UV–Vis, magnetic, SEM, TEM and X-ray diffraction analyses. Mg (II), Cu (II), Zn (II), Fe (III) and Se (VI) complexes form six-coordinate systems with a distorted octahedral geometry. The obtained results clarified that ceftriaxone metal complexes, especially (CFX/Mg, CFX/Zn and CFX/Se), greatly ameliorated hepatic enzyme functions and enhanced the antioxidant activities of hepatic tissues while reducing the excessive triggering of excessive reactive oxygen species (ROS) and the protection of hepatic structures as compared with ceftriaxone-treated groups alone. For antibacterial activity, the CFX/Zn complex was highly effective against *Streptococcus pneumoniae*, while CFX/Se was highly effective against *Staphylococcus aureus* and *Escherichia coli*. These results are very promising in providing protection for the hepatic tissues and reducing damaging effects and the severe oxidative stress induced by antibiotics on liver tissues, especially during the COVID-19 pandemic; in addition, it brings the recent research that correlated the damage of hepatic functions and severe instances of SARS-CoV-2 on the health of the body up to date.

Author Contributions: Conceptualization, S.M.E.-M., S.H.Q., F.S.A. and R.Z.H.; methodology, S.M.E.-M., S.H.Q. and R.Z.H.; validation, S.M.E.-M., S.H.Q., F.S.A. and R.Z.H.; formal analysis, S.M.E.-M., S.H.Q., F.S.A. and R.Z.H.; investigation, S.M.E.-M., S.H.Q., F.S.A. and R.Z.H.; resources, S.M.E.-M., S.H.Q., F.S.A. and R.Z.H.; data curation, S.M.E.-M. and R.Z.H.; writing—original draft preparation, S.M.E.-M., S.H.Q., F.S.A. and R.Z.H.; writing—review and editing, S.M.E.-M., S.H.Q., F.S.A. and R.Z.H.; supervision, S.M.E.-M., F.S.A., S.H.Q. and R.Z.H.; project administration, S.M.E.-M. and R.Z.H.; funding acquisition, S.M.E.-M., S.H.Q., F.S.A. and R.Z.H. All authors have read and agreed to the published version of the manuscript.

Funding: This work received no external funding.

Institutional Review Board Statement: The animal study was conducted in accordance with the Declaration of Helsinki, and approved by Ethics Committee of Taif University.

Informed Consent Statement: Not applicable.

Data Availability Statement: All the data are available inside the text.

Conflicts of Interest: The authors declare no conflict of interest.

References

1. Bilirubin, R.B.; Alex, R. Ceftriaxone binding to human serum albumin. *Mol. Pharmacol.* **1989**, *36*, 478–483.
2. Quaglia, M.G.; Bossu, E.; Dell'Aquila, C.; Guidotti, M. Determination of the binding of a β2-blocker drug, frusemide and ceftriaxone to serum proteins by capillary zone electrophoresis. *J. Pharm. Biomed. Anal.* **1997**, *15*, 1033–1039. [CrossRef]
3. Nerli, B.; Farruggia, B.; Pico, G. A Comparative study of the binding characteristics of ceftriaxone, cefoperazone and cefsudolin to human serum albumin. *Biochem. Mol. Biol. Int.* **1996**, *40*, 823–831. [CrossRef] [PubMed]
4. Kong, K.-F.; Schneper, L.; Mathee, K. Beta-lactam antibiotics: From antibiosis to resistance and bacteriology. *APMIS* **2010**, *118*, 1–36. [CrossRef] [PubMed]
5. Pacifici, G.M. Pharmacokinetics of cephalosporins in the neonate: A review. *Clinics* **2011**, *66*, 1267–1274. [CrossRef]
6. Anacona, J.R.; da Silva, G. Synthesis and antibacterial activity of cefotaxime metal complexes. *J. Chil. Chem. Soc.* **2005**, *50*, 447–450. [CrossRef]
7. Ebrahimi, S.; Farhadian, N.; Karimi, M.; Ebrahimi, M. Enhanced bactericidal effect of ceftriaxone drug encapsulated in nanostructured lipid carrier against gram-negative *Escherichia coli* bacteria: Drug formulation, optimization, and cell culture study. *Antimicrob. Resist. Infect. Control.* **2020**, *9*, 28. [CrossRef]
8. Kumar, S.; Bhanjana, G.; Kumar, A.; Taneja, K.; Dilbaghi, N.; Kim, K. Synthesis and optimization of ceftriaxone-loaded solid lipid nanocarriers. *Chem. Phys. Lipids.* **2016**, *200*, 126–132. [CrossRef]
9. Manten, A. Side effects of antibiotics. *Vet. Q.* **1981**, *3*, 179–182. [CrossRef]
10. Anacona, J.R.; Rodriguez, A. Synthesis and antibacterial activity of ceftriaxone metal complexes. *Trans. Met. Chem.* **2005**, *30*, 897–901. [CrossRef]
11. Ali, A.E. Synthesis, spectral, thermal and antimicrobial studies of some new tri metallic biologically active ceftriaxone complexes. *Spectrochim. Acta A* **2011**, *78*, 224–230. [CrossRef] [PubMed]
12. Anacona, J.R.; Estacio, J. Synthesis and antibacterial activity of cefixime metal complexes. *Trans. Met. Chem.* **2006**, *31*, 227–231. [CrossRef]
13. Arayne, M.S.; Sultana, N.; Khanum, F.; Ali, M.A. Antibacterial studies of cefixime copper, zinc and cadmium complexes. *Pak. J. Pharm. Sci.* **2002**, *15*, 1–8.
14. Pillai, M.S.; Latha, S.P. Designing of some novel metallo antibiotics tuning biochemical behaviour towards therapeutics: Synthesis, characterisation and pharmacological studies of metal complexes of cefixime. *J. Saudi Chem. Soc.* **2012**, *20*, S60–S66. [CrossRef]
15. Sodeifian, F.; Seyedalhosseini, Z.S.; Kian, N.; Eftekhari, M.; Najari, S.; Mirsaeidi, M.; Farsi, Y.; Nasiri, M.J. Drug-Induced Liver Injury in COVID-19 Patients: A Systematic Review. *Front. Med.* **2021**, *8*, 731436. [CrossRef] [PubMed]
16. Kulkarni, A.V.; Kumar, P.; Tevethia, H.V.; Premkumar, M.; Arab, J.P.; Candia, R.; Talukdar, R.; Sharma, M.; Qi, X.; Rao, P.N.; et al. Systematic review with meta-analysis: Liver manifestations and outcomes in COVID-19. *Aliment. Pharmacol. Ther.* **2020**, *52*, 584–599. [CrossRef] [PubMed]
17. Hoffmann, M.; Kleine-Weber, H.; Schroeder, S.; Krüger, N.; Herrler, T.; Erichsen, S.; Schiergens, T.S.; Herrler, G.; Wu, N.H.; Nitsche, A.; et al. SARS-CoV-2 cell entry depends on ACE2 and TMPRSS2 and is blocked by a clinically proven protease inhibitor. *Cell* **2020**, *181*, 271–280. [CrossRef] [PubMed]
18. Xu, X.W.; Wu, X.X.; Jiang, X.G.; Xu, K.J.; Ying, L.J.; Ma, C.L.; Li, S.B.; Wang, H.Y.; Zhang, S.; Gao, H.N.; et al. Clinical findings in a group of patients infected with the 2019 novel coronavirus (SARS-CoV-2) outside of Wuhan, China: Retrospective case series. *BMJ* **2020**, *368*, m606. [CrossRef]
19. Lykhin, A.O.; Novikova, G.V.; Kuzubov, A.A.; Staloverova, N.A.; Sarmatova, N.I.; Varganov, S.A.; Krasnov, P.O. A complex of ceftriaxone with Pb (II): Synthesis, characterization, and antibacterial activity study. *J. Coord. Chem.* **2014**, *67*, 2783–2794. [CrossRef]
20. Refat, M.S.; Altalhi, T.; Fetooh, H.; Alsuhaibani, A.M.; Hassan, R.F. In neutralized medium five new Ca(II), Zn(II), Fe(III), Au(III) and Pd(II) complexity of ceftriaxone antibiotic drug: Synthesis, spectroscopic, morphological and anticancer studies. *J. Mol. Liq.* **2021**, *322*, 114816. [CrossRef]
21. Hamza, R.Z.; Sheshah, Z.A.; Suleman, R.H.; Al-Juaid, N.F.; Hamed, N.A.; Al-Juaid, M.A. Efficacy of some antibiotics and some metal complexes (Nano-formula) that could increase their effectiveness during COVID-19. *Int. J. Biol. Pharm. Sci. Arch.* **2022**, *3*, 8–14. [CrossRef]
22. Novikova, G.V.; Tsyplenkova, D.I.; Kuzubov, A.A.; Kolenchukova, O.A.; Samoilo, A.S.; Vorobyev, S.A. Complex of Ca (II) with Ceftriaxone: Synthesis, Structure, Spectral and Antibacterial Properties. *J. Sib. Fed. Univ. Chem.* **2021**, *14*, 290–301. [CrossRef]
23. Alhumaidha, K.A.; El-Awdan, S.A.; El-Iraky, W.I.; Ezz-El-Din, S. Protective effects of ursodeoxycholic acid on ceftriaxone-induced hepatic injury in rats. *Bull. Fac. Pharm. Cairo Univ.* **2014**, *52*, 45–50. [CrossRef]
24. Mihara, M.; Uchiyama, M. Determination of malonaldehyde precursor in tissues by thiobarbituric acid test. *Anal. Biochem.* **1978**, *86*, 271–278.
25. Ellman, G.L. Tissue sulfhydryl groups. *Arch. Biochem. Biophys.* **1959**, *82*, 70–77. [CrossRef]

26. Marklund, S.; Marklund, G. Involvement of the superoxide anion radical in the autoxidation of pyrogallol and a convenient assay for superoxide dismutase. *Eur. J. Biochem.* **1974**, *47*, 469–474. [CrossRef]
27. Sinha, A.K. Colorimetric assay of catalase. *Anal. Biochem.* **1972**, *47*, 389–394. [CrossRef]
28. National Committee for Clinical Laboratory Standards. *Performance Vol. Antimicrobial Susceptibility of Flavobacteria*; National Committee for Clinical Laboratory Standards: Wayne, PA, USA, 1997.
29. National Committee for Clinical Laboratory Standards. *Methods for Dilution Antimicrobial Susceptibility Tests for Bacteria That Grow Aerobically*; Approved Standard M7-A3; National Committee for Clinical Laboratory Standards: Villanova, PA, USA, 1993.
30. Liebowitz, L.D.; Ashbee, H.R.; Evans, E.G.; Chong, Y.; Mallatova, N.; Zaidi, M.; Gibbs, D.; Global Antifungal Surveillance Group. A two year global evaluation of the susceptibility of Candida species to fluconazole by disk diffusion. *Diagn. Microbiol. Infect. Dis.* **2001**, *40*, 27–33. [CrossRef]
31. Bancroft, J.D.; Gamble, M. *Theory and Practice of Histological Techniques*; Elsevier Health Sciences: Amsterdam, The Netherlands, 2008.
32. Cotton, F.A.; Wilkinson, C.W. *Advanced Inorganic Chemistry*, 3rd ed.; Interscience Publisher: New York, NY, USA, 1972.
33. Nakamoto, K. *Infrared and Raman Spectra of Inorganic and Coordination Compounds*, 4th ed.; Wiley: New York, NY, USA, 1986.
34. Deacon, G.B.; Philips, R.J. Relationships between the carbon-oxygen stretching frequencies of carboxylato complexes and the type of carboxylate coordination. *Coord. Chem. Rev.* **1980**, *33*, 227. [CrossRef]
35. Zaman, R.; Rehman, W.; Hassan, M.; Khan, M.M.; Anjum, Z.; Shah SA, H.; Abbas, S.R. Synthesis, characterization and biological activities of cephalosporin metals complexes. *Int. J. Biosci.* **2016**, *9*, 163–172.
36. Lever, A.B.P. *Electronic Spectra of dn Ions Inorganic Electronic Spectroscopy*, 2nd ed.; Elsevier: Amsterdam, The Netherlands, 1984.
37. Franchini, G.C.; Giusti, A.; Preti, C.; Tosi, L.; Zannini, P. Coordinating ability of methylpiperidine dithiocarbamates towards platinum group metals. *Polyhedron* **1985**, *9*, 1553. [CrossRef]
38. Hadjikostas, C.C.; Katsoulos, G.A.; Shakhatreh, S.K. Synthesis and spectral studies of some new palladium (II) and platinum (II) dithiocarbimato complexes. Reactions of bases with the corresponding N-alkyldithiocarbamates. *Inorg. Chim. Acta* **1987**, *133*, 129. [CrossRef]
39. Cullity, B.D. *Elements of X-ray Diffraction*, 2nd ed.; Addision-Wesley Publishing Company: Boston, MA, USA, 1978.
40. Andrade, R.; Lopez-Vega, M.; Robles, M.; Cueto, I.; Lucena, M.I. Idiosyncratic drug hepatotoxicity: A 2008 update. *Expert Rev. Clin. Pharmacol.* **2008**, *1*, 261–276. [CrossRef] [PubMed]
41. Stine, J.; Lewis, J. Hepatotoxicity of antibiotics: A review and update for the clinician. *Clin. Liver Dis.* **2013**, *17*, 606–642. [CrossRef] [PubMed]
42. Raghunath, M.; Bakal, S. Formulation and evaluation of a fixed dose combination of ceftriaxone disodium and ornidazole. *Int. J. Pharm. Life Sci.* **2013**, *5*, 750–756.
43. Simmons, C. From Your Newsletter Beware: Antibiotic-induced hepatotoxicity is rare but deadly. *Hosp. Pharm.* **2002**, *37*, 326–333. [CrossRef]
44. Vial, T.; Biour, M.; Descotes, J.; Trepo, C. Antibiotic-associated hepatitis: Update from 1990. *Ann. Pharm.* **1997**, *31*, 204–220. [CrossRef]
45. Bell, M.; Stockwell, D.; Luban, N.; Shirey, R.; Shaak, L.; Ness, P.; Wong, E. Ceftriaxone-induced hemolytic anemia and hepatitis in an adolescent with hemoglobin SC disease. *Pediatr. Crit. Care Med.* **2005**, *6*, 363–366. [CrossRef]
46. Rivkin, A. Hepatocellular enzyme elevations in a patient receiving ceftriaxone. *Am. J. Health Syst. Pharm.* **2005**, *62*, 2006–2010. [CrossRef]
47. Elsayed, M.; Elkomy, A.; Aboubakr, M. Hepatotoxicity evaluation in albino rats exposed to ceftriaxone. *Asian J. Phar. Biol. Res.* **2011**, *1*, 145–150.
48. Malomo, S. Toxicological implications of ceftriaxone administration in rats. *Nig. J. Biochem. Mol. Biol.* **2000**, *15*, 33–38.
49. Bhamidimarri, K.; Eugene, S. Drug-Induced Cholestasis. *Clin. Liver Dis.* **2013**, *17*, 519–531. [CrossRef] [PubMed]
50. Al-Salmi, F.A.; Hamza, R.Z. Efficacy of Vanadyl Sulfate and Selenium Tetrachloride as Anti-Diabetic Agents against Hyperglycemia and Oxidative Stress Induced by Diabetes Mellitus in Male Rats. *Curr. Issues Mol. Biol.* **2021**, *44*, 94–104. [CrossRef]
51. El-Megharbel, S.M.; Al-Salmi, F.A.; Al-Harthi, S.; Alsolami, K.; Hamza, R.Z. Chitosan/Selenium Nanoparticles Attenuate Diclofenac Sodium-Induced Testicular Toxicity in Male Rats. *Crystals* **2021**, *11*, 1477. [CrossRef]
52. El-Megharbel, S.M.; Refat, M.S.; Al-Salmi, F.A.; Hamza, R.Z. In Situ Neutral System Synthesis, Spectroscopic, and Biological Interpretations of Magnesium(II), Calcium(II), Chromium(III), Zinc(II), Copper(II) and Selenium(IV) Sitagliptin Complexes. *Int. J. Environ. Res. Public Health* **2021**, *18*, 8030. [CrossRef]
53. Gupta, S.P. Roles of metals in human health. *MOJ Biorg. Org. Chem.* **2018**, *2*, 221–224. [CrossRef]
54. Daniel, R.; Urmesh, K.O.; Sangeeta, J.; Chandrashekhar, P.; Mrutyunjay, S. The Small RNA DsrA Influences the Acid Tolerance Response and Virulence of Salmonella enterica Serovar Typhimurium. *Front. Microbiol.* **2016**, *7*, 599.

MDPI
St. Alban-Anlage 66
4052 Basel
Switzerland
Tel. +41 61 683 77 34
Fax +41 61 302 89 18
www.mdpi.com

Antibiotics Editorial Office
E-mail: antibiotics@mdpi.com
www.mdpi.com/journal/antibiotics

www.ingramcontent.com/pod-product-compliance
Lightning Source LLC
Chambersburg PA
CBHW061519210926
51850CB00008B/1970